POSTHARVEST BIOLOGY AND

BIOTECHNOLOGY

F
N
P

BOOKS AND JOURNALS
IN
FOOD SCIENCE AND TECHNOLOGY

THE SCIENCE OF MEAT AND MEAT PRODUCTS—
J.F. Price and B.S. Schweigert

PRINCIPLES OF FOOD SCIENCE — Georg Borgstrom
VOLUME 1 — FOOD TECHNOLOGY
VOLUME 2 — FOOD MICROBIOLOGY AND BIOCHEMISTRY

FOOD SCIENCE AND TECHNOLOGY, THIRD EDITION —
Magnus Pyke

FOOD POISONING AND FOOD HYGIENE, FOURTH EDITION —
Betty C. Hobbs

JOURNAL OF FOOD BIOCHEMISTRY — Herbert O. Hultin
Norman F. Haard and John R. Whitaker

JOURNAL OF FOOD PROCESS ENGINEERING —
Dennis R. Heldman

JOURNAL OF FOOD PROCESSING AND PRESERVATION —
Theodore P. Labuza

JOURNAL OF FOOD QUALITY — Amihud Kramer and
Mario P. DeFigueiredo

JOURNAL OF FOOD SAFETY — M. Solberg and Joseph D. Rosen

JOURNAL OF TEXTURE STUDIES — P. Sherman and
Alina S. Szczesniak

Library of Congress Catalog Card Number: 78-61444
ISBN: O-917678-05-2

Printed in the United States of America

POSTHARVEST BIOLOGY
AND
BIOTECHNOLOGY

Edited by

Herbert O. Hultin, Ph.D.
UNIVERSITY OF MASSACHUSETTS
DEPARTMENT OF FOOD SCIENCE AND NUTRITION
AMHERST, MASSACHUSETTS 01003

Max Milner, Ph.D.
MASSACHUSETTS INSTITUTE OF TECHNOLOGY
DEPARTMENT OF NUTRITION AND FOOD SCIENCE
CAMBRIDGE, MASSACHUSETTS 02139

FOOD & NUTRITION PRESS, INC.
WESTPORT, CONNECTICUT 06880 USA

CONTRIBUTORS

D.B. BECHTEL, U.S. Department of Agriculture, Manhattan, Kansas, 66502.

R.J. BOTHAST, U.S. Department of Agriculture, Northern Regional Research Center, Agricultural Research Service, Peoria, Illinois, 61604.

W.G. BURTON, Formerly Food Research Institute, Norwich, England.

W. BUSHUK, Department of Plant Science, University of Manitoba, Winnipeg, Manitoba, R3T 2NT, Canada.

JOHN P. CHERRY, U.S. Department of Agriculture, Southern Regional Research Center, Agricultural Research Service, New Orleans, Louisiana 70179.

MILDRED CODY, Department of Home Economics and Nutrition, New York University, New York, New York 10003.

RODNEY CROTEAU, Department of Agricultural Chemistry and Program in Biochemistry and Biophysics, Washington State University, Pullman, Washington 99164.

JOSEPH W. ECKERT, Department of Plant Pathology, University of California, Riverside, California 92502.

K.F. FINNEY, U.S. Department of Agriculture, North Central Region, Agricultural Research Service, U.S. Grain Marketing Research Laboratory, Manhattan, Kansas, 66502.

CHAIM FRENKEL, Department of Horticulture and Forestry, Cook College, Rutgers University, New Brunswick, New Jersey 08903.

NORMAN F. HAARD, Department of Biochemistry, Memorial University of Newfoundland, St. John's, Newfoundland, A1C 5S7, Canada.

J.W. LEE, CSIRO, Wheat Research Unit, North Ryde, N.S.W., Australia.

IRVIN E. LIENER, Department of Biochemistry, College of Biological Sciences, University of Minnesota, St. Paul, Minnesota 55108.

W.B. McGLASSON, Plant Physiology Unit, Division of Food Research, CSIRO, and School of Biological Sciences, Macquarie University, North Ryde, N.S.W. 2113.

Y. POMERANZ, U.S. Department of Agriculture, U.S. Grain Marketing Research Center, Agricultural Research Service, Manhattan, Kansas 66502.

JOSEPH J. RACKIS, U.S. Department of Agriculture, Northern Regional Research Center, Agricultural Research Service, Peoria, Illinois 61604.

LOUIS B. ROCKLAND, U.S. Department of Agriculture, Western Regional Research Center, Agricultural Research Service, Berkeley, California 94710.

SIGMUND SCHWIMMER, U.S. Department of Agriculture, Western Regional Research Center, Agricultural Research Service, Berkeley, California 94710.

IKUZO URITANI, Laboratory of Biochemistry, Faculty of Agriculture, Nagoya University, Chikusa, Nagoya, 464, Japan.

PREFACE

This is the second in a series of basic symposia on a topic of major importance to food scientists and food technologists sponsored by the Institute of Food Technologists and the International Union of Food Science and Technology. The subjects covered by the Symposium on "Postharvest Biology and Biotechnology" are extremely complex from a scientific point of view and yet critical to solutions of global food problems. It was clear from the start that not all aspects of the subject could be covered. It was felt nevertheless that the most useful approach would be to emphasize the fundamental aspects involved and hopefully to point out how our elucidation of the basic science of a system could be utilized to improve the quality attributes and storage stability of plant tissues. This approach was based on a firm belief that major advances in processing and preserving edible plant tissues will come from more fundamental understanding of the nature of the tissues involved.

Thus, this Symposium focused on the relationship between the basic biological, chemical and physical phenomena which affect postharvest plant tissues and maintenance of nutritional, esthetic, and keeping qualities of their products. Since it was not possible to cover all important subject areas, an attempt was made to solicit papers which would be representative of the types of problems that are of general importance in the field. The participants were asked to give a brief review of the subject matter, to include some current work, and to give their interpretation of future directions that should be developed to most fully utilize our dwindling resources. It was hoped that the Symposium would serve as a focal point to not only bring people up-to-date but also to orient them in the directions which are most important and where progress is most likely to come. Due to the large number of potential topics the decision was made not to deal with processing problems *per se*. Hopefully this important area might be the subject of a future Basic Symposium of the Institute of Food Technologists.

A second major thrust of the symposium was to take a phenomenological approach to postharvest biology and biotechnology. This was done to emphasize the similarities between edible plant tissues rather than to stress the differences. This should not be construed as indicating that the organizing committee was ignorant of the very important and significant differences between edible plant tissues, but to emphasize the desirability, yes even necessity, of beginning to cross some traditional boundaries to find, if not common solutions, at least common approaches. In general, researchers working with fruits and vegetables seem to go one way, and investigators dealing with cereals and legumes another. Hopefully, putting subject matter together dealing with both wet and dry tissues would at least begin to get people who work in the different areas talking together. At the time the planning of these sessions began, it became evident that as two individuals with at least nominal qualifications in food science and technology, individually we were not aware that this title had a somewhat different meaning for each of us. One (H.O.H.) had in mind the physiological, physical and biochemical phenomena in what might be called the wet-tissue food crops, while the other (M.M.) reacted to this designation from a principal background in the area of dry-tissue crops, that is, the grains and seeds. We were pleased to discover that a term which had been largely preempted by the grain and cereal specialists, indeed had relevance to broader areas of food science. Perhaps there is unity in science after all, if only in semantics!

Certain fundamental aspects are common to both types of edible plant tissues, e.g., many of the same stress conditions are deleterious, and to a significant degree, disruptive procedures have similar, if not identical effects. Therefore, without categorization as to whether the plant tissue contained high or low moisture, the Symposium was divided into four areas: (1) Biochemical Changes in Edible Plant Tissue During Maturation and Storage; (2) Response of Edible Plant Tissues to Stress Conditions; (3) Native Structure of Edible Plant Tissues and Effects of Disruption; and (4) Quality Attributes of Edible Plant Tissues.

Another purpose of this Symposium was to stress the problems of postharvest biology and biotechnology. In our opinion, there has not been enough emphasis on this subject area in the past by the IFT membership. We are not exactly sure of the reason why this is so; it may be in part due to the fact that often these products constitute a small proportion, in an economic sense, of the final processed food supply. Where there has been an important interest in postharvest plant tissues, such as in cereal crops, societies have developed to serve as the medium for information exchange. In areas where there has been a keen academic interest, such as ripening and senescent processes in fruits, workers have outlets in societies such as those dealing with plant physiology. In addition, the handling of fresh fruits and vegetables is often by relatively small operators who cannot afford to support major research projects. We hope that this Symposium will serve to kindle greater interest in the biology, biochemistry and pathology of postharvest plant tissues among the membership of the Institute of Food Technologists. We cannot forget that the majority of the calories, proteins, vitamins and minerals for most of the people of the world are derived from these types of products. A major objective of this Symposium was to provide a better understanding of the problem involved in the utilization of these basic raw materials which often do not get sufficient attention by the food processor.

Obviously all science, and particularly that affecting human welfare, must in these terms respond more urgently to the critical issues surrounding the global food supply. At no other time in the history of North America as a surplus food producer, have we been forced to recognize as we must today, the increasing constraints on our food production capabilities of decreasing availability of land, water and energy, as well as the environmental implications of all applied science and technology. Most importantly, we must be sensitive to the impact of all these problems on food costs. It is not difficult to make a prediction that the major focus of future research in food science, in the face of these constraints, will be precisely on the need to increase the productivity and efficiency of our food systems, and also to find effective means to reduce the waste which regrettably characterizes some sectors of our consumer food economy and national life styles. The need for dietary changes in order to improve our national health status is also now widely recognized.

Coincident with the implications of these problems to the responsibilities of the food scientist, our attention is increasingly diverted to what appears to be a growing concern among a highly vocal group of consumers that all is not well with the quality and safety of our food supply. While these convictions are usually expressed in terms not entirely meaningful to the food scientist, they are nevertheless generating growing pressures on the food industry and on governments, which in turn may place additional stress on our entire food chain. Specifically, for example, the apparently growing consumer demand for "fresh" foods, but with corresponding critical attitudes toward the use of processes or adjuncts that would improve or retain freshness, suggests that food scientists will be facing some increasingly perplexing issues and challenges. Clearly, it will not be easy to produce the greater quantities of crops and from them the food products which are needed for distribution through an increasingly complex food chain, without the use of processing and processing adjuncts. The scope for a trade-off or for freedom of action between these opposing pressures will, we believe, become narrower; which brings us back full circle to the implication of our Symposium. Clearly, if at least some of these antithetical demands are to be satisfied, food scientists will come under urgent pressure to gain more information about the fundamental physiological and biochemical processes in food crops which affect not only their yield but also their physical properties, stability and nutritional quality in storage, distribution and processing.

Many people, in addition to the speakers, contributed in a significant way to the development and successful completion of this Symposium. In particular, we wish to thank the chairmen of the various sessions who did an outstanding job in maintaining and controlling the scientific pace of the sessions and were able to zero in on the important problems.

Therefore, we extend our sincere gratitude to William J. Hoover, D.K. Salunkhe, Autar K. Mattoo, Henry H. Kaufmann, David A. Fellers, T. Solomos, Owen Fennema and Joseph J. Jen. We would like to recognize the contribution of Dr. Ernest Briskey, past President of IFT, who originated the idea of the Basic Symposia and, indeed, suggested the topic of this particular one. Also, to Dr. John Ayres, who as President of IFT the year the Symposium was given, thanks are due for his support and encouragement throughout the development of the Symposium. We want also to thank the members of the Basic Symposium Committee who helped with suggestions, advice and criticism and hope that their valuable ideas were not distorted too severely in formulating the final program.

The success of the second Basic Symposium was the result also of the expert assistance of Calvert L. Willey, Executive Director of IFT, and John B. Klis, Director of Publications, and their staff for providing publicity for the symposium, coordinating registration and taking care of the many details of arranging meeting rooms, hotel reservations and the numerous other details that go into a successful meeting. John Klis has served as coordinator and Anna May Schenck (JFS Assistant Scientific Editor) as copy editor for publication of the manuscripts presented at the Symposium and it is through their concerted efforts that this monograph has being.

HERBERT O. HULTIN

MAX MILNER

CONTENTS

Section III. Native Structure of Edible Plant Tissues and Effects of Disruption

Section IV. Quality Attributes of Edible Plant Tissues

Biochemical Changes in Edible Plant Tissue During Maturation and Storage

BIOCHEMICAL AND FUNCTIONAL CHANGES IN CEREALS: MATURATION, STORAGE AND GERMINATION

W. BUSHUK

Dept. of Plant Science
University of Manitoba
Winnipeg R3T 2N2, Canada

and

J. W. LEE

C.S.I.R.O. Wheat Research Unit
North Ryde, N.S.W., Australia

ABSTRACT

Grain development and maturation are discussed in terms of changes in dry weight, grain yield, and starch and protein synthesis. The main factor that controls grain size appears to be the duration of grain filling and not the availability of the primary metabolites for starch and protein synthesis.

Under normal commercial storage conditions, cereal grains can be stored for many years without deterioration. Rapid deterioration that is sometimes encountered in practice results from (1) infestation by parasites such as fungi, molds, insects and rodents; (2) unsound condition of the grain; (3) abnormally high moistures; and (4) high temperatures. If the grain is to be used for seeding, a certain period of storage is required to eliminate dormancy. During prolonged storage, seed viability decreases gradually. However, grain stored under optimum conditions can retain its viability for several decades.

Of the grain constituents, the lipids are the first to undergo changes that indicate deterioration during storage. Changes in carbohydrates and pro-

1

teins occur at slower rates. Relative to nutritive quality, the major change during post harvest storage of cereals is the degradation of thiamine, and vitamin A in yellow corn.

Storage of freshly harvested wheat produces an initial improvement in breadmaking quality. This is followed by gradual deterioration at a rate dependent upon storage conditions. With malting barley, storage eliminates dormancy and facilitates uniform germination.

Germination is important to propagation of all cereal grains and to utilization of some (e.g. malting barley). Germination is characterized by de novo synthesis of a number of enzymes and hydrolysis of starch and proteins to produce metabolites for the new plant. With some cereals (rye and wheat) premature germination can occur in the field before harvesting and thereby lower the end-use quality of the grain. The high α-amylase activity that develops during germination is particularlyy detrimental to breadmaking quality.

INTRODUCTION

Cereal grains contribute about 70% of calories and 50% of the protein in human nutrition. Wheat, rice and corn are the most important cereals. A significant increase in world food supply, required to meet the rising population, will require a major increase in the total amount of cereal grains available for human consumption. This can be achieved in two ways: first, by increasing production through an increase in seeded area and yield; and second, by decreasing post maturity losses. This chapter deals with the biochemical and functional changes in cereal grains during maturation, storage and germination and is, therefore, a relevant part of the story on the world food supply.

GRAIN DEVELOPMENT AND MATURATION

A seed is a self-perpetuating plant organ and thus seed development may be considered as part of the postharvest biological processes of the previous generation. Not only does seed development play an obvious role in ensuring survival, but variations in development are often of great importance in determining the value of the mature seed for food and feed and its suitability for processing. Perhaps the only significant example of an immature cereal seed being used as food is sweet corn, and here stage of maturity is obviously important in determining the relative levels of sugar and starch in the seed. For all these reasons, discussion of cereal seed development is included here, and while most of the material presented concerns wheat, much of it is relevant to other members of the family *Gramineae.*

Dry Weight

There are two controlling factors which determine the way in which a fertilized embryo develops into a mature seed; the genotype and the environmental conditions during growth. The latter include soil structure and nutrients, water supply, temperature, irradiation, presence of pathogens or insects and the state of the vegetative parts of the plant at anthesis. After an initial lag of a few days, grain dry weight in wheat and other cereals increases essentially linearly to a maximum and often shows a small decline as desiccation leading to full maturity proceeds. High temperatures cause lower grain dry weights largely through decreased times, rather than rates, of grain filling (Marcellos and Single 1972; Sofield *et al.* 1974). Wheat genotype may also influence final grain size as shown in Fig. 1.1 for *Triticum aestivum* cvs. Timgalen and Heron. Again the duration of grain filling appears a more significant determinant of grain size than its rate. The identity of the factors controlling the rate and duration of grain filling are not known but it is likely that they are hormonal. Under optimum growing conditions, the supply of primary substrates for starch and protein synthesis, sucrose and amino acids respectively, do not appear limiting but the capacity to utilize them is (Jenner 1976; Donovan *et al.* 1977).

Yield

The yield components of a cereal crop are in broad terms, the product of grain size and grain number per unit area. The factors affecting grain number are determined before, during and after anthesis. Ear number and spikelets per ear are established relatively early in plant development by such factors as solar radiation, nutrient and water availability, plant density and genetic capabilities of the cultivar. The number of mature grains available for harvest is, in turn, influenced by the number which abort or do not reach an acceptable size. These effects, like tillering and spikelets per ear are strongly influenced by environmental conditions and by cultivar. The various yield components depend very much on the sequence of environmental effects during the plant's growth. There are often strong interactions or compensations between yield factors; for example, high grain numbers or high tiller numbers may be associated with small grain size. Grain size is a heritable character but it is still very much influenced by environmental conditions during the post-anthesis period.

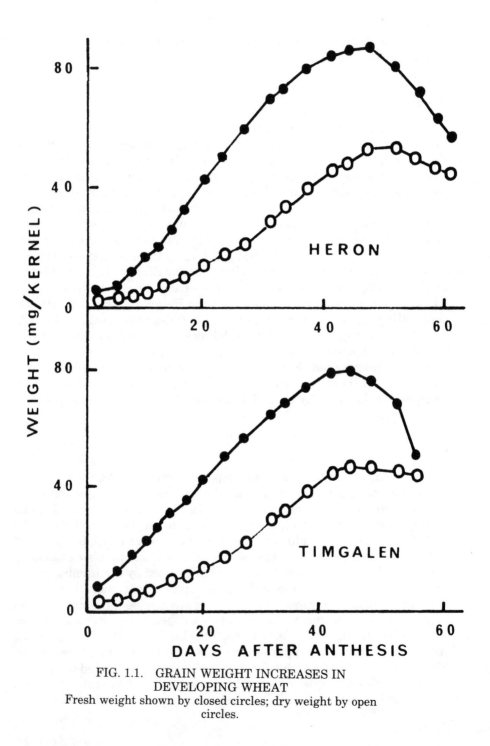

FIG. 1.1. GRAIN WEIGHT INCREASES IN
DEVELOPING WHEAT
Fresh weight shown by closed circles; dry weight by open
circles.

Starch Synthesis

Starch is quantitatively the most important component of all mature cereal grains and it is found almost entirely in the endosperm where its function is presumably a source of carbon for germinating grain. Starch granules appear in endosperm cells from approximately six days after anthesis. Early in seed development starch occurs in all endosperm cells including those of the aleurone but at maturity the aleurone is devoid of starch. In wheat, primary starch granules (A-type) are formed in plastids (amyloplasts) and ultimately reach diameters of up to 35 μm. At about 14 days post-anthesis, small granules (B-type) appear in the stroma space of the plastid and these are extruded by first stretching then constricting the plastid membrane (Buttrose 1963). Starch granule size varies considerably in different cereals and the distribution of granule size within many of the cereal genera has been studied extensively. Evers and Lindley (1977) using two analytical methods following microsieving showed that between one-third and one-half of the weight of mature wheat endosperm starch consisted of granules of less than 10 μm diameter. He also pointed out that a wheat kernel containing 35 mg starch would contain approximately 7 $\times$ 10^8 starch granules which collectively would have a surface area of about 350 cm^2.

The primary substrate for starch synthesis is sucrose which is translocated from other parts of the plant to the endosperm. Jenner (1974) showed that sucrose hydrolysis was not a necessary pre-requisite for its transport into the wheat endosperm cell and that the rate of arrival of sucrose within the cell probably controlled the rate of starch synthesis. In contrast Shannon and Dougherty (1972) presented evidence which they believed showed that sucrose hydrolysis actually controlled the entry of sugar into the corn endosperm cell. During the period of rapid wheat grain growth, the enzymes sucrose synthase, ADPG- or UDPG-pyrophosphorylase and starch synthase do not appear limiting (Donovan et al. 1977). In the later stages of grain development where starch synthesis ceases, however, sucrose is present in adequate concentrations to maintain synthesis, so presumably the enzymes of the amyloplast are incapable of carrying out all the reactions necessary for the production of starch (Jenner and Rathjen 1975).

Protein Synthesis

While cereals represent primarily an energy source in food and feed, their constituent proteins often have an overriding importance in food processing and sometimes in nutrition. Functions have been somewhat arbitrarily ascribed to the various classes of protein in both developing and mature seeds of the *Gramineae*. Thus, the albumins and globulins have been named as metabolic or cytoplasmic proteins and the prolamines and glutelins are thought to be storage proteins. While there is little firm evidence that these are indeed functions performed by the different protein classes, the generalizations are probably substantially correct and serve as a useful basis for study of protein synthesis in the developing grain. It should be pointed out that the proteins are placed in classes on the basis of their solubility and there is considerable evidence of class overlap where solubility is used as the sole criterion. Here we will consider the albumins and globulins as those proteins extractable by dilute salt solution and serving a metabolic function. The storage proteins occur in the seed in the form of protein bodies and are not soluble in dilute salt solutions, but are soluble in disaggregating solvents such as aqueous alkali, urea, etc. (Jennings *et al*. 1963). The prolamines which form part of the storage protein in cereals are soluble in 70% aqueous ethanol, a solvent which extracts virtually none of the albumins, globulins or glutelins.

In the mature wheat grain, all but approximately 5% of the nitrogen present occurs as protein so that analysis of the mature wheat seed for nitrogen gives a good measure of its protein content ($N \times 5.7$). As will be discussed later, nonprotein nitrogen (NPN) levels in developing grain can be very significant but from a practical viewpoint the accumulation of protein follows closely the accumulation of nitrogen. Total nitrogen increases, during wheat grain development, parallel closely dry weight increases (Fig. 1.2) and protein accumulation ceases when dry weight reaches its maximum. Data given in Fig. 1.2 are for an Australian "high protein" wheat cv. Timgalen and a "low protein" variety cv. Heron. The higher percentage nitrogen in Timgalen can be accounted for by two factors: a greater net synthesis of protein (nitrogen) per grain and a smaller grain weight. This observation emphasizes the point that the protein or nitrogen content (percent) of mature grain essentially represents a ratio between the amount of protein and the amount of starch per grain. High protein percentage in mature grain does not necessarily reflect efficient utilization of soil nitrogen; it may be due to reduced starch synthesis. It should be emphasized that low grain weights or low weights of starch per grain do not always lead to high protein percentages.

Accumulation of protein nitrogen during cereal seed development is the net result of synthesis and cycling through turnover or degradation. There has been no real study of the significance of turnover in the developing seed, but it is known that it does occur in detached wheat leaves (Brady and Tung 1975) and that there are proteases in both developing (Bushuk *et al*. 1971; Dalling *et al*. 1976) and mature (Kaminski and Bushuk 1969) wheat grain. Both metabolic and storage protein synthesis proceeds concurrently in the developing wheat grain and both prolamines and glutelins are present from 14 days post-anthesis or earlier (Bushuk and Wrigley 1971). Storage protein synthesis continues for a longer period than synthesis of the albumins and globulins (Fig. 1.3). The amount of NPN per kernel remains relatively constant during wheat grain development (Jennings and Morton 1963) which is perhaps not unexpected when it is considered that it represents a continuing source of substrate for protein synthesis. A similar pattern of accumulation of metabolic and storage protein has been reported for corn (*Zea mays*) by Moureaux and Landry (1972). They also reported that NPN levels per kernel remained relatively constant after reaching a maximum early in development.

As mentioned earlier, storage protein in seeds is deposited in the form of protein bodies. Protein bodies are first visible in cereal endosperm at from 10–14 days post-anthesis and remain visible and can be isolated from mature grain of millet, barley, corn and sorghum but not from wheat (Adams *et al.* 1976). Early in wheat endosperm development, the protein bodies, which may or may not be surrounded by a membrane, are commonly found in vacuoles (Graham *et al.* 1963; Barlow *et al.* 1974). That wheat protein bodies do indeed consist largely of prolamine (gliadin) and glutelin (glutenin) was shown by Graham *et al.* (1963) both by direct chemical analysis and by radioincorporation studies. A number of enzymic activities has been shown to be associated with protein bodies but their contribution to the total protein is probably quite small. After wheat endosperm cell division has ceased at 12–14 days post-anthesis, existing protein bodies may range in size up to 20 μm while larger aggregates are clearly visible under the light microscope (Fig. 1.4A, 1.4B). There is a progressive increase in the amount of rough endoplasmic reticulum (RER) visible under the electron microscope as development proceeds and at around day 24 there is evidence of enlargement or distension of the lumen of the RER (Fig. 1.4C). Barlow *et al.* (1974) have suggested that the protein synthesized by the ribosomes of the RER is secreted into the lumen which causes swelling. There is evidence (Fig. 1.4C) of "rounding up" of some of the membrane of the RER giving rise to a second type of membrane enclosed protein body with ribosomes still attached to the outer surface of the membrane. At the same time as storage protein bodies are growing in number and size, starch granule synthesis is continuing. Towards maturity at the onset of desiccation the cell contents have become crowded and the relatively rigid starch granules distort the more aqueous and softer protein bodies and membranes. At maturity, the starch granules still retain the shape they had at the completion of synthesis while the protein bodies and remnants of membranes occupy whatever space is available between the starch granules (Fig. 1.4D).

The metabolic proteins of cereal endosperm are probably largely as-

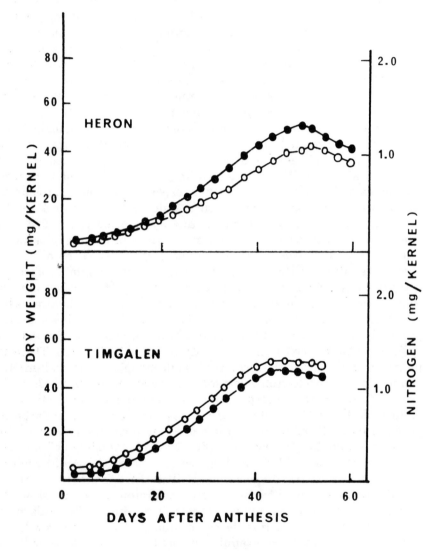

FIG. 1.2.   NITROGEN CHANGES IN DEVELOPING WHEAT
GRAIN
Grain dry weight is shown by closed circles; total nitrogen per
grain open circles.

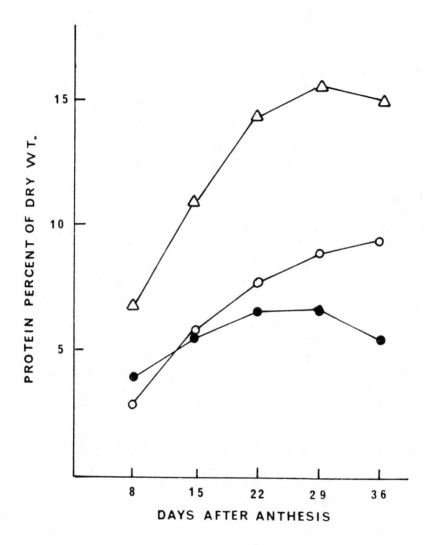

FIG. 1.3. CHANGES IN PROTEIN IN DEVELOPING
WHEAT GRAIN

Data are for *Triticum aestivum* cv. Timgalen. Closed circles
show sodium pyrophosphate-soluble protein (albumins and
globulins). Open circles show storage protein. Open triang-
les show total protein.

sociated with cell organelles such as the amyloplasts, mitochondria, RER, dictyosomes, lysosomes and the nucleus, although some are almost certainly present in the cytosol and mention has already been made of their association with protein bodies. Small amounts of protein are also associated with the cell wall but whether these are enzymically active is not clear. Immune fluorescence microscopic studies by Barlow *et al.* (1973) have shown that the major area of concentration of buffer soluble protein in wheat endosperm is around the starch granules where the protein is presumably that of the amyloplast membrane. It has been postulated (Simmonds *et al.* 1973) that these proteins play a role as a cementing material influencing the hardness of mature wheat grain.

There is a considerably higher concentration of protein in the subaleurone cells compared with the central endosperm of the mature wheat grain. Kent (1966) found that in a mature Hard Red Winter wheat with an overall protein content of 13.7%, the subaleurone cells contained approximately 50% protein while in the central endosperm the level was about 8%. These observations which are supported by direct microscopic examination, show that there is a gradient of protein concentration from the outer to the central endosperm. Evers (1970) found that the younger subaleurone cells actually contain approximately the same weight of protein as the central endosperm cells, but the former are smaller and contain less starch. There are reports of differences in this gradient of protein concentration in some samples of different wheat varieties (Farrand and Hinton 1974), but whether these differences are attributable to differences in the amounts of protein per cell, to differences in the gradient of cell size, or to varying states of maturity in the different samples, is not clear. This gradient in protein concentration is a feature of the wheat grain which has not yet been fully exploited by millers but with newer techniques it may become feasible to produce flours with a wide range of protein content.

The proteins of cereal grains are often of primary importance in determining the cereal's usefulness as food or feed and particularly its suitability for processing. Conventional breeding techniques and agronomic practice have been used to vary the level of protein and, to some extent, the quality of that protein in cereals. It may well be that with the increasing amount of knowledge of the factors controlling the conversion of soil nitrogen to cereal grain protein that new cereals can be produced with much higher protein contents and improved protein composition. Some success has already been achieved with the "high lysine" cereals following the discovery of the *opaque-2* and *floury-2* corn mutants (Mertz *et al.* 1964; Nelson *et al.* 1965). High lysine genotypes of both barley (Ingversen *et al.* 1973) and sorghum (Singh and Axtell 1973) have also been found. In each

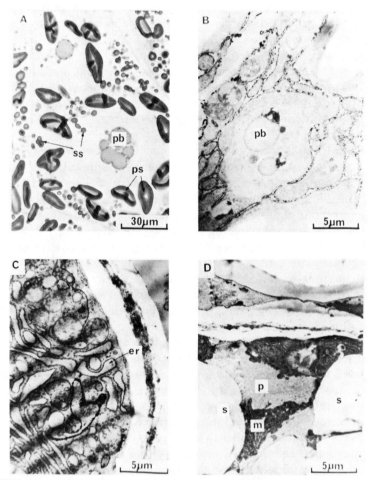

FIG. 1.4. INTERNAL MORPHOLOGY OF DEVELOPING WHEAT
ENDOSPERM

A. Light micrograph of 20-day endosperm stained with Periodic Acid-Schiff
reagent and Coomassie Blue.

B. Transmission electron micrograph of 20-day endosperm.

C. Transmission electron micrograph of 30-day endosperm.

D. Transmission electron micrograph of mature wheat endosperm.

All plates are of *Triticum aestivum* cv. Timgalen; er, endoplasmic reticulum; m,
membrane residues; p, protein; pb, protein body; ps, primary starch granule; s,
starch; ss, secondary starch granule.

of these cereals the high lysine mutants show reduced prolamine synthesis and somewhat enhanced synthesis of the other classes which are richer in lysine than the prolamines. Increased levels of lysine in corn, barley and sorghum proteins have been achieved by suppression of synthesis of a complete protein class. There may be scope for seeking mutants lacking individual proteins or groups within a class. Huebner *et al.* (1974) have shown that there is a fourfold difference between the highest and the lowest lysine contents in wheat glutenin fractions and it may be possible to exploit this character.

Although there is a considerable amount of data on the way that agronomic and environmental factors influence the quantity of nitrogen removed from the soil by cereals and the pattern of redistribution in the plant, little is known of the control of the biochemical processes involved. It now seems certain that no one identifiable biochemical factor can be used by the cereal breeder in selecting genotypes for efficient nitrogen utilization. There have been suggestions that the enzyme nitrate reductase (Croy and Hageman 1970; Deckard *et al.* 1973) could be used as an aid to selection but practice has shown that this enzyme alone is not an adequate criterion. A complication of this type of approach was illustrated by Rao *et al.* (1977) who when comparing nitrate absorption from two wheat cultivars, found significant differences in *in vitro* nitrate reductase activity but little difference in *in vivo* leaf nitrate reduction. The final step in nitrogen assimilation by cereals is the process of protein synthesis in the grain itself. In the intact plant, studies on protein synthesis are complicated by an absence of information on the rate of arrival and identity of substrate. Studies with detached wheat ears (Donovan and Lee 1977) grown in defined media make interpretation easier. Using this technique, it has been shown that changes in grain constituents during development are similar when ears are fed nitrogen either as a complete mixture of amino acids, glutamine, asparagine or ammonium nitrate. It has also been established that grain in detached ears from some high-protein wheat cultivars can incorporate labelled amino acids more rapidly than low-protein types (Donovan *et al.* 1977). The identity of the control factors of the extremely complex series of physiological and biochemical processes involved in converting soil nitrogen to plant protein is unlikely to be solved by studies on whole plants alone. By using results obtained from live plant organs such as roots, leaves, ears and grains, it may be possible to assemble a model which illustrates where control is exerted.

The modification of the composition of cereal grains by conventional breeding techniques, by the production of new genera like X *Triticosecale* Wittmack (triticale) and by selection from existing or new mutants or species, has considerable scope. Examples of practical success in this area are now reasonably numerous. The high-lysine cereals, corn and rice with different amylose to amylopectin ratios, cereals with high- or low-protein content are but a few examples. If such cereals are to be used for feed or subjected to little processing, the modifications in composition will probably not result in any major problems. If, however, processing requires chemical and physical properties to be held within narrow limits, it may not be possible to use modified cereals in existing processes. The high lysine barley mutants may be unsuitable for beer production because of excessive browning caused by the additional amino groups reacting with reducing sugars. It is highly likely that major modifications in the composition of wheat protein would render the visco-elastic properties unsuitable for use in bakeries using conventional techniques. It can only be guessed as to whether the development of radically different cereals or modifying processes to use them will be the more difficult problem to solve.

CHANGES DURING POST-HARVEST STORAGE

Post-harvest losses of grain is a significant factor in the world food supply. Estimates of total losses as high as 50% for India have been reported (Parpia 1976). Most of these result from infestation and consumption (followed by deterioration) by rodents, insects, fungi, molds and other microorganisms. Losses and deterioration from those external factors are outside the scope of this review and will be referred to only when it is not possible to separate them from those of specific concern here. A much smaller, but not insignificant, proportion of the total losses results from the gradual deterioration of viability, nutritive quality, and utilization quality during storage under normal commercial conditions. These are the changes that will be discussed in this section.

As in previous sections of this chapter, emphasis will be on wheat because the authors are most familiar with this crop and because most of the published work has been on this grain.

Chemical and physiological changes in seeds occur progressively during storage. Some of these changes are beneficial to subsequent use of the seed but many are detrimental. The rates at which these changes occur depend on the composition of the seed (i.e. species) and on a large variety of external factors. Seeds that contain large percentages of oil (i.e. oats) compared with wheat, deteriorate much faster due primarily to the changes in the oil component which occur much more rapidly than the changes in other components. Of the external factors, the most important are humidity, temperature and composition of interstitial atmosphere excluding parasites.

Respiration

Respiration is a natural phenomenon that occurs in all grain during storage. It can be either aerobic and anaerobic. It is usually measured by the respiratory quotient (R.Q.) which is the ratio of the amount of carbon dioxide produced to that of oxygen consumed. R.Q.'s are useful as indicators of the relative extent of aerobic and anaerobic respiration in grain.

Most of the respiration in grain in normal commercial storage can be attributed generally to the respiration of the parasitic microflora such as fungi and molds. This is especially true in grain with high moisture contents. Hummel *et al.* (1954) showed that in mold-free wheat containing 16–31% moisture, respiration at 35° C occurred at a low constant rate.

The rate of respiration, which usually determines the rate of deterioration, depends on a variety of internal and external factors. Of the internal factors, the most important are the soundness and extent of damage, if any, of the grain. Bailey and Gurjar (1918) showed that under controlled conditions, respiration in wheat was definitely greater for unsound grain than for sound grain. In commercial practice, the potential detrimental effects of those internal factors are taken into account in the grading systems by minimizing the percentages of unsound or damaged kernels in the top grades.

The principal external factors that affect grain respiration are moisture, temperature and oxygen tension. Of these, moisture is the most important in commercial storage.

Moisture.—Grain with moisture contents below the established "safe" minimums can be stored under normal commercial conditions for 2–3 years with very little deterioration. Respiration rate, due primarily to fungal proliferation, increases sharply with moisture content above the safe minimum with subsequent development of heat spots which can eventually raise the grain temperature to sufficiently high levels to denature proteins or inactivate enzymes that are important to functional quality. It has been generally assumed that the safe minimum moisture contents are the equilibrium moisture contents of the grain at 75% relative humidity (RH). Since the equilibrium moisture content at a specific relative humidity depends on grain composition, the equilibrium moisture contents at 75% RH differ widely among different species (Table 1.1). The safe minimums adopted by different countries differ somewhat (compare Canadian and U.S. data in Tables 1.2 and 1.3). The main reason for this difference is the ambient temperature. Higher moistures are considered as safe for storage at lower ambient temperatures.

The main factor that controls equilibrium moisture content of grain is the relative humidity of the interstitial atmosphere. The relationship between moisture content and relative humidity, the so-called sorption isotherm, is sigmoid in shape. Different grain species give slightly different isotherms (Coleman and Fellows 1925). Furthermore, most biological materials show a hysteresis effect with the adsorption (increasing relative humidity) isotherm being slightly lower than the desorption (decreasing relative humidity) isotherm. With grains, the equilibrium moisture contents remain relatively low until the isotherm reaches the point of about 80% RH. Beyond this point, the moisture content increases exponentially with increasing relative humidity.

On occasion it is expedient or even essential to harvest grain with moisture contents well above the levels that are considered as safe for storage. Such grain must be artificially dried before it is placed into storage. The commercial value of the grain in terms of germination capacity, nutritive quality and utilization quality (e.g. breadmaking) can be drastically reduced if proper drying conditions are not used.

The most important factor in drying damp grain is the temperature of the air that comes in contact with the grain. Stansfield and Cook (1932) were the first to publish data on the "critical" or maximum temperature

TABLE 1.1

EQUILIBRIUM MOISTURE CONTENTS OF GRAINS AT 75% RH AND 25° C[1]

Grain	Moisture (%)
Barley	14.4 (25–28° C)
Buckwheat	15.0 (25–28° C)
Corn	14.3
Cottonseed	11.4
Flaxseed	10.3
Oats	13.4
Peanuts	10.5
Rice	14.0
Rye	14.9 (25–28° C)
Sorghum	15.3 (25–28° C)
Soybeans	14.0; 14.4
Sunflower	10.4; 11.7
Wheat	
Durum	14.1 (25–28° C)
Hard red spring	14.8 (25–28° C)
Hard red winter	14.6 (25–28° C)
Soft red winter	14.7 (25–28° C)
White	15.0 (25–28° C)

[1] Adapted from Pomeranz (1974).

TABLE 1.2

MAXIMUM MOISTURE CONTENTS FOR STRAIGHT GRADE IN CANADIAN GRADING SYSTEM[1]

	%
Barley	14.8
Corn	14.0
Oats	14.0
Rye	14.0
Wheat	14.5
Buckwheat	16.0
Flaxseed	10.5
Mustard seed	11.0
Rapeseedd10.5	
Safflower seed	9.6
Sunflower seed	9.6
Peas	16.0
Pea beans	19.9
Soybeans	14.0

[1]Source: Grain Grading Handbook, Canadian Grain Commission, Winnipeg, Canada.

TABLE 1.3

MAXIMUM MOISTURE CONTENTS OF GRAINS FOR NORMAL U.S. GRADES[1]

	%
Barley	14.5
Corn	14.1[2]
Oats	14.0
Rice	
Rough	14.0
Brown	14.5
Milled	15.0
Rye	16.0
Wheat[3]	—
Sorghum	13.0[2]
Flax seed	9.5
Soybeans	13.0[2]
Peas	15.0
Beans	18.0[4]

[1] Source: USDA Agricultural Marketing Service Grain Division Standards.
[2] U.S. No. 1 grade only.
[3] No statutory maximum; actual moisture shown on grade certificate.
[4] 17.0% for lima beans.

that can be used for drying wheat of different moisture content. The temperatures of the drying air ranged from 50–133° C and the grain moistures from 14–21%. Wheat of any moisture can be dried without loss of loaf volume so long as the air temperature does not exceed 83° C and the rate of drying is not excessive. Cashmore (1942) showed that the germination capacity of wheat was retained when grain with moisture content of 18% was dried with air at temperatures below 67° C and that with 30% moisture at 43° C. Finney *et al.* (1962) reported that somewhat higher air temperatures can be used for drying high moisture immature wheat to be used for breadmaking quality tests. Linko (1966) published a comprehensive review of the published work on the drying damage of functional properties of wheat.

Milner and Woodforde (1965) found no differences in the feeding value for young chicks of wheat samples harvested at 22% moisture content and dried at 80° C for 37 min, and at 104° C for 120 min, compared with the control sample dried at 27° C. According to the official recommendation of the Canadian Grain Commission, tough (14.6–17.5% moisture) and damp (above 17.5% moisture) wheat should be dried with air temperatures below 57° C in order to retain the native breadmaking quality. It is common knowledge that the maximum temperature of drying air can be increased as the grain dries.

With malting barley, the critical temperatures are generally lower than those used to dry wheat for milling purposes (Pomeranz 1974).

Corn is often harvested at moisture contents well above the safe minimum for storage and must be dried to prevent deterioration. Emerick *et al.* (1961) showed that corn that was air-dried in thin layers at temperatures from 10–177° C retained its nutritive value for chicks and rats. A significant decrease in nutritive value was obtained for samples dried at 232° C for 2½ hr. Cabell *et al.* (1958) showed that extended heating above 57° C decreased the nutritive value of corn protein for rats but drying at 116° C for 1.5 hr had no effect.

Brown rice dried in the sun (*ca.* 30° C) or in storage at 90° C gave milled and cooked rice that contained more soluble matter and higher starch-iodine blue value than the samples dried at 10° C (Sato *et al.* 1971).

Temperature.—Temperature can affect respiration directly by its effect on the rate of the enzyme reactions involved in the process, or indirectly by influencing the equilibrium moisture content of the grain. Respiration of sound grain increases with increase in temperature until the rate of thermal inactivation of the enzymes involved in the respiration becomes significant and limiting.

Temperature has a minor effect on the equilibrium moisture content. Because sorption of water by biological materials has a negative temperature coefficient, the moisture content at a specific relative humidity increases slightly with decreasing temperature. Milner and Geddes (1954) reported that for wheat, this effect amounts to 0.6–0.7% increase in moisture content for each 10° C drop in temperature.

A related factor that is important to commercial storage is the interchange in moisture in the vapor phase due to temperature gradients produced by fluctuating external temperatures. Under extreme conditions, relatively humid warm air reaching a cold region (e.g. outside bin walls) may be cooled below its dew point whereby droplets of water will condense on the cool grain adjacent to the bin walls. This, in turn, can raise the respiratory rate to a level that could lead to "pocket" heating and subsequent deterioration of quality.

Oxygen.—Aerobic respiration involves direct consumption of oxygen and production of carbon dioxide. The process is, therefore, inhibited by lack of oxygen. Storage under inert atmospheres is used commercially (e.g. wheat in Australia) to extend the safe storage period.

Dormancy and Seed Viability

Seeds of many plant species will not germinate immediately after harvest even under ideal germination conditions. This natural phenomenon to resist germination is called dormancy. Dormancy in seeds has been defined (Roberts 1972) as being either innate, induced or enforced. Innate dormancy is present in the seed immediately; the embryo ceases to grow while still attached to the parent plant. Induced dormancy is encountered in seeds which have lost their innate dormancy but do not germinate because of some factor in the environment such as oxygen or carbon dioxide levels. Induced dormancy is characterized by its persistence after the inhibitory factor has been removed. Enforced dormancy is normally confined to the condition where some limitation in the environment prevents germination but on transfer to favorable conditions, the dormancy is immediately removed. The period of dormancy may range from several days to years depending on the species and cultivar. During post-harvest storage, under normal conditions, a series of complex biochemical and physiological changes occur that eliminate dormancy. This disappearance of dormancy is called after-ripening. In some cases, dormancy can also be eliminated relatively quickly by external factors (e.g. light, heat treatment or chemical treatment) (MacLeod 1967).

According to Brown et al. (1948) barley, oats and sorghum lost their dormancy after storage for 1–6 months at 40° C. Barley and oats stored at 2° C at relatively high humidity retained its dormancy for 3 yr. The degree of dormancy in sorghum was less than in barley and oats.

While dormancy is essential to the survival of wild plants, prolonged dormancy creates problems in end-use applications that require uniform germination of freshly harvested seed (e.g. malting of barley). With wheat and rye, it is the lack of post-harvest dormancy that presents greater problems in commercial grain production than unduly long dormancy. Premature germination lowers the grade and milling quality of the grain, and increases α-amylase activity in the flour to undesirably high levels.

Dormancy in most species is a cultivar characteristic. Accordingly, cultivars with improved dormancy can be developed by plant breeding.

Within cultivars, dormancy can be markedly affected by environmental factors. Belderok (1968) in his extensive studies of the dormancy of wheat cultivars grown in The Netherlands showed that dry and sunny weather during the period when the grain is in the dough stage of development markedly shortens the dormancy. From these studies, he developed an "early warning" system that is used to warn farmers of the potential danger of sprouting. The system, which advises the farmers when they should harvest their crop, has been applied successfully to commercial production in The Netherlands and Western Germany.

Viability of seeds deteriorates slowly but continuously with time under commercial storage conditions. Seeds that do not germinate under favorable growth conditions are either dormant or dead. The maintenance of viability during storage and under a wide range of environmental conditions is of primary importance with cereals as not only is it of importance for growth into a new plant, but also it is commonly a good index of the seed's suitability for processing. Some cereal seeds, particularly some with abnormal chromosome complements, are inherently nonviable but from a practical viewpoint the loss of viability caused by environmental conditions during maturation, harvest or storage are of most significance.

Cereal seeds can remain viable for long periods if stored under favourable conditions. Roberts (1972) discussed an earlier observation that barley and oat seed found in glass tubes in the foundation stone of the Nuremberg City Theatre where it had lain for 123 years, still retained some viability. Haferkamp et al. (1953) showed that barley stored for 32 years under dry conditions had 96% germination capacity. It is not known how long cereal seed will remain viable if stored optimally but there is interest in very long term storage of plant breeders' and geneticists' material at very low temperatures in an atmosphere low in oxygen and where the seed is protected from radiation.

Of the external factors, temperature and humidity are the most important. According to Bakke and Noecker (1933) viability of oats was reduced to 6% after 3 days storage at 40° C of grain containing 26.4% moisture. Carter and Young (1945) stored wheat of 12.2–18.6% moisture in sealed containers at different temperatures. Loss of viability was significant in samples stored at 40° C for 279 days, but not in those stored at lower temperatures.

In most studies using relatively extreme storage conditions, loss of viability can be generally attributed to the development of storage molds or fungi. If storage conditions are satisfactory and the grain is sound, small grains, especially cereals, can retain their viability for a long time.

Chemical and Nutritional Changes

Carbohydrates and Related Factors.—Under some post-harvest storage conditions, cereal grains show a net gain in dry weight (Pomeranz 1974). This has been attributed to the addition of water in the hydrolysis of starch by endogenous amylases. However, under conditions of high respiration there is generally a loss of dry weight due to conversion of starch to carbon dioxide. Total sugars content normally increases during storage at a rate that depends strongly on storage conditions, especially moisture content. Glass et al. (1959) showed that under anaerobic conditions the increase in reducing sugars content of wheat could be accounted for by the decrease in the content of nonreducing sugars (Table 1.4). Somewhat different trends in sugars content during 8 wk of storage were obtained by Lynch et al. (1962) (Table 1.5). The reducing sugars content remained unchanged in the sample stored in air but increased substantially in the samples stored under carbon dioxide and nitrogen. A much greater decrease in nonreducing sugars content occurred in the sample stored in air than in the samples stored under carbon dioxide or nitrogen. According to Iwasaki and Tani (1967), rice containing 16% moisture showed a marked increase in reducing sugars content during storage for 1 yr at −2° C to 33° C.

Changes in grain carbohydrates during storage are related to concomitant changes in the activity of endogenous amylases. Studies of Popov and Timofeev (1933) showed a slight increase in amylase activity during the early post-harvest storage. On the other hand, Loney and Meredith (1974) showed that amylase activity of wheat flour decreased continually with storage time at room temperature. The activity remained essentially constant over 2½ yr when the flour was stored at −25° C. The changes in grain carbohydrates during storage are obviously highly dependent on the storage conditions and on grain species.

TABLE 1.4
EFFECT OF ATMOSPHERE AND MOISTURE LEVEL ON MOLD COUNT AND ON REDUCING AND NONREDUCING SUGAR CONTENT OF WHEAT AFTER STORAGE FOR 16 WK AT 30° C[1]

	Moisture Content of Wheat, and Atmosphere							
	15%		16%		17%		18%	
	Air	Nitrogen	Air	Nitrogen	Air	Nitrogen	Air	Nitrogen
Mold count[2]	0.7	0	1.4	0	10.6	0	40	0
Nonreducing sugars[3]	186	188	133	176	127	144	98	129
Reducing sugars[4]	32	38	44	52	44	83	48	108

[1] Data of Glass et al. (1959); taken from Pomeranz (1974).
[2] Mold count per g × 10)[6]
[3] Expressed as mg sucrose per 10g wheat (dry-matter basis). Initial value for wheat before storage trial was 232 mg sucrose per 10g.
[4] Expressed as mg maltose per 10g wheat (dry-matter basis). Initial value for wheat before storage trial was 37 mg maltose per 10g.

TABLE 1.5

CHANGES IN THE MONO AND DISACCHARIDES OF WHEAT STORED
UNDER VARIOUS ATMOSPHERES FOR 8 WK AT 30° C AND 20%
MOISTURE[1]

Sugar[2]	Control	Stored in Air	Stored in Nitrogen	Stored in Carbon Dioxide
Fructose	6	5	18	16
Glucose	8	7	24	23
Galactose	2	3	9	9
Sucrose	54	21	39	36
Maltose	5	1	4	3
Total reducing sugars, as maltose	41	41	117	117
Total nonreducing sugars, as sucrose	190	43	100	115

[1] Data of Lynch et al. (1962); taken from Pomeranz (1974).
[2] Concentration of sugars, mg per 10g dry-matter basis.

Proteins, Amino Acids and Enzymes.—Wheat stored for 8 yr under commercial storage conditions showed no change in protein content (Pixton and Hill 1967). Slight increases in total nitrogen have been obtained for wheat that had been stored under conditions of high respiration. This change has been attributed to a loss of carbohydrate (Daftary et al. 1970). Jones and Gersdorff (1941) carried out an extensive study of the changes in various protein fractions of wheat stored in closed jars at −1° C and 24° C for 9 and 24 months. The changes were small; however, there was a decreasing trend in the proportion of soluble protein. A decrease in the amount of water-soluble protein was reported by Kozlova and Nekrasov (1956) under long term (to 16 yr) commercial storage.

The free amino acids content of sound wheat is very low (less than 1% of total nitrogen). These values remain unchanged during storage until extensive deterioration occurs. In such grain, it is difficult to separate the contributions of endogenous and exogenous (from molds and fungi) proteases. It is generally presumed that the increase in the free amino acids observed for some storage conditions is from microorganisms associated with grain spoilage.

Changes in several enzymes have been used as indicators of onset of deterioration in storage. Linko and Sogn (1960) showed that for wheat and corn the percentage of germination correlated positively with the activity of glutamic acid decarboxylase (GADA). Linko (1961) developed a manometric technique for GADA activity which can be used as a measure of seed viability. The well-known tetrazolium test for seed viability depends on the activity of dehydrogenases found in the seed embryo. The activity of these enzymes decreases rapidly as grain deteriorates in storage.

Lipids.—Two types of reactions can occur in grain lipids during storage, oxidation and hydrolysis. Oxidative reactions produce a variety of products that lead to rancid taste and odor. Under normal storage conditions these reactions occur to a negligible extent, because most grains contain active antioxidants which inhibit them.

Hydrolytic breakdown of grain lipids is catalyzed by natural lipases and yields free fatty acids. In sound grain, lipid hydrolysis occurs much faster than the hydrolysis of carbohydrates or proteins (Pomeranz 1974). A rise in free fatty acid content is commonly used as an index of onset of storage deterioration. A 3-yr survey by the U.S. Department of Agriculture (1954) showed that freshly harvested sound wheat samples had fat acidity values under 20. Wheat that had deteriorated in storage because of mold growth, high respiratory activity or spontaneous heating had fat acidity values of 100 or higher (Baker *et al.* 1959).

Changes in lipids of soft and hard wheats with high moisture contents stored at high temperatures for up to 132 days were examined by Daftary and Pomeranz (1965). They found that the nonpolar lipid content decreased by about 25% and there was a marked breakdown of polar lipids in samples that showed extensive deterioration. Those samples also showed a rapid rise in mold count; it is quite likely that the changes in the lipids were due to mold growth.

Nutritional Changes.—Major changes in the nutritive quality of grains occur with extensive deterioration arising from growth of bacteria, fungi or mold. In some cases, loss of nutritive quality results indirectly from the various toxic compounds produced by those microorganisms. Infestation of grain by insects or rodents usually makes the grain unsuitable for human food although some of it can be occasionally salvaged for animal feed. Since changes under abnormal storage conditions are outside the scope of this article, secondary effects due to various infestations will not be discussed further.

There is no change in mineral matter under suitable commercial storage (Pomeranz 1974). In flour and to a lesser extent in wheat, phosphate content increased slightly from the decomposition of phytin by the enzyme phytase (Greaves and Hirst 1925). Sreenivasan (1939) reported that digestibility of rice improved with storage. This was attributed to a decrease in amylase activity.

The palatability (for rats) of wheat and corn decreased with storage (Jones and Gersdorff 1939, 1941); this change was attributed to gradual decrease in protein solubility. Prolonged storage decreased the *in vitro* susceptibility of the protein of barley, buckwheat, corn, millet, peas, rice and wheat to pepsin and trypsin (Salun and Nadezhnova 1971). On the other hand, Mitchell and Beadles (1949) did not detect any changes in the nutritive quality for rats of corn and wheat protein of grain stored for up to 3 yr. Similar results were obtained with rice and wheat stored under controlled conditions (Yannai and Zimmerman 1970) and with barley (Koch and Meyer 1957; Lund *et al.* 1971).

Some workers focused their attention on the changes in lysine, the limiting amino acid in wheat. Dobczynska (1966) noted decreases of 6–8% in wheat under normal storage conditions; losses were greater in grain of higher moisture content stored at higher temperatures. Losses of lysine during drying of damp grain were significant and increased with increase in air temperature.

Cereal grains can contribute to the diet significant proportions of the following vitamins: thiamine, niacin, pyridoxine, inositol, biotin and vitamin E. Yellow corn contains significant quantities of vitamin A.

There have been a few studies on the changes in vitamin content of grain during storage. Bayfield and O'Donnell (1945) showed that wheat at 12% moisture lost about 12% of its thiamine during 5 months of storage. On the other hand, the work of Fifield and Robertson (1945) showed that under ideal conditions wheat lost very little of its thiamine during storage for up to 21 yr.

Extensive losses of vitamin A in yellow corn occurred during storage (Jones *et al.* 1943). Most rapid losses occurred during the first year of storage. Changes in other vitamins in grain during storage have not been investigated.

Changes in Properties Related to Utilization Quality

Of the various uses of cereal grains for food and feed, breadmaking has received the greatest attention in grain storage studies. In general, breadmaking quality of wheat tends to improve during immediate post-harvest storage. This level of quality can be retained for relatively long periods under ideal storage conditions (Greer *et al.* 1954). Under normal commercial storage conditions, there is a gradual loss of breadmaking quality during storage after the initial period of improvement (Pomeranz 1971A, B). The more drastic the storage conditions, the greater will be the rate of loss in quality.

Improvements in breadmaking quality of wheat during immediate post-harvest storage is generally attributed to the so-called improver effect of oxygen through its oxidation of sulfhydryl groups of the endosperm proteins (Yoneyama *et al.* 1970A, B). Wheat that has been stored for prolonged periods (or under adverse conditions) has poor baking quality mainly due to the detrimental effect of free unsaturated fatty acids (which increase during storage, see above) on gluten formation. The type of changes in breadmaking quality that might occur in wheat can be interpolated from the results on flour and farina (coarse flour) obtained by Larmour *et al.* (1961). This study covered a period of 5 yr and a variety of packaging materials and storage conditions. Changes in fat acidity, fat content, farinograms, and loaf volume obtained in this study are shown in Fig. 1.5. For each set of conditions, changes in the farina were definitely lower than those in flour. Accordingly, one would expect that those changes in wheat would be slower than in farina.

The numerous studies of wheat indicate that the extent of deterioration is strongly dependent on storage conditions and infestation of the grain by microorganisms or insects. Under ideal storage conditions, wheat stored for up to 22 yr showed satisfactory breadmaking quality (Fifield and Robertson 1945).

Much less has been published on functional changes during storage of grains other than wheat. Cox *et al.* (1947) showed that artificial drying of damp corn decreased the starch yield in wet milling by 45%.

Yasumatsu and Moritaka (1964) showed that the fatty acids released in rice during storage produced an increase in amylograph viscosity and decreased the quality of cooked rice. Rice stored at room temperature was inferior in cooked form to the rice stored at 4° C (Suzuki and Matsumoto 1971). Oats develop a bitter taste during storage which apparently derives from substances formed by certain oxidizing enzymes such as peroxidase (Pomeranz 1974).

In summary, grain can undergo many biochemical, physiological and physical changes during storage. The rates at which those changes occur depend on the composition of the grain and on external factors such as relative humidity, temperature, oxygen tension and parasites. The changes can lead to significant economic losses due to deterioration of viability, nutritive quality, and utilization quality. However, the detrimental changes occur very slowly when dry grain is stored at ambient temperatures in storages that prevent infestation. Under such conditions, grain can be stored for long periods without significant deterioration.

GERMINATION

Interest in the germination of cereal seeds covers three main areas. First and of obvious importance is the role of germination in the survival of the

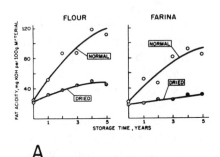

A

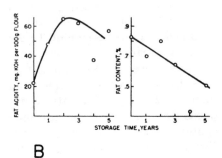

B

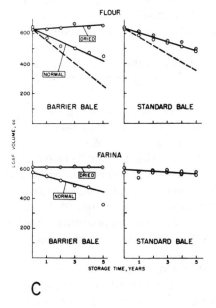

C

D

From Larmour et al. (1961)

FIG. 1.5. CHANGES DURING STORAGE OF WHEAT FLOUR AND FARINA
UNDER VARIOUS CONDITIONS

A. Fat acidity at two moisture contents stored outdoors in moisture barrier-type packages.

B. Fat acidity (left) and fat content (right) in one flour maintained at near the normal moisture level in a barrier package.

C. Loaf volume during outdoor storage for normal and dried flour and farina in barrier and standard bales (dashed line shows maximum initial loss for normal flour).

D. Farinograms of flour and farina after storage under various conditions.

species. Secondly, germination or some of the biochemical reactions involved in it are essential in malting to provide a ready source of hydrolytic enzymes and fermentable carbohydrate and thirdly, the early stages of germination are responsible for considerable damage to some crops through "weather damage." The starchy endosperm of mature cereal seed is dead tissue and represents only a food source to sustain the early development of the plant but it is in this starchy endosperm that the consequences of germination are of most significance in food and beverage processing.

Germination of seeds is normally considered from three points of view: viability, dormancy and vigor. Changes in viability and dormancy during storage were discussed in the previous section. Some of the implications of those factors in germination capacity were already considered.

Seed vigor is perhaps most usefully described as the ability to produce plants of maximum performance even under unfavorable germination conditions. Seed vigor is influenced by factors such as genotype (e.g. heterosis exhibited by hybrids), stage of seed development at harvest, storage and seed size.

The detailed physiological and biochemical events involved in cereal germination are now reasonably well understood, at least for barley (Briggs 1973). The two principal areas of the grain showing synthetic activity during early germination are the embryo including the scutellum and the aleurone while it is the starchy endosperm which is first degraded. Cell division occurs, however, only in the embryo. While all the morphological parts of the seed mentioned above are of importance for the development of a new plant, it is simpler to consider the parts separately when considering the effects of germination on seed which is to be processed for food use.

In traditional malting a high viability is essential to ensure adequate synthesis of hydrolytic enzymes by the embryo and aleurone. The embryo, after damping, produces both gibberellins and enzymes and it is believed that the gibberellins induce de novo protein synthesis in the aleurone resulting in the production of a range of hydrolases. Visual evidence of attack by these enzymes on the starchy endosperm can be seen first immediately below the scutellum with which part of the aleurone is contiguous. Not all enzymes active after germination arise from de novo protein synthesis. Phytase, for example, is believed to become activated after reaction with some material which reaches the aleurone from the embryo (Eastwood and Laidman 1971). More recent malting and brewing techniques involving the use of exogenous enzymes, gibberellins and adjuncts like starch have made total grain viability of less importance. De-embryonized grains are still capable of producing large amounts of hydrolases when gibberellins are added. Exogenous gibberellin is used even with viable grains to increase enzyme production.

The early events of germination which are sought after in the malting process, are generally undesirable where the seed is to be processed into baked goods or pasta or which is to be used for prime starch manufacture. Rice, which is normally not subjected to extensive processing, has been under considerable natural and human selection pressure so that preharvest sprouting is not normally a problem (MacKey 1976). It is with rye, durum wheat and particularly with bread wheat that so-called "weather damage" is of most significance. The increases in the levels of hydrolases, particularly alpha-amylase, which occur during the early stages of germination can cause serious problems in the production of baked goods and pasta. Processing normally requires that starch hydrolysis and its gelation on heating should be held within narrow limits and consequently excessive amounts of alpha-amylase may render the seed or its milled product unfit for food use.

There is a continuing search for new cereal genotypes, particularly wheat or species which can be crossed with wheat, with increased innate dormancy. One of the costs of development of free-threshing wheats is the loss of dormancy regulated by hulls or glumes. It is unlikely, however, that growers or processors would welcome the reintroduction of wheat or rye where the husk remained attached during harvesting. Belderok (1976) distinguished seed coat dormancy and embryo dormancy. In the former case germination does not occur or is very slow when the seeds are placed in optimal water, light and temperature conditions even though the embryo is ready for growth. Embryo dormancy is due to inhibitory factors in the embryo itself and no germination occurs when the seed coat is removed or damaged. It is assumed by most that the causes of dormancy (germination resistance) in wheat lie largely in the seed coat. This assumption is based on evidence obtained from studies on red and white grained wheats. According to Belderok (1976) no white-grained germination resistant wheats have been found although Bhatt et al. (1976) have reported on breeding white-grained spring wheat for low alpha-amylase synthesis and insensitivity to gibberellic acid. Kruger (1976) suggested three ways in which the problem of germination (sprout) damage might be attacked. Firstly, breeding directly for sprout resistance and slow alpha-amylase release is obviously the most practical approach until all the factors controlling dormancy are understood. Secondly, it may be possible selectively to inhibit such enzymes as alpha-amylase without affecting the overall quality of the cereal for processing. The final approach is to learn enough about the biochemical factors controlling dormancy and the onset of germination so that breeding for specific inhibitors may become practical.

REFERENCES

ADAMS, C. A., NOVELLIE, L. and LIEBENBERG, N.v.d.W. 1976. Biochemical properties and ultra structure of protein bodies isolated from selected cereals. Cereal Chem. *53*, 1–12.

BAILEY, C. H. and GURJAR, A. M. 1918. Respiration of stored wheat. J. Agr. Res. *12*, 685–713.

BAKER, D., NEUSTADT, M. H. and ZELENY, L. 1959. Relationships between fat acidity values and types of damage in grain. Cereal Chem. *36*, 308–311.

BAKKE, A. L. and NOECKER, N. L. 1933. The relation of moisture to respiration and heating in stored oats. Iowa Res. Bull. *165*, 320–336.

BARLOW, K. K., LEE, J. W. and VESK, M. 1974. Morphological development of storage protein bodies in wheat. *In* Mechanisms of Regulation of Plant Growth. R. L. Bielski, A. R. Ferguson and M. M. Creswell (Editors). Bull. 12, The Royal Society of New Zealand, Wellington, 793–797.

BARLOW, K. K., SIMMONDS, D. H. and KENRICK, K. G. 1973. The localization of water-soluble proteins in wheat endosperm as revealed by fluorescent antibody techniques. Experientia *29*, 229–231.

BAYFIELD, E. G. and O'DONNELL, W. W. 1945. Observations on the thiamine content of stored wheat. Food Res. *10*, 485–488.

BELDEROK, B. 1968. Seed dormancy problems in cereals. Field Crop Abstr. *21*, 203–211.

BELDEROK, B. 1976. Physiological-biochemical aspects of dormancy in wheat. Cereal Res. Commun. *4*, 133–137.

BHATT, G. M., DERERA, N. F. and McMASTER, G. J. 1976. Breeding white-grained spring wheat for low alpha-amylase synthesis and insensitivity to gibberellic acid in grain. Cereal Res. Commun. *4*, 245–249.

BRADY, C. J. and TUNG, H. F. 1975. Role of protein synthesis in senescing, detached wheat leaves. Australian J. Plant Physiol. *2*, 163–176.

BRIGGS, D. E. 1973. Hormones and carbohydrate metabolism in germinating cereal grains. *In* Biosynthesis and Its Control in Plants. B. V. Milborrow (Editor). Academic Press Inc., London, England.

BROWN, E., STANTON, T. R., WIEBE, G. A. and MARTIN, J. H. 1948. Dormancy and the effect of storage on oats, barley, and sorghum. U.S. Dep. Agr. Tech. Bull., 953.

BUSHUK, W., HWANG, P. and WRIGLEY, C. W. 1971. Proteolytic activity of maturing wheat grains. Cereal Chem. *48*, 637–639.

BUSHUK, W. and WRIGLEY, C. W. 1971. Glutenin in developing wheat grain. Cereal Chem. *48*, 448–455.

BUTTROSE, M. S. 1963. Ultrastructure of the developing wheat endosperm. Australian J. Biol. Sci. *16*, 305–317.

CABELL, C. A., DAVIS, R. E. and SAUL, R. A. 1958. Some effects of variation in drying temperature, heating time, air flow rate, and moisture content on nutritive value of field shelled corn. J. Animal Sci. *17*, 1204–1207.

CARTER, E. P. and YOUNG, G. Y. 1945. Effect of moisture content, temperature, and length of storage on the development of "sick" wheat in sealed containers. Cereal Chem. *22*, 418–428.

CASHMORE, W. H. 1942. Temperature control of farm grain driers. Agriculture (London) *3*, 144–149.

COLEMAN, D. A. and FELLOWS, H. C. 1925. Hygroscopic moisture of cereal grains and flaxseed exposed to atmospheres of different relative humidity. Cereal Chem. 2, 275–287.

COX, J. J., MacMASTERS, M. M. and RIST, C. E. 1947. Laboratory processing studies on soft corn. Paper presented at the 32nd AACC annual meeting, Abstracts Program.

CROY, L. I. and HAGEMAN, R. H. 1970. Relationship of nitrate reductase activity to grain protein production in wheat. Crop Sci. 10, 280–285.

DAFTARY, R. D. and POMERANZ, Y. 1965. Changes in lipid composition in wheat during storage deterioration. Agr. Food Chem. 13, 442–446.

DAFTARY, R. D., POMERANZ, Y. and SAUER, D. B. 1970. Changes in wheat flour damaged by mold during storage. Effects on lipid, lipoprotein, and protein. Agr. Food Chem. 18, 613–616.

DALLING, M. J., BOLAND, G. and WILSON, J. H. 1976. Relation between acid proteinase activity and redistribution of nitrogen during grain development in wheat. Australian J. Plant Physiol. 3, 721–730.

DECKARD, E. L., LAMBERT, R. J. and HAGEMAN, R. H. 1973. Nitrate reductase activity in corn leaves related to yields of grain and grain protein. Crop Sci. 13, 343–350.

DOBCZYNSKA, D. 1966. Bull. Inform. 40(4), 73–83; cited by Pomeranz (1974).

DONOVAN, G. R. and LEE, J. W. 1977. The growth of detached wheat heads in liquid culture. Plant Sci. Lett. 9, 107–113.

DONOVAN, G. R., LEE, J. W. and HILL, R. D. 1977. Compositional changes in the developing grain of high- and low-protein wheats. II. Starch and protein synthetic capacity. Cereal Chem. 54, 646–656.

EASTWOOD, D. and LAIDMAN, D. L. 1971. The mobilization of macronutrient elements in the germinating wheat grain. Phytochem. 10, 1275–1284.

EMERICK, R. J., CARLSON, C. W. and WINTERFELD, H. L. 1961. Effect of heat drying upon the nutritive value of corn. Poultry Sci. 40, 991–995.

EVERS, A. D. 1970. Development of the endosperm of wheat. Ann. Bot. 34, 547–555.

EVERS, A. D. and LINDLEY, J. 1977. The particle-size distribution in wheat endosperm starch. J. Sci. Food Agr. 28, 98–102.

FARRAND, E. A. and HINTON, J. J. C. 1974. Study of relationships between wheat protein contents of two U.K. varieties and derived flour protein contents at varying extraction rates. II. Studies by hand dissection of individual grains. Cereal Chem. 51, 66–74.

FIFIELD, C. C. and ROBERTSON, D. W. 1945. Milling, baking, and chemical properties of Marquis and Kanred wheat grown in Colorado and stored 14 to 22 years. J. Amer. Soc. Agron. 37, 233–239.

FINNEY, K. F. et al. 1962. Chemical, physical, and baking properties of preripe wheat dried at varying temperatures. Agron. J. 54, 244–247.

GLASS, R. L., PONTE, J. G., JR., CHRISTENSEN, C. M. and GEDDES, W. F. 1959. Grain storage studies. XXVIII. The influence of temperature and moisture level on the behavior of wheat stored in air or nitrogen. Cereal Chem. 36, 341–356.

GRAHAM, J. S. D., MORTON, R. K. and RAISON, J. K. 1963. Isolation and characterization of protein bodies from developing wheat endosperm. Australian J. Biol. Sci. 16, 375–383.

GREAVES, J. E. and HIRST, C. T. 1925. The influence of storage on the composition of flour. Utah Agr. Exp. Sta. Bull. 194.

GREER, E. N., JONES, C. R. and MORAN, T. 1954. The quality of flour stored for periods up to 27 years. Cereal Chem. *31*, 439–450.

HAFERKAMP, M. E., SMITH, L. and NILAN, R. A. 1953. Studies of age of seed. I. Relation of age of seed to germination and longevity. Agron. J. *45*, 434–437.

HUEBNER, F. R., DONALDSON, G. L. and WALL, J. S. 1974. Wheat glutenin subunits. II. Compositional differences. Cereal Chem. *51*, 240–249.

HUMMEL, B. C. W., CUENDET, L. S., CHRISTENSEN, C. M. and GEDDES, W. F. 1954. Grain storage studies. XII. Comparative changes in respiration, viability, and chemical composition of mold-free and mold-contaminated wheat upon storage. Cereal Chem. *31*, 143–150.

INGVERSEN, J., KØIE, B. and DOLL, H. 1973. Induced seed protein mutant of barley. Experientia *29*, 1151–1152.

IWASAKI, T. and TANI, T. 1967. Effect of oxygen concentration on deteriorative mechanisms of rice during storage. Cereal Chem. *44*, 233–237.

JENNER, C. F. 1974. An investigation of the association between the hydrolysis of sucrose and its absorption by grains of wheat. Australian J. Plant Physiol. *1*, 319–329.

JENNER, C. F. 1976. Physiological investigations on restrictions to transport of sucrose in ears of wheat. Australian J. Plant Physiol. *3*, 337–347.

JENNER, C. F. and RATHJEN, A. J. 1975. Factors regulating the accumulation of starch in ripening wheat grain. Australian J. Plant Physiol. *2*, 311–322.

JENNINGS, A. C. and MORTON, R. K. 1963. Changes in carbohydrate, protein and non-protein nitrogenous compounds of developing wheat grain. Australian J. Biol. Sci. *16*, 318–331.

JENNINGS, A. C., MORTON, R. K. and PALK, B. A. 1963. Cytological studies of protein bodies of developing wheat endosperm. Australian J. Biol. Sci. *16*, 366–374.

JONES, D. B., FRAPS, G. S., THOMAS, B. H. and ZELENY, L. 1943. The effect of storage of grains on their nutritive value. Nat. Res. Council U.S., Reprint and Circ. Ser. No. 116.

JONES, D. B. and GERSDORFF, C. E. F. 1939. The effect of storage on the proteins of seeds and their flours. Soybeans and wheat. J. Biol. Chem. *128*, XLIX-1.

JONES, D. B. and GERSDORFF, C. E. F. 1941. The effect of storage on the protein of wheat, white flour, and whole wheat flour. Cereal Chem. *18*, 417–434.

KAMINSKI, E. and BUSHUK, W. 1969. Wheat proteases. I. Separation and detection by starch gel electrophoresis. Cereal Chem. *46*, 317–324.

KENT, N. L. 1966. Sub-aleurone cells of high protein content. Cereal Chem. *43*, 585–601.

KOCH, B. A. and MEYER, J. H. 1957. Effects of storage upon nutrient value of barley grains as a source of protein. J. Nutr. *61*, 343–356.

KOZLOVA, L. T. and NEKRASOV, B. P. 1956. Changes in wheat quality during prolonged storage. Trudy Tsentr. Nauch.-Issledovatel. Lab. Glavnoe Upravlenie Gosudarst. Material. No. *4*, 60–80. (Chem. Abstr. *55*, 9715a).

KRUGER, J. E. 1976. Biochemistry of pre-harvest sprouting in cereals and practical applications in plant breeding. Cereal Res. Commun. *4*, 187–194.

LARMOUR, R. K., HULSE, J. H., ANDERSON, J. A. and DEMPSTER, C. J. 1961. Effect of package type on stored flour and farina. Cereal Sci. Today *6*, 158, 160–164.

LINKO, P. 1961. Quality of stored wheat. Simple and rapid manometric method for determining glutamic acid decarboxylase activity as quality index of wheat. Agr.

Food Chem. *9*, 310–313.

LINKO, P. 1966. Beobachtungen über Hitzeschäden bei Getreide. Getreide Mehl 16, 86–90; cited by Pomeranz (1974).

LINKO, P. and SOGN, L. 1960. Relation of viability and storage deterioration to glutamic acid decarboxylase in wheat. Cereal Chem. *37*, 489–499.

LONEY, D. P. and MEREDITH, P. 1974. Note on amylograph viscosities of wheat flours and their starches during storage. Cereal Chem. *51*, 702–705.

LUND, A., PEDERSEN, H. and SIGSGAARD, P. 1971. Storage experiments with barley at different moisture contents. J. Sci. Food Agr. *22*, 458–463.

LYNCH, B. T., GLASS, R. L. and GEDDES, W. F. 1962. Grain storage studies. XXXII. Quantitative changes occurring in the sugars of wheat deteriorating in the presence and absence of molds. Cereal Chem. *39*, 256–262.

MAC KEY, J. 1976. Seed dormancy in nature and agriculture. Cereal Res. Commun. *4*, 83.

MAC LEOD, A. M. 1967. The physiology of malting—a review. J. Inst. Brew. (London) *73*, 146–162.

MARCELLOS, H. and SINGLE, W. V. 1972. The influence of cultivar, temperature and photoperiod of post-flowering development of wheat. Australian J. Agric. Res. *23*, 533–540.

MERTZ, E. T., BATES, L. S. and NELSON, O. E. 1964. Mutant gene that changes protein composition and increases lysine content of maize endosperm. Science *145*, 279–280.

MILNER, M. and GEDDES, W. F. 1954. Respiration and heating. *In* Storage of Cereal Grains and Their Products. J. A. Anderson and A. W. Alcock (Editors). Monograph series, Vol. II. American Association of Cereal Chemists: St. Paul, MN.

MILNER, C. K. and WOODFORDE, J. 1965. The effect of heat in drying on the nutritive value of wheat for animal feed. J. Sci. Food Agr. *16*, 369–373.

MITCHELL, H. H. and BEADLES, J. R. 1949. The effect of storage on the nutritional qualities of the proteins of wheat, corn, and soybeans. J. Nutr. *39*, 463–484.

MOUREAUX, T. and LANDRY, J. 1972. The maturation of the grain of corn. Qualitative and quantitative distribution of nitrogenous substances. Physiol. Veg. (Paris) *10*, 1–18.

NELSON, O. E., MERTZ, E. T. and BATES, L. S. 1965. Second mutant gene affecting the amino acid pattern of maize endosperm proteins. Science *150*, 1469–1470.

PARPIA, H. A B. 1976. Postharvest losses—impact of their prevention on food supplies, nutrition, and development. *In* Nutrition and Agricultural Development. N. S. Scrimshaw and M. Béhar (Editors). Plenum Press, New York, NY.

PIXTON, S. W. and HILL, S. T. 1967. Long-term shortage of wheat. II. J. Sci. Food Agr. *18*, 94–98.

POMERANZ, Y. 1971A. A review of some recent studies on biochemical and functional changes in mold-damaged wheat and wheat flour. Cereal Sci. Today *16*, 119–122, 131.

POMERANZ, Y. 1971B. Biochemical and functional changes in stored cereal grains. Critical Rev. Food Technol. *2*, 45–60.

POMERANZ, Y. 1974. Biochemical, functional, and nutritive changes during storage. *In* Storage of Cereal Grains and Their Products. Y. Pomeranz (Editor). American Association of Cereal Chemists Inc., St. Paul, MN.

POPOV, N. F. and TIMOFEEV, L. I. 1933. Some data on the chemistry of wheat ripened after harvesting in storage, silos, or elevators. Sci. Inst. Cereal Res. (USSR) *11*, 59–83. (Chem. Abstr. *29*, 2607).

RAO, K. P., RAINS, D. W., QUALSET, C. O. and HUFFAKER, R. C. 1977. Nitrogen nutrition and grain protein in two spring wheat genotypes differing in nitrate reductase activity. Crop Sci. *17*, 283–286.

ROBERTS, E. H. 1972. Dormancy: a factor affecting seed survival in soil. *In* Viability of Seeds. E. H. Roberts (Editor). Syracuse University Press.

SALUN, I. P. and NADEZHNOVA, L. A. 1971. Changes in the *in vitro* digestibility of proteins contained in groats during storage. Voprosy Pitaniya *30*(3), 70–73. (Food Sci. Technol. Abstr. 10M1116).

SATO, T., TACHIBANA, T. and ITO, K. 1971. Quality change of rice during storage. II. Changes of cooking characteristics and susceptibility of cooked and steamed rice to amylase action according to their drying conditions. Nippon Jozo Kyokai Zasshi *65(3)*, 266–268. (Chem. Abstr. *75*, 150553v).

SHANNON, J. C. and DOUGHERTY, C. T. 1972. Movement of ^{14}C-labelled assimilates into kernels of *Zea mays* L. II. Invertase activity of the pedicel and placento chalazal tissues. Plant Physiol. *49*, 203–207.

SIMMONDS, D. H., BARLOW, K. K. and WRIGLEY, C. W. 1973. The biochemical basis of grain hardness in wheat. Cereal Chem. *50*, 553–562.

SINGH, R. and AXTELL, J. D. 1973. High lysine mutant gene (hl) that improves protein quality and biological value of grain sorghum. Crop Sci. *13*, 535–539.

SOFIELD, I., EVANS, L. T. and WARDLAW, I. F. 1974. The effects of temperature and light on grain filling in wheat. *In* Mechanisms of Regulation of Plant Growth. R. L. Bielski, A. R. Ferguson and M. M. Creswell (Editors). Bulletin 12, The Royal Society of New Zealand, Wellington, 909–915.

SREENIVASAN, A. 1939. Studies on quality in rice. IV. Storage changes in rice after harvest. Indian J. Agr. Sci. *9*, 208–222.

STANSFIELD, E. and COOK, W. H. 1932. The drying of wheat (2nd report). Dominion Can. Natl. Res. Council Rep. No. 25.

SUZUKI, Y. and MATSUMOTO, F. 1971. Storage temperature of rice and eating quality of cooked rice. Kaseigaku Zasshi *22*(5), 288–295. (Chem. Abstr. *75*, 150568d).

U.S. DEPARTMENT OF AGRICULTURE. 1954. Sound grain fat acidity survey—1953 crop. Agr. Marketing Service. Grain Division Publication.

YANNAI, S. and ZIMMERMANN, G. 1970. Influence of controlled storage of some staple foods on their protein nutritive value in lysine-limited diets. I. Protein nutritive value of defatted milk powder, wheat, and rice. J. Food Sci. Technol. 7, 179–184.

YASUMATSU, K. and MORITAKA, S. 1964. Fatty acid compositions of rice lipid and their changes during storage. Agr. Biol. Chem. *28*, 257–264.

YONEYAMA, T., SUZUKI, I. and MUROHASHI, J. 1970A. Natural maturing of wheat flour. I. Changes in some chemical components and in farinograph and extensigraph properties. Cereal Chem. *47*, 19–26.

YONEYAMA, T., SUZUKI, I. and MUROHASHI, M. 1970B. Natural maturing of wheat flour. II. Effect of temperature on changes in soluble SH content, and some rheological properties of doughs obtained from the flour. Cereal Chem. *47*, 27–33.

CHAPTER 2

BIOCHEMICAL CHANGES IN SOYBEANS: MATURATION, POSTHARVEST STORAGE AND PROCESSING, AND GERMINATION

JOSEPH J. RACKIS

Northern Regional Research Center
Agricultural Research Service
U.S. Department of Agriculture
Peoria, Illinois 61604

ABSTRACT

Mature soybeans as flours, concentrates, and isolates are a primary source of protein for world needs. Recent reports imply that development of acceptable foods from green, immature, and germinated soybeans may further increase their importance and versatility as a protein resource. The reliance on soybeans hinges upon continued research on problems associated with productivity, protein quality, antinutritional factors, flavor and flatulence in all soybean forms. Trypsin inhibitors and other antinutritional factors are readily destroyed by moist heat treatment, and protein efficiency ratios of the toasted immature, mature, and germinated soybeans range from 2.0–2.2 based on a value of 2.5 for casein. Depending upon variety, soybeans contain 34–49% protein and 14–25% oil. Methionine content varies from 1.0–1.9 g/16g N.

The direct use of green immature soybeans holds promise because their protein content is twice that of other food legumes. Long cooking times are required to improve texture and palatability of mature soybeans whereas immature soybeans cook to a tender nutlike texture much more quickly. The primary difficulty with immature soybeans is the need for special harvesting and shelling equipment to prevent damage. Mature soybeans are practically devoid of ascorbic acid and β-carotene, whereas, in immature and germinated soybeans, ascorbic acid and β-carotene content is increased severalfold; however, most of the ascorbic acid is destroyed by cooking needed to increase palatability.

Raffinose and stachyose in mature soybeans cause flatulence. These oligosaccharides are completely hydrolyzed during germination for 4–6 days. In immature soybeans, raffinose and stachyose content is very low and starch content is high. Regardless of variety or type, the organoleptic qualities of immature, mature, and germinated soybeans are very low because of the presence of grassy/beany and bitter flavors. Special processes are needed to improve flavor. Accumulated breakage during handling of soybeans from harvest time, postharvest storage, and subsequent shipment to domestic and foreign processors reduces quality of the oil and protein, affects flavor qualities and lysine availability.

It is important to minimize loss of lysine availability in soy proteins because these proteins are essential ingredients for enhancing the nutritive value of cereals deficient in lysine.

INTRODUCTION

The status and future potential of various protein resources in order to expand their availability have been evaluated (National Science Foundation 1976). Mature soybeans processed into flours, concentrates, and isolates are a primary international protein resource. Development of acceptable food products from green immature and germinated soybeans may be another way to further increase the versatility of soybeans. The reliance on soybeans for production of animal protein and for increasing use in human foods hinges around continued research on problems associated with productivity and acceptance in food systems. These problems and biochemical changes that occur in soybeans during maturation, postharvest storage, and subsequent germination will be evaluated with respect to nutritional quality, elimination of physiological and biological substances, and improvement of organoleptic quality.

Thermostable and heat-labile substances in soybeans cause flatulence, inhibit growth, enlarge the pancreas, and reduce mineral bioavailability. Weakly estrogenic and goitrogenic constituents are also present. As a result, precise control of moist-heat treatment and special extraction processing technology is required to improve protein quality. Preparation and use of fermented soybean products have been reviewed elsewhere (Hesseltine and Wang 1972) and will not be discussed here.

POTENTIAL, PROBLEMS AND PRODUCTS

Projections

In discussing world protein resources, Milner (1975) has cited U.S. Department of Agriculture projections that estimated demand for U.S. soybeans will increase 73% by 1985 over the 1970–1972 production of about 35 million metric tons. Projected percentage increases for other U.S. protein crops during the same time span are: peanuts, 58; chickens, 39;

beef, 33; corn, 30. Soybean production may increase to 85 million metric tons by the year 2000. Anton (1975) and Johnson (1976) predict that 1974 production of 0.71 and 1.13 billion pounds of edible soy protein products, respectively, could more than double by 1980 and 1985, respectively. (Table 2.1).

TABLE 2.1
ESTIMATED PRODUCTION OF EDIBLE SOY PROTEIN PRODUCTS IN 1974
AND FUTURE PROJECTIONS IN MILLIONS OF POUNDS

Products	1974		1980[1]	1985[2]
	I[1]	II[2]		
Soy flour and grits	300	900	600	2000
Textured soy proteins	160	100	1080	450
Soy protein concentrate	175	70	350	600
Soy protein isolate	75	60	400	450
Soy milk-type products	—	nil	—	200
Total	710	1130	2430	3700

[1]Anton (1975).
[2]Johnson (1976).

Problems

Thus, the reliance on soybeans as a primary protein resource may continue to increase provided that research on bioagronomic constraints on productivity and on acceptance problems with soy protein foods is intensified (National Science Foundation 1976; National Soybean Research Coordinating Committee 1977; Hill 1976). It is concluded that first priority in utilization research on soybeans should be assigned to edible protein products. Some of the more important problems that limit the potential use of soybeans for food are summarized in Table 2.2.

The mature soybean itself is rarely eaten; it does not fit into the general category of a ready-to-eat food because of the long cooking times required to improve texture and palatability. Long cooking times increase energy consumption and decrease the availability of nutrients. Special processes are also needed to remove or modify physiological-biological factors that affect organoleptic and nutritional quality (Rackis 1974; Rackis *et al.* 1975B) and to modify functional properties (Rackis 1977). In addition, more research is needed to develop control data for monitoring quality factors during all phases of postharvesting, preprocessing, postprocessing, and subsequent storage of manufactured soy products.

TABLE 2.2
SOYBEAN RESEARCH PROBLEMS

Production	Flavor
Detection methods for soy in foods	Color
Labeling requirements	Flatulence
Antinutritional factors	Methionine-cystine content
Functionality in foods	Cookability
Post-harvest losses in quality	

Soybean Food Products

The use of soybeans at the household level has been limited to certain countries, mainly in Southeast Asia, where up to 77g of soybeans are consumed per capita per day (Protein Advisory Group 1973). In Japan, some 12–13% of the dietary intake of protein per capita is derived from soybeans. To achieve such a relatively high intake of soy protein, a great deal of technology is needed to convert whole soybeans into a variety of nonfermented and fermented foods. The more traditional foods of Southeast Asia include: soy milk, tofu, yuba, kinako, miso, shoyu, natto, sufu, tempeh, and fermented milks (Smith and Circle 1972).

Soybeans are also valued for their oil and, as a consequence, major processors of soybeans have built large solvent extraction plants for the production of soybean oil and defatted soybean meal and flour. The use of soybean oil is very important to the potential use of soy protein products, in that the more economic return from the oil, the less expensive protein needs to be to make soybeans a profitable crop. Of the approximately 18–19 million metric tons of defatted soybean meal produced annually in the United States, only about 3% is used directly in human food.

Soy flour, concentrates, and isolates are the basic forms of processed protein products, which are further modified into a large number of other products for specific nutritional and functional uses (Smith and Circle 1972) and into meatlike extenders and analogs (American Soybean Association 1974). The relative cost of soy protein compared with that of animal protein indicates that soy protein will continue to expand its present markets and penetrate new ones (U.S. Department of Agriculture 1976).

Research has advanced soy protein technology to a sophisticated level. Some other workers visualize that soybeans as a vegetable could be an even more universal and versatile food resource. Smith and Van Duyne (1951) listed soybean varieties most suitable for use as a fresh vegetable or for freezing and canning. More recently, there has been renewed interest in dried whole soybeans (Dougherty and Knapp 1972); green soybeans

(Collins and Sanders 1973; Bates and Matthews 1975); pickled soybeans (Bates *et al.* 1977); and sprouts (Fordham *et al.* 1975). Home use of soybeans has been described (Mueller *et al.* 1974; Sinclair *et al.* 1974). Green immature soybeans, known as Endamame, and sprouted soybeans are important vegetables in the Orient but have not found ready acceptance by most other people.

Dal, a Hindi term for both dry dehulled split pulses and a gruel type of food made from the pulses, are eaten in many countries. A process to produce a soybean dal similar in appearance and texture to indigenous dals, but with a much higher protein content, has been developed (Spata *et al.* 1974). Cost of production was calculated to be competitive with dals prepared from several varieties of Indian pulses.

CHARACTERISTICS OF MATURE SOYBEANS

Soybean Types

Garden-type varieties are soybeans which the people in the Orient use during the summer as green beans for the table. The garden types, which are also referred to as vegetable soybeans, were the seeds first introduced into the United States (Smith and Van Duyne 1951). Eventually, through subsequent breeding, field-type varieties were developed for processing into oil and meal. At the present time, little differentiation can be made between garden and field varieties on the basis of chemical composition (Gupta and Deodhar 1975), protein quality, and digestibility (Gupta *et al.* 1976). Generally, the pods and seeds of garden varieties are much larger and easier to shell in the green immature stage. Also, the seed coat is looser and cracks more easily, which probably accounts for their increased cookability compared to the field varieties. Regardless of variety and type, raw soybeans have grassy/beany and bitter flavors at about the same intensity level, but some workers claim that a few garden varieties have somewhat better organoleptic qualities after cooking.

Cookability

Certain varieties of soybeans, primarily the vegetable type, were found to cook to a more "tender, nutlike" texture more quickly (U.S. Department of Agriculture 1971). Two vegetable varieties of soybeans required shorter cooking times than did field-type beans (boiling time: 80 min vs 115 min) based on subjective evaluations of texture and flavor (Perry *et al.* 1976). Boiling times were decreased substantially for all varieties when the beans were soaked and cooked in sodium bicarbonate solutions. The recommended amounts of baking soda varies between 0.07% (Perry *et al.* 1976; U.S. Department of Agriculture 1971) and 0.5% (Nelson *et al.* 1976). Cooking times were reduced by more than 50% when soaked soybeans were fried in oil prior to boiling (Perry *et al.* 1976). Soybeans cooked (30–60 min of boiling) by this latter procedure had the highest flavor scores. Even shorter cooking times can be achieved with a patented Hydravac process (Rockland and Metzler 1967; Rockland *et al.* 1967). The process consists of intermittent vacuum treatment to facilitate hydration and infusion of salt mixtures into whole dry beans.

Oil and Protein

Oil and protein content and fatty acid composition of the oil in soybeans are affected by environmental conditions and genetic influences. Soybeans grown under warmer temperatures tend to produce beans with a higher oil and protein content, a lower degree of polyunsaturation, and with lesser amounts of linoleic and linolenic fatty acids (Smith and Circle 1972; Chapman *et al.* 1976). Results of environmental and genetic influences on oil content and fatty composition of oil in 24 soybean genotypes grown at two locations are given in Table 2.3.

A summary of the range in oil, protein, methionine and cystine content of several soybean varieties and strains is presented in Table 2.4. When the soybean breeding program began in the United States, the high price of oil encouraged breeders to develop high oil varieties. This, in turn, has made it possible to formulate low-cost animal rations with defatted soybean meal. From 1947–1973, the meal has accounted for 47–67% of the total value of a bushel of soybeans processed (Horan 1976). Plant breeders should consider the development of high-protein soybeans because: (a) there is a need for more protein and (b) there are ample supplies of vegetable oil owing to dramatic increases in palm oil as well as increases in other oilseeds.

Essential amino acid composition of various soybean meal fractions is given in Table 2.5. A valuable characteristic of all soy protein products is that they have a much higher lysine content than most plant proteins. The amino acid composition of many different varieties and strains of soybeans has been evaluated, particularly in regard to methionine, the first limiting amino acid of soybean protein; but, thus far, few significant differences have been found (Krober 1956; Kakade *et al.* 1972). There is an indication that methionine content in the seed increases proportionately with protein content (Krober and Cartter 1966). Tryptophan content varies widely (Gupta *et al.* 1976). Depending upon variety, yield of soy protein isolate ranges from 34–56g per 100g dehulled, defatted soy flakes (Smith *et al.* 1966). Generally, the amino acid availability in processed soy flour is high

TABLE 2.3

RANGE IN OIL AND FATTY ACID COMPOSITION OF 24 SOYBEAN GENOTYPES GROWN AT TWO LOCATIONS[1]

Soybean Genotype	Number in Group	Oil (%)	Fatty Acid Composition (% of Total Fatty Acid)				
			Palmitic	Stearic	Oleic	Linoleic	Linolenic
Group IV-S	12						
Tifton, Georgia		20.9–24.2	9.7–12.2	3.0–4.8	18.4–40.3	42.5–58.9	3.4–8.0
Blairsville, Georgia		19.0–23.0	9.6–11.7	3.5–4.9	20.7–36.8	45.3–57.3	3.4–7.6
Mean value		21.4	10.4	4.3	26.8	52.1	5.8
Group V	12						
Tifton, Georgia		21.4–24.0	10.8–12.8	2.7–3.9	18.9–31.8	48.9–58.6	4.4–7.3
Blairsville, Georgia		20.4–31.4	10.5–12.7	2.9–4.1	18.0–27.4	51.1–58.3	7.2–8.5
Mean value		22.7	11.5	3.4	23.5	54.7	6.9

[1]Dry-basis, Chapman et al. (1976).

TABLE 2.4
OIL, PROTEIN, METHIONINE AND CYSTINE CONTENT OF SEVERAL
SOYBEAN VARIETIES AND STRAINS[1]

Item	Range in Values (Dry Wt Basis)
Oil, %	14–25
Protein (N × 6.25), %	34–49
Methionine, g/16g N	1.0–1.9
Cystine, g/16g N	1.6–3.5
Methionine + cystine, g/16g N	3.1–4.3

[1]Source: Data compiled from the following: Krober (1956); Krober and Cartter (1966); Smith and Circle (1972); Kakade et al. (1972); Deodhar et al. (1973); Chapman et al. (1976); Hill (1976); Kapoor et al. (1975); Gupta and Deodhar (1975).

and remains fairly constant. Burgos et al. (1973) suggest that difficulties experienced in rupturing cell structures in some soybean varieties can affect amino acid availability.

Other Constituents

Vitamin and mineral content of all kinds of soy protein products has been summarized (Liener 1972; American Soybean Association 1974). The wide variation in the values clearly indicates that better methods are needed to obtain more reliable data, particularly in different soybean varieties. Vegetable-type soybeans may contain more thiamine than the field type. However, 50–70% of the thiamine is destroyed by various cooking methods (Perry et al. 1976). Changes in chemical composition of soybeans during maturation and germination will be discussed in appropriate sections below.

Protein Quality

In the raw form, nutritive value of soybeans and of protein products prepared from them is very low. By live steam treatment, a process referred to as toasting, nutritive value of soy flour can be raised to values nearly equal to those of meat and milk (Fig. 2.1 and 2.2). Proper heat treatment is an absolute requirement for all soy protein products if essential nutrients are to be utilized maximally (Rackis 1974; Rackis et al. 1975B). Heat-labile factors in soybeans are listed in Table 2.6. As shown in Table 2.7, associated with the increase in nutritive value is the inactivation of trypsin inhibitors (TI), elimination of pancreatic hypertrophy, and the conversion of raw refractory proteins to more readily digestible forms

TABLE 2.5
ESSENTIAL AMINO ACID COMPOSITION OF SOYBEAN MEAL FRACTIONS[1]

Amino Acid	Whole Meal	Residue	Acid-precipitated Protein	Whey Protein	Hulls	Hypocotyl Meal	Acid-Precipitated Protein of Hypocotyl
			(Grams of amino acid/16g N)				
Arginine	8.42	7.44	9.00	6.64	4.38	8.32	6.38
Histidine	2.55	2.70	2.83	3.25	2.54	2.60	2.65
Lysine	6.86	6.14	5.72	8.66	7.13	7.45	7.80
Tyrosine	3.90	3.30	4.64	4.67	4.66	3.48	3.78
Tryptophan	1.28	—	1.01	1.28	—	—	—
Phenylalanine	5.01	5.24	5.94	4.46	3.21	3.88	4.22
Cystine	1.58	0.71	1.00	1.82	1.66	1.24	—
Methionine	1.56	1.63	1.33	1.92	0.82	1.72	1.79
Threonine	4.31	4.67	3.76	6.18	3.66	4.00	3.82
Leucine	7.72	8.91	7.91	7.74	5.93	6.62	7.22
Isoleucine	5.10	6.02	5.03	5.06	3.80	4.11	4.53
Valine	5.38	6.37	5.18	6.19	4.55	4.82	5.28

[1]Rackis et al. (1961).

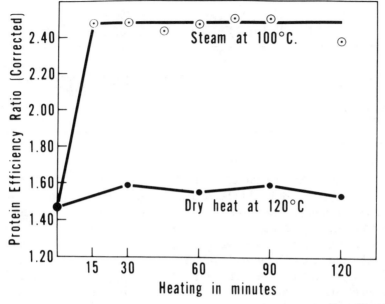

From Rackis (1974)

FIG. 2.1.   EFFECT OF TYPE AND EXTENT OF HEATING ON NUTRITIONAL
VALUE OF SOYBEAN MEAL

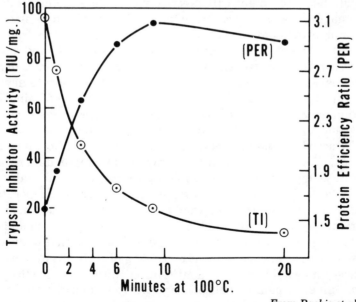

From Rackis et al. (1975A)

FIG. 2.2.   EFFECT OF STEAMING ON TRYPSIN INHIBITOR ACTIVITY (TI)
AND PROTEIN EFFICIENCY RATIO (PER) OF SOY MEAL

TABLE 2.6
HEAT-LABILE FACTORS IN SOYBEANS

Property	
Inhibit growth	Stimulate pancreatic enzyme secretion
Reduce protein digestibility	Stimulate gallbladder activity
Increase sulfur amino acid requirements	Reduce metabolizable energy
Enlarge pancreas	Inhibit proteolysis

(Rackis *et al.* 1975A; Kakade *et al.* 1973). This subject is discussed in greater detail elsewhere in this book. Kakade *et al.* (1972) analyzed over 100 varieties and experimental strains of soybeans to determine whether breeding can eliminate such antinutritional factors. Results showed that raw flours prepared from these varieties had very low protein efficiency ratios (PER) and caused pancreatic hypertrophy. As indicated in Table 2.8, PER of all soybean varieties were markedly improved after autoclaving. PER values ranged from 78–106% of that for casein. Pancreas weights of the rats fed heated soybeans were not significantly different from those receiving casein. There was little correlation between PER and level of sulfur amino acids.

Although heat treatment will effectively eliminate the heat labile antinutritional factors, careful control of processing conditions is essential to prevent both nutritional and functional damage to the protein from excessive heat treatment. Overcooking, browning reactions, and high pH treatment can destroy essential amino acids and vitamins and can reduce mineral bioavailability. Gould and MacGregor (1977) have reviewed the relationship of alkali-treated proteins and the formation of lysinoalanine to nutritional quality. The PER values of soy beverages drop sharply when processed for 10 min at 121° C at pH above 8, conditions that readily destroy TI activity (Robinson *et al.* 1971). Labuza (1972) points out that determination of reaction kinetics can provide meaningful data regarding proper processing conditions that result in maximum protein quality of soy milks and other products. For example, an increase in temperature to double the rate of TI destruction would increase the destruction of essential nutrients by four- to five-fold. Therefore, investigators should determine rate constants of various deteriorative reactions that occur during postharvest storage and subsequent processing of soybeans into food products. These data would result in identifying conditions needed to optimize nutritive value, calculate shelf-life, and provide information for nutritional labeling. Processing that maximizes nutrient retention would, in turn, increase our protein resources.

TABLE 2.7

NUTRITIVE VALUE OF DEFATTED SOY FLOUR CONTAINING GRADED LEVELS OF TRYPSIN INHIBITOR (TI)[1]
WHEN FED TO RATS

Diet No.	Dietary Protein[2]	TI Content mg/100g diet	Mean Body Weight (g) ± std. dev.	PER[3]	Nitrogen Digestibility[4] (%)	Pancreas Weight ± std. Dev. g/100 GBW[5]
13	Casein (0)	0	157 ± 16ab[6]	2.50	92	0.48 ± 0.03c
14	Soy (0)	1001	84 ± 4f	1.13	74	0.68 ± 0.11a
15	Soy (1)	774	94 ± 8ef	1.35	78	0.58 ± 0.01b
16	Soy (3)	464	123 ± 5d	1.75	77	0.51 ± 0.06c
17	Soy (6)	288	141 ± 12c	2.07	83	0.52 ± 0.04c
18	Soy (9)	212	146 ± 11bc	2.19	84	0.48 ± 0.06c
19	Soy (20)	104	139 ± 13c	2.08	83	0.49 ± 0.05c
LSD[7]			11	0.17		0.06

[1]Rackis et al. (1975A).
[2]Time (min) of heat treatment at 100° C is given in parenthesis.
[3]PER = Protein efficiency ratio corrected on a basis of a PER = 2.50 for casein.
[4]Digestibility = intake-fecal nitrogen/intake × 100.
[5]GBW = grams body weight.
[6]Letters not in common denote statistical significance (P<0.05).
[7]LSD = Least significant difference at the 95% confidence level.

TABLE 2.8
EFFECT OF AUTOCLAVING ON NUTRITIVE VALUE AND THE PANCREAS
WEIGHTS OF RATS FED DIFFERENT VARIETIES AND STRAINS OF
SOYBEANS[1]

Sample	Protein (%)	Total s-Amino Acids (g/16g N)	PER[2] Raw	Heated	Pancreas Weights[3] Raw	Heated[4]
Disoy	39.6	3.1	1.47	2.66	E	N
Provar	41.2	3.4	1.60	2.20	E	N
PI 15319	36.6	3.6	0.88	2.46	E	N
Harck	39.1	3.9	1.21	2.32	E	N
PI 153206	37.5	4.2	1.21	1.95	E	N

[1]Kakade et al. (1972).
[2]Protein efficiency ratio corrected on a basis of a PER = 2.50 for casein.
[3]E = enlarged; N = normal compared to casein.
[4]Autoclaved at 15 lb/in.^{2} (120° C) for 30 min.

COMPOSITIONAL CHANGES DURING MATURATION

The objective of this section is to describe the changes in content of protein, oil and fatty acids as well as other nutrients in developing soybean seeds. The importance of immature soybeans as a protein source, their nutritional and antinutritional properties, flavor, and flatulence qualities in relation to soy protein products prepared from mature soybeans will be discussed in the appropriate sections.

Degree of Maturity

As shown in Table 2.9, fresh weight, percent dry matter, along with days after flowering can be used as indices of maturity. Developing soybeans reach maximum fresh weight when the seeds are still green and the pods begin to yellow. Bates et al. (1977) refer to seeds at this stage of development as green-mature soybeans. The percentage of oil and protein and the acid fatty composition reach nearly maximum values at the green-mature stage (Rubel et al. 1972). During the final stages of maturation, there is a rapid loss of moisture and the seeds become hard to chew because of a tough buff-brown or yellow seedcoat. Some soybean varieties maintain a green color even at full maturity. The primary problem in harvesting green immature soybeans for use as a green vegetable is the difficulty of removing the tender green seeds from the tough pod. Mechanical equipment for harvesting and shelling green soybeans has been devised (Collins and McCarty 1969; Collins et al. 1971). Cytological and physiological changes occurring in soybeans during maturation have been described (Bils and Howell 1963).

TABLE 2.9
FRESH WEIGHT, DRY MATTER AND COLOR CHARACTERISTICS OF
MATURING SOYBEANS (1969 CROP YEAR)[1]

Days after Flowering[2]	Average Fresh Wt. (mg/seed)	Dry Matter (%)	Color of beans	
			Pods	Seeds
22	30[3]	16.0	Green	Green
24	59	22.2	Green	Green
27	131	23.0	Green	Green
29	220	23.2	Green	Green
29	295	27.6	Green	Green
31	313	28.0	Green	Green
35	384	30.0	Green	Green
40	498	32.4	Green	Green
44	568[4]	39.0	Yellow	Light-green
49	523	42.7	Yellow	Yellow-green
52	440	49.6	Brown	Yellow
55	331	71.2	Brown	Buff-brown
59[5]	253	84.1	Brown	Buff-brown
64[5]	209	91.6	Brown	Buff-brown

[1]Rackis *et al.* (1972).
[2]Flowering date, July 14, 1969; Hawkeye soybeans.
[3]Range in weight within ± 5%.
[4]Maximum fresh weight occurs during the yellowing stage.
[5]Mature beans after harvest.

Lipid Composition

Lipids undergo virtually complete transformation in composition during maturation (Roehm and Privett 1970). As shown in Table 2.10, about 46 days after flowering the fatty acid composition of oil from green immature soybeans is the same as that of oil extracted from mature soybeans. During the final days of maturation, oil content increased from 20.7–23.6%. At 63 days after flowering, the maximum fresh weight stage, 90% of the total oil found in mature beans was synthesized. These results agree closely with those of Singh and Privett (1970) and Privett *et al.* (1973). There are no significant differences in the saturate:polysaturate fatty acid ratio and nutritive value of soybean oil in either green-mature or mature soybeans. The major lipid groups that increase during maturation include neutral lipid, phospholipid, and glycolipid (Privett *et al.* 1973).

Protein Composition

As shown in Table 2.11, when Acme soybeans reached maximum fresh weight at 50 days after flowering (yellowing stage), only 61% of the protein found in mature seeds had been synthesized. On a dry weight basis,

TABLE 2.10

OIL AND FATTY ACID COMPOSITION OF HAROSOY 63 SOYBEANS DURING MATURATION[1]

Days after Flowering	Weight (mg/seed)		Oil[2] (%)	mg/Seed	Oil Composition (%)				
	Fresh	Dry			Palmitic	Stearic	Oleic	Linoleic	Linolenic
24	5.5	0.9	3.5	0.03	19.0	6.2	7.5	35.0	30.0
32	146.0	30.8	15.5	4.8	14.5	3.2	24.0	46.2	12.1
39	214.1	64.8	19.5	12.6	10.1	2.8	36.3	51.0	9.2
46	324.0	100.5	20.1	20.7	10.6	3.2	25.8	52.7	7.7
63	452.0	179.0	22.3	39.9	10.4	3.1	26.1	54.1	6.3
72 (maturity)	311.8	186.6	23.6	44.0	10.4	2.9	26.4	54.1	5.8

[1]Rubel et al. (1972).
[2]Dry basis.

TABLE 2.11

NITROGEN COMPOSITION OF ACME SOYBEANS DURING MATURATION[1]

Days After Flowering	Weight (mg/seed)		Protein[2]	
	Fresh	Dry	(%)	mg/Seed
25	22.2	3.8	32.2	1.2
32	124.2	27.3	31.8	8.7
41	266.3	75.7	34.1	25.6
50	370.8	122.2	34.8	42.5
60	359.4	165.6	36.0	59.6
74	238.5	200.6	35.0	70.2
(maturity)				

[1]Rubei et al. (1972).
[2]Dry basis.

percentage of crude protein increased slightly in going from the green mature to dry mature soybeans. There are other reports of small increases in crude protein and lipid during the final stages of maturation (Bates et al. 1977; Deodhar et al. 1973). The Chippewa and Harosoy 63 varieties analyzed for nonprotein and protein nitrogen showed the same protein trends as reported for Acme (see Table 2.11).

Although the percent of oil and protein content in green mature soybeans and in those harvested at full maturity is nearly comparable, the yield of oil and particularly of protein per acre would be greatly increased if soybeans were harvested at full maturity. Therefore, mature soybeans would be a much better protein resource.

Other Constituents

Carbohydrates.—During the development of the soybean, large amounts of starch accumulate and reach a maximum while in the green-mature stage. During subsequent development to full maturity, starch practically disappears and there is a corresponding increase in oligosaccharides and polysaccharides (Bils and Howell 1963). Mature soybeans contain about 0.5% starch (Boonvisut and Whitaker 1976).

The carbohydrate content of dehulled, defatted soybean meal is approximately 30%. The carbohydrate constituents are given in Table 2.12. The actual percentage of total carbohydrate which is available for energy ranges from 14% in chicks to 40% in rats. In the absence of conclusive data in humans, the latter figure is taken by the Food and Agriculture Organization of the United Nations to be the digestibility of carbohydrates in soybeans. On this basis, the caloric value becomes 1.68 cal per g of carbohydrate, rather than the factor of 4 cal per g for starch and other highly digestible sugars.

TABLE 2.12
CARBOHYDRATE CONTENT AND CARBOHYDRATE CALORIC VALUE OF
DEHULLED, DEFATTED SOYBEAN MEAL

Constituent	Meal (%)
Polysaccharide content, total[1]	15–18
Acidic polysaccharides	8–10
Arabinogalactan	5
Cellulosic material	1–2
Starch[2]	0.5
Oligosaccharide content, total[3]	15
Sucrose	6–8
Stachyose	4–5
Raffinose	1–2
Verbascose	Trace
Caloric value[4]	1.68 cal/g

[1]Aspinall et al. (1967).
[2]Boonvisut and Whitaker (1976).
[3]Kawamura (1967).
[4]Liener (1972).

Vitamins.—Ascorbic acid and β-carotene content of green mature soybeans, hand harvested when the pods were green and full size and the lower leaves were beginning to yellow, is shown in Table 2.13. There was a 2.5- and 1.8-fold variation of these two nutrients, respectively, over the entire season, representing five green mature harvests of comparable maturities. Changes in ascorbic acid and β-carotene during maturation and subsequent sprouting is illustrated in Figure 2.3. Ascorbic acid content of green mature, dry mature, and sprouted soybeans was 30, 2 and 11 mg/100g, respectively; β-carotene content was 0.35, 0.12 and 0.2 mg/100g, respectively. Deodhar et al. (1973) report that green soybeans grown in India and harvested about 2 weeks before maturity contained 1.52–6.25 mg ascorbic acid per 100g seeds. However, 50–70% of ascorbic acid is destroyed by cooking and canning, whereas β-carotene losses were only 10–20% (Bates and Matthews 1975). Tocopherol and sterols increase and then decrease during maturation (Vorob'ev 1967).

BIOCHEMICAL CHANGES OCCURRING DURING GERMINATION

General Aspects

Food legumes are probably the most important and economical source of protein that can be consumed directly by humans. The Protein Advisory Group of the United Nations (1973), now referred to as the Protein Calorie

TABLE 2.13
ASCORBIC ACID AND β-CAROTENE CONTENT OF TWO
VEGETABLE-TYPE SOYBEANS FROM FIVE HARVESTS AT THE GREEN
MATURE STAGE[1]

Variety	Ascorbic acid (mg/100 g)		B-Carotene (mg/100 g)	
	Fresh Wt	Dry Wt	Fresh Wt	Dry Wt
Early green				
Range	14.1–31.6	—	0.21–0.35	—
Mean[2]	30.0	100.0	0.29	0.97
Verde				
Range	18.3–45.9	—	0.20–0.32	—
Mean[2]	27.0	90	0.27	0.90

[1]Bates and Matthews (1975).
[2]Mean value on a dry basis was calculated by assuming a moisture content of 70% fresh green mature soybeans (see Fig. 2.3).

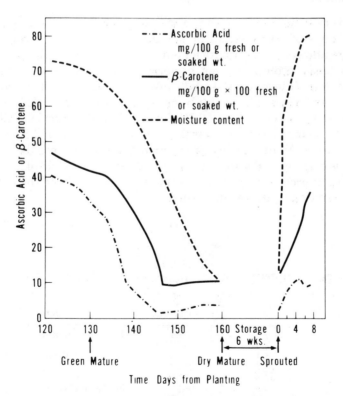

FIG. 2.3.   CHANGES IN ASCORBIC ACID, β-CAROTENE, AND MOISTURE
CONTENT OF SOYBEANS DURING MATURATION AND GERMINATION

Advisory Group, places great emphasis on food legumes, including soybeans, for use as a fresh low-cost vegetable for home use. Germinated legumes (sprouts) constitute a good portion of the total consumption of food legumes in Asia and the Orient. Sprouted soybeans, also referred to as Moyami, were developed as food by the Chinese centuries ago, but their acceptance is much lower than for sprouts prepared from other seeds.

After several years of neglect, sprouts, as well as green-mature soybeans, are again receiving attention because of the probability that flavor, nutritive value, and protein digestibility may be improved. Commercial equipment for preparing bean sprouts has been described (Chen 1970). Other methodology for sprouting soybeans has been reported (Anon. 1974; Fordham *et al.* 1975; Smith and Van Duyne 1951; Whyte 1973). Studies on the nutritive value of sprouted soybeans in relation to green-mature soybeans and processed soy protein products will be discussed below.

Protein Content

There is very little change in amino acid composition of soybeans during the first 5 days of germination or sprouting. For use as a vegetable, soybeans are allowed to germinate for only 4–6 days. When calculated on a dry weight basis, protein content of sprouted soybeans was somewhat higher than that of the original seeds (Table 2.14) because of the removal of the seed coat and a loss of leachable nonprotein constituents. McKinney *et al.* (1958) reported a weight loss of 0.7% during the soaking period. After 48 hr germination, dry matter loss was 0.8%, with a total loss of 2.6% after 6 days germination. Additional losses occur when sprouts are cooked and the cook water is discarded. Fordham *et al.* (1975) reported a yield of only 60% sprouts; in their process only the hypocotyl was recovered and the remainder of the sprouted bean was discarded.

TABLE 2.14
PROTEIN CONTENT OF SOYBEAN SEEDS AND SPROUTS

Sample	Moisture (%)	Protein (N × 6.25) (%) Dry Basis	Reference
Seed	8.0	39.4	Bates *et al.* (1977)
Sprouts	72.6	41.7	
Seed	8.4	41.9	Kylen and McCready (1975)
Sprouts (raw)	73.2	44.8	
Sprouts (cooked)	67.2	39.9	
Seed	5.7	45.3	Fordham *et al.* (1975)
Sprouts (Hypocotyl only)	92.3	49.4	

Proteolytic activity increases markedly during the initial stages of germination and then decreases. However, as protein hydrolysis continues, there is a compensatory increase in free amino acids and peptides so that crude protein content (N × 6.25) remains essentially constant during the first 4–5 days of germination (Adjei-Twum *et al*. 1976). Catsimpoolas *et al*. (1968A, B), using disc and immuno electrophoresis techniques, observed that the major reserve proteins in germinating soybeans were being degraded at a slow rate. Nonprotein nitrogenous constituents also increase during germination (Thapar *et al*. 1974). That germination does not improve the biological value of the proteins in soybeans will be discussed in another section.

Carbohydrates and Lipids

Changes in carbohydrate constituents during germination will be discussed in the section dealing with the problem of flatulence. Adjei-Twum *et al*. (1976) report lipid reserve decreased 26% after 3 days and decreased slightly thereafter.

Minerals and Vitamins

As shown in Tables 2.15 and 2.16, only limited data are available concerning the effect of sprouting on the mineral and vitamin content of soybeans. The variability of calcium levels in sprouted soybeans is probably due to the absorption of calcium from tap water and the frequent use of chlorinated lime in the sprouting medium. Manganese content of hypocotyl sprouts was greatly increased.

As shown previously in Figure 2.3 and Table 2.16, ascorbic acid and β-carotene content increased during sprouting; however, more than 65% of the ascorbic acid in soybean sprouts was destroyed after 10 min boiling (Bates and Matthews 1975). Therefore, it is questionable whether soybean sprouts are a good source of ascorbic acid. The data compiled in Table 2.16 from a number of sources also reflect an incomplete assessment of the value of soybean sprouts as a source of vitamins. A wide range of values has been reported. Except for increases in riboflavin and pyridoxine, changes in the content of the other B vitamins during sprouting were small. Vitamin content was greatly increased in sprouts prepared by the procedure of Fordham *et al*. (1975) (see footnote d, Table 2.15). However, unless some food use is found for the large amount of material that was discarded, this method has little practical significance.

TABLE 2.15

CONTENT OF SELECTED MINERALS IN MATURE SOYBEANS AND SPROUTS

Soybean Product	Moisture (%)	Ash (%)	Mineral content, mg/100g (dry basis)				
			Calcium	Iron	Zinc	Magnesium	Manganese
Seeds[1]	8.4	5.3	240	1.6	6.8	—	—
Sprouts							
Raw	73.2	11.2	280	1.5	6.0	—	—
Cooked[2]	67.2	9.8	249	1.2	6.4	—	—
Mature beans[3]	—	—	160–470	9–15	3.7	220–240	3.2
Sprouts	—	—	400	10	—	—	—
Seeds[4]	5.7	4.7	337	10.1	—	212	19.3
Sprouts	92.3	5.2	416	7.8	—	195	132.5

[1]Kylen and McCready (1975).
[2]Stir fry technique for 2 min.
[3]Liener (1972), range of values compiled from several other sources.
[4]Fordham et al. (1975), only the hypocotyl was analyzed; the remainder of the germinated seed was discarded; yield of sprouts was only 60%.

TABLE 2.16

VITAMIN CONTENT OF MATURE SOYBEANS AND SPROUTS

Soybean Product	Moisture (%)	Content, mg/100g (dry basis)						
		Thiamine	Riboflavin	Niacin	Pantothenic acid	Pyridoxine	Biotin	Ascorbic acid
Seeds[1]	8.4	1.30	0.25	3.3	—	—	—	—
Sprouts[1]								
Raw	73.2	1.19	0.60	4.1	—	—	—	44.8
Cooked	67.2	1.28	0.58	3.4	—	—	—	36.6
Mature beans[2]	—	1.10–1.75	0.23	2.0–2.6	1.20	0.64	0.06	20.0
Sprouts[2]	—	1.19–2.19	0.48–0.70	1.9–4.0	1.88–3.44	1.11–1.77	0.11–0.17	40.0
Seeds[3]	5.7	1.09	0.34	2.33	—	—	—	8.0
Sprouts[3]	92.3	2.34	1.95	35.3	—	—	—	274

[1]Kylen and McCready (1975).
[2]Liener (1972), range of values compiled from several other sources.
[3]Fordham et al. (1975), only the hypocotyl was analyzed; the remainder of the germinated seed was discarded.

EFFECT OF MATURATION AND GERMINATION
ON PROTEIN QUALITY

Very few studies have been carried out on the nutritive value of food legumes during maturation and germination. Most reports have described the germination process, and the results are contradictory and inconclusive (Elias *et al*. 1973; Palmer *et al*. 1973; Jaya *et al*. 1976; Venkataraman *et al*. 1976).

The nutritive value of immature, mature and sprouted soybeans is shown in Table 2.17. In the raw form, protein quality was very low for all three stages of maturity. After heat treatment, PER values were greatly increased to values 80% of that for casein. Although protein content was increased slightly with maturation and in sprouting, sulfur-amino acid content and protein quality (PER) of the heat-treated products were independent of maturity. Based on the net protein utilization and PER as parameters of protein quality, Standel (1963) reported that nutritive value of green soybeans (Endamame) was superior to most Oriental soybean foods in rats. Smith and Van Duyne (1951) cite a human experiment in which cooked green soybeans had a biological value of 65, a value that appears to be low compared to most soybean products (Liener 1972). This difference, however, may reflect differences in sulfur-amino acid content among soybean varieties. Some of the earlier investigators found an increase in nutritive value of raw soybeans during germination (Everson *et al*. 1944; Desikachar and De, 1947). But when both raw mature soybeans and raw germinated soybeans are properly processed by moist heat treat-

TABLE 2.17

COMPARATIVE PROTEIN QUALITY OF IMMATURE, MATURE, AND
SPROUTED SOYBEANS IN THE RAT[1]

Diet	Protein content (% of diet)	Methionine content of diet (g/16g N)	Cystine content of diet (g/16g N)	PER[2]
A Casein	10.4	2.46	0.21	2.50
B Green mature, raw[3]	10.6	1.22	0.59	0.77
C Green mature, heated	9.1	1.24	0.59	2.05[4]
D Dry mature, raw	10.5	1.18	0.61	0.75
E Dry mature, heated	9.9	1.22	0.61	2.11[4]
F Sprouts, raw	9.5	1.19	0.63	0.64
G Sprouts, heated	9.7	1.21	0.58	2.02[4]
H Early green variety, cooked[5]	9.5	1.33	0.59	2.14[4]

[1]Bates *et al*. (1977).
[2]Protein efficiency ratio corrected on a basis of PER = 2.50 for casein.
[3]Diets B through G, Bragg soybeans, field type.
[4]No significant difference in PER in respect to stage of maturity or variety.
[5]Diet H, early green soybeans, vegetable type.

ment, protein quality does not differ. The major nutritional change in soybeans during maturation and germination is an increase in content of certain B-vitamins, ascorbic acid and β-carotene.

In studies with other food legumes, protein digestibility was not significantly altered during maturation and germination. Elias *et al.* (1973) suggest that in *Phaseolus vulgaris*, the presence of toxic constituents at different physiological stages of maturation may affect nutritive value. Most food legumes contain a number of heat-labile and heat-stable substances (National Academy of Science 1973).

BIOLOGICAL AND PHYSIOLOGICAL FACTORS

"Nutritive value" and "biological-physiological effects" can be differentiated by restricting the former to the ability of soy products to supply amino acids and other essential nutrients, and the latter to substances that interfere with the utilization of essential nutrients or cause adverse reactions. Chemical composition and nutritive value of green, mature, and sprouted soybeans and of other protein products with respect to protein, fat, vitamins, and minerals have been discussed earlier. Table 2.6 listed the biological-physiological effects that occur in animals fed raw soybean meal. These heat-labile effects are interrelated and represent an animal's inability to digest protein and utilize the amino acids in the most effective and efficient manner rather than an irreversible response to a toxic substance. There are, however, qualitative differences in the responses of various species of animals fed raw soybeans (Table 2.18). A number of

TABLE 2.18
BIOLOGICAL EFFECTS OF RAW SOYBEAN MEAL IN VARIOUS ANIMALS[1]

Species	Growth Inhibition	Pancreas	
		Size	Enzyme Secretion
Rat[2]	+[3]	+	+
Chicken[2]	+	+	+
Pig[2]	+	−	±
Calf	+	−	±
Dog	−	−	[4]
Human	[5]	[5]	[5]

[1]Rackis (1974).

[2]Although adult animals maintain body weight, pancreas effects still occur.

[3]+ = growth inhibition and pancreatic hypertrophy and hypersecretion; − = no effect; ± = hyposecretion.

[4]Hyposecretion initially; normal after continued feeding.

[5]Unknown. However, two adults, in a 9-day feeding trial, had positive nitrogen balance for both raw and autoclaved soy flour.

other biological-physiological factors are also present in various soy products which are listed in Table 2.19. Very little information is available concerning their presence in immature and sprouted soybeans.

Trypsin Inhibitors and Hemagglutinins

Mature beans.—The increase in growth and the decrease in pancreatic hypertrophy in rats fed soy flour parallels the destruction of TI activity (see Fig. 2.2 and Table 2.7). The destruction of hemagglutinin by moist heat treatment activity parallels the inactivation of TI. Although earlier reports indicated that soybean hemagglutinins could account for 25% of the growth-inhibitory effect of raw soybean meal, more recent research now shows that the hemagglutinins in soybeans have no nutritional significance (Turner and Liener 1975). However, phytohemagglutinins in other food legumes inhibit growth of rats at levels as low as 0.5% of diet (Liener 1974).

TI and hemagglutinin activities in 108 varieties and strains of soybeans are summarized in Table 2.20. Gupta and Deodhar (1975) reported that TI activity in 16 field-type soybeans grown in India ranged from 33–86 specific activity units per mg protein (mean 57 units), whereas in vegetable soybeans the range was 21–66 units (mean 41 units).

Maturation

TI activity of vegetable- and field-type soybeans generally increase during maturation (Collins and Sanders 1976). As shown in Table 2.21, TI activity in immature, mature and germinated soybeans is readily destroyed by autoclaving for 15 min.

Germination

As shown in Table 2.22, TI activity in water extracts of germinated soybeans decreased very little; whereas Bates et al. (1977) report that TI

TABLE 2.19
BIOLOGICAL-PHYSIOLOGICAL FACTORS PRESENT IN SOYBEANS[1]

Property	
Trypsin inhibition	Estrogenicity
Hemagglutinating activity	Lowered bioavailability of minerals and vitamins
Allergenicity	
Flatulence	Mineral chelators (phytic acid)
Goitrogenicity	Amino acid availability

[1]Rackis (1974).

TABLE 2.20

TRYPSIN INHIBITOR AND HEMAGGLUTINATING ACTIVITIES OF
SEVERAL VARIETIES AND STRAINS OF MATURE SOYBEANS

Parameters	Range of Values[1]	Number of Strains or Varieties	Reference
Trypsin inhibitor activity	66–233[2]	108	Kakade et al. (1972)
Trypsin inhibitor activity	33–86[3]	16	Gupta and Deodhar (1975)
Trypsin inhibitor activity	21–66[4]	16	Gupta and Deodhar (1975)
Hemagglutinating activity	60–426[5]	108	Kakade et al. (1972)

[1]Values represent activity extracted under conditions employed and may not represent total activity of the whole seed.
[2]Expressed as trypsin units/mg protein.
[3]Specific activity units/mg protein, field-type varieties.
[4]Same as c above, vegetable-type varieties.
[5]Expressed as hemagglutinating units/mg protein.

TABLE 2.21

EFFECT OF MATURATION AND GERMINATION ON TRYPSIN INHIBITOR
ACTIVITY IN SOYBEANS[1]

Stage of Maturity	Trypsin Inhibitor Activity Reaction Rate, [2,3]	
	Raw	Heated[4]
Immature	49.0	1.5
Mature	52.2	0.6
Sprouts	17.8	1.7

[1]Bates et al. (1977).
[2]Change in absorbance at 257 nm/min/g protein compared to control with no trypsin inhibitors, Bragg soybeans.
[3]Activity in extracts under conditions employed.
[4]Autoclaving 121° C for 15 min.

TABLE 2.22

TRYPSIN INHIBITOR[1] ACTIVITY OF GERMINATING SOYBEANS[2]

Treatment	Days of Germination	Variety		
		Kanrich	Soylima	Dare
Mature	0	22.8	17.2	26.9
Germination	1	22.6	16.5	26.9
Germination	2	21.4	16.8	26.4
Germination	3	20.5	16.2	25.2
Germination	4	19.8	16.9	24.7

[1]Trypsin inhibitor (mg/g) dry basis, values represent activity extracted under conditions employed and may not represent total activity of the sample.
[2]Collins and Sanders (1976).

activity decreased about 70% in a 4-day germination period (see Table 2.21). Chen and Pan (in press, A) report that hemagglutinating activity decreased 50–87% during 5-day germination.

Significance of TI.—Special mention should be made concerning the TI activity values reported for immature, mature and germinated soybeans. Values given in Tables 2.20–2.22 represent activity extracted under the conditions employed and may not represent total TI activity of the intact bean (Rackis *et al*. 1974; Kakade *et al*. 1974). The amount of TI activity destroyed depends upon temperature, duration of heating, and moisture conditions. Initial moisture content of the intact beans is another major factor that influences inactivation of TI (Rackis 1974). For example, only 10% of the total TI activity of whole soybeans containing about 10% moisture was destroyed during steaming at 100° C for 20 min; but, when the mature soybeans were first soaked or tempered to 25%, over 97% TI activity was destroyed during heat treatment for 20 min.

Maximum growth occurs in rats fed soy diets in which only 79% of the TI activity was eliminated by live steam treatment (Rackis *et al*. 1975A). Residual TI activities in properly processed soy products are below biological threshold levels and do not have nutritional significance (Rackis *et al*. 1975A; Churella *et al*. 1976). Apparently, no heat stable TI's are present in soybeans of varying maturity. The practical significance of these findings is that TI and other antinutritional substances in immature, mature, and sprouted soybeans (regardless of variety) can be readily eliminated by ordinary cooking and moist heat treatment.

Estrogenic Activity.—Substances exhibiting estrogenic activity are distributed widely in animals and plants. In soybeans, the major estrogenic substance is the isoflavone, genistin. However, the amount present in soy protein products probably has no nutritional significance in man (Rackis 1974). The presence of coumesterol, an estrogenic phenolic compound, in soybeans is still a matter of conjecture. Knuckles *et al*. (1977) report that soybean products contain 0.2–1.2 μg/g of coumesterol; soybean sprouts contain 71.1 μg/g. Its biological significance, at such a level, is unknown.

Phytic Acid and Mineral Bioavailability

Rackis and Anderson (1977) last summarized the literature covering the relationship between phytic acid and mineral bioavailability in soy protein products. A detailed discussion will not be presented here.

Suffice it to say that various soy protein isolates, when fed as the sole source of protein in the diet, require 0–100 ppm of added zinc to maintain rat growth, while soy flours do not. These differences in zinc requirements are attributed to processing conditions that promote the formation of phytate-protein-mineral complexes during the manufacture of protein isolates. It is of special importance to cite the recent report of Churella and Vivian (1976), who found no differences in bone or carcass ash, calcium, phosphorus, or zinc between rats fed a soy isolate containing reduced levels of phytate and those fed an isolate with normal levels of phytate. The significance of these data is that the isolates used in their studies are manufactured for formulation of infant formulas. Because of their high iron content and the high biological availability of iron, whole soybeans and soy flours are recommended for programs oriented to prevent iron-deficiency anemia.

Maturation and Germination.—The iron in ^{59}Fe-labeled mature soybeans is more available to iron-deficient rats than the iron in immature green soybeans, even though mature soybeans contain three times more phytic acid (Table 2.23). Absorption of ^{59}Fe in mature soybeans compares favorably with that from ^{59}FeCl$_3$. During germination, phytic acid content decreased which correlates with an increase in phytase activity (Chen and Pan, in press, B). There are no reports on the biological availability of other minerals in immature and germinated soybeans.

Other Aspects

Soybeans contain almost twice as much phosphorus as most cereals, but approximately 70–80% is present as phytic acid. Although no experiments

TABLE 2.23

ABSORPTION OF ^{59}Fe BY RATS FED ^{59}Fe-LABELED IMMATURE AND
MATURE SOYBEANS AND ^{59}Fe-LABELED FeCl$_2$[1]

Sample[2]	Fe Consumed in Dose (μg)	Absorption (% of dose)	Fe Absorbed (μg)	Phytic Acid Content (%)[3]
Immature soybeans				0.61
Low iron	32.1	34.4	11.0	
High iron	51.5	30.2	15.6	
Mature soybeans				1.71
Low iron	46.9	48.4	22.7	
High iron	37.9	55.5	21.0	
FeCl$_2$				
Low	14.5	59.8	8.7	
High	29.6	50.7	15.0	

[1]Welch and Campen (1975).

[2]Low or high iron refers to the level of iron supplied to the soybean plants in the nutrient solution.
[3]Dry basis.

with humans have been conducted to determine the availability of phosphorus from soybeans, phytic acid phosphorus is well utilized by the rat but not by the chick (Liener 1972). Phytase activity in soybeans is low. Kenkey and Ogi are sour fermented maize foods that are consumed in large amounts by people in many African countries. Extensive phytic acid hydrolysis occurs during fermentation (Amoa and Muller 1976). Soy-Ogi is produced in Nigeria (Rackis and Akers 1976). Such combinations may be a practical way to increase phosphorus availability in soybeans.

FLATULENCE

Raffinose, stachyose and verbascose cause flatulence in man and animals (Rackis 1975, 1976; Cristofaro et al. 1974). These α-D-galactooligosaccharides escape digestion and are not absorbed into the blood. Consequently, the bacteria in the lower intestinal tract metabolize them to form large amounts of carbon dioxide and hydrogen. Flatus activity of soy protein products in man is shown in Table 2.24. Full-fat and defatted soy flour cause flatulence. Soybean oil does not. When soy flour is extracted with 80% ethyl alcohol, the flatulence effects are reduced. Alcohol extractives produced very high amounts of flatus. Protein isolates and high-molecular-weight polysaccharides (water-insoluble residue) are devoid of flatus activity. Soy whey solids, which contain most of the oligosaccharides, produced large amounts of flatus.

TABLE 2.24
EFFECTS OF SOY PRODUCTS ON FLATUS IN MAN[1]

Product[2]	Daily Intake (g)	Flatus Vol (cc/hr) Average	Flatus Vol (cc/hr) Range
Full-fat soy flour	146	30	0–75
Defatted soy flour	146	71	0–290
Soy protein concentrate	146	36	0–98
Soy proteinate	146	2	0–20
Water-insoluble residue[3]	146	13	0–30
Whey solids[4]	48	300[5]	—
80% Ethanol extractives[4]	27	240	220–260
Navy bean meal	146	179	5–465
Basal diet	146	13	0–28

[1]Data from Steggerda et al. (1966), except for whey solids and 80% ethanol extractives from Rackis et al. (1970).
[2]All products were toasted with live steam at 100° C for 40 min.
[3]Fed at a level three times higher than that present in the defatted soy flour diet.
[4]Amount equal to that present in 146g of defatted soy flour.
[5]One subject, otherwise four subjects per test.

Elimination of Flatulence

Raffinose and stachyose can be removed by aqueous alcohol extraction and by the process of preparing soy protein concentrates and isolates (Table 2.24). Enzyme processes have been patented (Rackis 1975). Although the raffinose and stachyose contents of soybean varieties differ, elimination of these oligosaccharides by genetic means does not look promising (Hymowitz and Collins 1974).

The amount of raffinose and stachyose that can be removed from mature whole soybeans by leaching and soaking is summarized in Table 2.25. Oligosaccharide content can also be reduced by germination (Table 2.26). Calloway *et al.* (1971) did not observe any change in flatus activity of germinated white beans, other beans, and soybeans when fed to humans. Becker *et al.* (1974) observed that the disappearance of raffinose and stachyose during autolysis of California white beans reduces flatulence, as

TABLE 2.25

PROTEIN LOSS AND OLIGOSACCHARIDE REMOVAL FROM WHOLE SOYBEANS BY VARIOUS TREATMENTS

Treatment	Bean-to-Water Ratio	Protein Loss (%)[1]	Removal Oligosaccharide (%)[2]	Reference
Soak, rt, 15 hr	—		8	Kim *et al.* (1973)
Boil, 20 min, water	1:10	1.0	33	Ku *et al.* (1976)
Boil, 60 min, water	1:10	2.6	59	Ku *et al.* (1976)
Boil, 20 min, 0.5% NaHCO₃	1:10	1.3	21	Ku *et al.* (1976)
Boil, 60 min, 0.5% NaHCO₃	1:10	6.8	60	Ku *et al.* (1976)
Boil, 60 min, pH 4.3	1:10	2.0	46	Ku *et al.* (1976)

[1]g Protein/100g protein in original dry bean.
[2]g Oligosaccharide/100g oligosaccharide in original dry bean.

TABLE 2.26

EFFECT OF GERMINATION ON AUTOLYSIS OF SUCROSE, RAFFINOSE AND STACHYOSE

Time of Germination (hr)	% Loss of Oligosaccharides			
	Sucrose	Raffinose	Stachyose	Reference
48	0	30	50	East *et al.* (1972)
96	80	100	96	
96	—	80	80	Adjei-Twum *et al.* (1976)
120	—	100	90	

measured by hydrogen production in the rat, and that other components may also cause flatulence. In respect to soybeans, these results would appear to indicate that the high-molecular-weight polysaccharides, which normally do not cause flatulence (see Table 2.24), may have been partially hydrolyzed during germination. These modified polysaccharides may now become substrates for the formation of flatus by the intestinal microflora and, thereby, compensate for the loss in raffinose and stachyose. On the other hand, the failure to eliminate flatus activity in the mung beans and other food legumes during germination may be related to the presence of undigested starch granules that enter the lower intestinal tract. Flatulence problems with food legumes have been reviewed (Jaffe 1973). Immature soybeans contain numerous starch granules which disappear during final days of maturity (Bils and Howell, 1963). But whether immature soybeans cause flatulence is not known.

FLAVOR

Mature Soybeans

Because of their functional and nutritional properties, soy protein products find outlets in a variety of foods. Rising prices forecast for animal protein have provided economic incentives to use less expensive vegetable proteins. To achieve an increased market for soy products, disagreeable flavors must be eliminated.

The major objectionable flavors of raw, full-fat, and defatted soy flours prepared from mature soybeans representing several varieties, in order of decreasing intensity are: beany, bitter and green (Moser *et al.* 1967). After raw soy flour was steamed 10–40 min, flavor scores increased from 1.5 to a maximum of about 6.0–6.3. Our laboratory studies indicate that the flavor scores of flours prepared from vegetable- and field-type soybeans are not significantly different. An organoleptic evaluation in 1971 of commercial soy protein products confirms that they are not bland and that some of the beany and bitter flavors of mature whole soybeans remain (Kalbrener *et al.* 1971). A combination of toasting and hexane:ethanol azeotrope extraction can be used to produce soy flours and concentrates approaching the blandness of wheat flour (Table 2.27). Flavor scores of azeotrope-extracted flakes and proteinates prepared from these extracted flakes are significantly higher than those prepared by present commercial practices. Azeotrope extraction effectively removes residual lipids which are primarily responsible for the undesirable flavors in soy products. Oxidized phosphatidylcholines may well contribute to the bitter taste of soybeans (Sessa *et al.* 1976).

TABLE 2.27

FLAVOR OF SOY PROTEIN PRODUCTS

Product	Flavor Score[2]	Flavor Intensity Value[3]		
		Grassy/beany	Bitter	Astringent
A Raw, defatted flakes	4.0	2.9	1.0	0.9
B Toasted, defatted flakes	6.6	0.6	0.8	0.6
C Azeotrope-extracted defatted flakes, toasted[4]	7.8	0.4	0.3	0.1
D Soy protein concentrate prepared from C	7.9	0.2	0.4	—
E Wheat flour	8.1	—	—	—
F Soy protein isolate prepared from A	5.2	1.8	0.9	0.8
G Soy protein isolate prepared from C	7.3	0.2	0.6	0.2
H Sodium caseinate	8.0	0.1	0.2	0.1
I Commercial products[5]	4.2–6.6			
J LSD[6]	0.67			

[1]Honig et al. (1976).
[2]Strong = 1, bland = 10.
[3]FIV, based on a value of 1 for weak, 2 for moderate, 3 for strong. Rackis et al. (1972).
[4]Hexane:ethanol azeotrope (82:18 v/v b.p. 59° C).
[5]Flours, concentrates and isolates (Kalbrener et al. 1971).
[6]Least significant difference at the 95% level.

Immature Soybeans

Flavor intensity values (FIV) with respect to beaniness varied from 1.6–2.7 during maturation for Hawkeye soybean (Table 2.28); no significant trends were noted. The range for the Amsoy variety was 2.0–2.7. The average FIV for bitter for the two varieties increased three- to four-fold during maturation. A correlation ($r = 0.73$) exists between lipoxygenase activity and the increase in FIV for bitter as soybeans mature.

Effect of Variety

Specific activity of lipoxygenase in several soybean varieties ranged from 0.28 units/mg protein to 1.02 units/mg protein (Chapman et al. 1976). Most likely, even the lower levels of lipoxygenase activity are sufficient to generate objectionable flavors, since no significant differences in flavor scores were observed for soy flours prepared from several varieties (Moser et al. 1967). Bates and Matthews (1975) also found little significant difference in acceptance of vegetable- and field-type soybeans either in the green-mature or dry-mature stage. Other reports suggest that vegetable types are preferred to field varieties (Dougherty and Knapp 1972). Cook-

TABLE 2.28
FLAVOR EVALUATION OF MATURING SOYBEANS (HAWKEYE VARIETY, 1969 CROP)[1]

Days After Flowering	Dry Matter Tasted (mg)	Beany[2] FIV[3]	Bitter %[4]	Bitter FIV
24	21	1.6	25	0.40
24	26	1.7	33	0.42
27	60	2.0	21	0.29
27	102	2.1	36	0.43
29	104	2.5	29	0.57
29	152	2.4	29	0.43
31	175	2.5	42	0.65
33	115[5]	2.0	25	0.50
35	232[5]	2.4	43	0.57
40	169	2.4	36	0.46
44	222	2.2	38	0.54
49	223	2.0	69	1.10
52	219	2.0	50	0.94
55	236	2.1	57	1.10
59	216	2.7	77	1.60

[1]Rackis et al. (1972).
[2]Includes all "beaniness" responses: green-beany, beany, and raw beany. Except for one taster, all the others recorded a positive beaniness response.
[3]Flavor intensity value (FIV) = [(number of weak responses) + 2 (number of moderate responses) + 3 (number of strong responses)]/n, where n is the number of tasters.
[4]Percent of tasters giving a positive bitter response; total number of tasters ranged between 12 and 15.
[5]Repeat of taste test at twice the level of intake of beans at the same mautrity level.

ing procedures have a large effect on the various parameters used to determine total palatability scores of cooked mature soybeans. Mean scores for individual palatability factors between bean varieties did not differ too much, although Amsoy, a field variety, tended to have lower values (Perry et al. 1976).

POSTHARVEST STORAGE

This section will examine the storage stability of soybean protein products with the realization that postharvest losses or reduction in quality of essential nutrients in these products can lead to food shortages. Another reason for the need for greater information on the storage stability of soy products is the Food and Drug Administration (1973) announcement of a nutrition labeling program. Nutrition labeling will result in more attention being given to the maximum retention of essential nutrients and maintenance of protein quality and to factors that affect storage stability.

Whole Soybeans

Foster and Holman (1973) reported that the repeated commercial handling of soybeans will produce splits or foreign matter exceeding limits for U.S. No. 1 grade soybeans. The amounts of breakage that can occur are summarized in Tables 2.29 and 2.30. Breakage will lower market grade, reduce the quality of soybean oil, and lower flavor acceptability and nutritive value of soy protein products. List *et al.* (1977) have described the problems associated with the refining of soybean oil from damaged soybeans. Studies have been reported on biochemical reactions occurring in stored soybeans (Edje and Burris 1970; Friedlander and Navarro 1972) and on changes in nutritive value of various soy protein products (Zimmerman *et al.* 1969; Ben-Gera and Zimmerman 1972; Yannai and Zimmerman 1970). These workers found that the net protein utilization of the protein in defatted soybean meal decreased with prolonged storage time at temperatures of 20° and 40° C. Isolated soybean protein, likewise, underwent a loss in available lysine.

TABLE 2.29

SOYBEAN BREAKAGE WITH FOUR HANDLING METHODS[1]

Method	When Used	Breakages (%)
Free-fall drop (ft)	Storage bins	
100		4.5
70		2.1
40		1.1
Sprouting	Railcars	1.0
Grain thrower	Railcars, barges, ships	0.7
Bucket elevator	All above	0.3

[1]Foster and Holman (1973).

TABLE 2.30

EFFECT OF REPEATED HANDLING ON SOYBEAN BREAKAGE
(100 FT/FREE-FALL DROP ON CONCRETE)

| Test Condition | | Breakage in Each Run | | | | Cumulative |
| Moisture of Beans | Temperature | (%) | | | | Breakage |
(%)	(° F)	1	2	3	4	(%)
10.7	46	3.2	3.0	2.5	2.0	10.7
11.0	32	4.0	2.5	2.0	1.6	10.1
12.6	50	1.4	1.2	1.4	1.2	5.2

[1]Foster and Holman (1973).

Kwolek and Bookwalter (1971) developed several models for determining acceptable conditions for product storage and predicting product stability in respect to different criteria of quality. Labuza (1972) points out that determination of reaction kinetics can provide meaningful data relating to the effect of drying and subsequent storage of soy products. Much more data are needed to determine rate constants of various deteriorative reactions so that the best processing methods can be selected to optimize nutritive value. Similar data can also be used to calculate shelf-life and provide information for nutrition labeling of soy products. One of the effects of storage of cereal-soybean foods is the loss of available lysine (Bookwalter 1977). It is important to minimize loss of lysine availability in soy proteins because these proteins are essential ingredients for enhancing the nutritive value of cereals deficient in lysine. Furthermore, processed cereal-soybean foods are used in large amounts throughout the world to combat malnutrition (Agency for International Development 1976).

Control of moisture content is critical in maintaining good storage stability in respect to protein quality, flavor acceptability, functionality and microbial growth. The importance of the moisture content/equilibrium relative humidity relationships has been discussed (Pixton and Warburton 1971A). Data for whole soybeans (Pixton and Warburton 1971B) and defatted soybean meal (Pixton and Warburton 1975) have been reported. Bean et al. (1976) report that deterioration of soy-wheat flour blends as measured by baking performance, bread quality, and flavor occurs when the blends are stored for 24 weeks at 100° F and 13% moisture. As shown in Table 2.31, a 3% lowering of the moisture content appears to retard development of off-flavors more than a 10° F lowering of tempera-

TABLE 2.31
ORGANOLEPTIC EVALUATION OF BREAD FROM SOY FLOUR-WHEAT FLOUR BLENDS STORED 24 WEEKS[1]

| | Flavor Ranking[2] | | | | Acceptability Scores[3] Yes, Response (%) | |
| | Defatted Soy | | Full-Fat Soy | | Defatted | Full-Fat |
Storage Conditions	Test 1	Test 2	Test 1	Test 2	Soy	Soy
−10° F, 13% moisture	37[4]	38[4]	26[4]	29[4]	86	86
100° F, 10% moisture	43	45	46	44	83	79
90° F, 13% moisture	54	49	61[5]	27	63	68
100° F, 13% moisture	66[4]	68[4]	67[4]	70[4]	40	41

[1]Bean et al. (1976).
[2]Lower numbers represent least off-flavors.
[3]Percent of total number of panel members giving an acceptable yes response.
[4]Significant difference at 1% level.
[5]Significant difference at 5% level.

ture. These workers discovered in a subsequent paper (Mecham *et al*. 1976) that the wheat flour was the principal source of off-flavors that developed at 100° F. No adverse soy-wheat interactions, in respect to off-flavor, occurred in the blend during storage. Whether such blends are in fact stable enough for use in overseas aid programs should be determined by acceptability tests after distribution through regular channels.

RESEARCH NEEDS

Food quality guidelines are required for upgrading human nutrition through the improvement of soybeans. The problem of soybean quality is a complex area which requires the input of various disciplines including genetics, plant physiology and biochemistry, analytical chemistry and processing technology. While increasing crop yield and improving consumer acceptance are primary objectives, research to increase protein content, elevate methionine-cystine levels, improve protein digestibility and to improve functionality should also have high priority. To be successful, rapid analytical procedures are needed for use by plant breeders. Other research priorities include: expanded knowledge in cellular and subcellular structure of soybeans, improved technologies to recover oil and protein bodies and to eliminate flavors, flatus and antinutritional factors.

Regardless of the progress made in breeding and processing, greater attention should be directed toward preserving soybean quality during postharvest storage of mature soybeans and processed soy protein products and utilizing green, immature and germinated soybeans.

BIBLIOGRAPHY

ADJEI-TWUM, D. C., SPLITTSTOESSOR, W. E. and VANDEMACK, J. S. 1976. Use of soybeans as sprouts. HortScience *11*, 235–236.

AGENCY FOR INTERNATIONAL DEVELOPMENT. 1976. Food for Peace, PL-480, Title II, Commodities Reference Guide, U.S. Department of State, Washington, D.C. Nov. 15.

AMERICAN SOYBEAN ASSOCIATION. 1974. Proceedings of the World Soy Protein Conference, Munich, Germany, Nov. 11–14, 1973. J. Am. Oil Chem. Soc. *51*(1), 49A–207A.

AMOA, B. and MULLER, H. G. 1976. Studies on Kenkey with particular reference to calcium and phytic acid. Cereal Chem. *53*, 365–375.

ANONYMOUS. 1974. Sprouts in your kitchen. *In* Sunset, Ed. Central, Vol. 152, p. 64. Lane Magazine and Book Company, Menlo Park, Ca.

ANTON, J. J. 1975. Good market climate nurtures soy industry growth. Food Prod. Dev. *9*(8), 96–99.

ASPINALL, G. O., BEGBIE, R. and McKAY, J. E. 1967. Polysaccharide components of soybeans. Cereal Sci. Today *12*: 223, 226–228, 260–261.

BATES, R. P., KNAPP, F. W. and ARAUJO, P. E. 1977. Protein quality of green-mature, dry mature and sprouted soybeans. J. Food Sci. 42, 271–272.

BATES, R. P. and MATTHEWS, R. F. 1975. Ascorbic acid and β-carotene in soybeans as influenced by maturity, sprouting, processing and storage. Proc. Fla. State Hort. Soc. 88, 266–271.

BATES, R. P., WEISS, D. D. and MATTHEWS, R. F. 1977. Pickled soybeans as a nutritious snack. Proc. Fla. State Hort. Soc. 89, 210–213.

BEAN, M. M., HANAMOTO, M. M., MECHAM, D. K., GUADAGNI, D. G. and FELLERS, D. A. 1976. Soy-fortified wheat flour blends. 2. Storage stability of complete blends. Cereal Chem. 53, 397–404.

BECKER, R., OLSON, A. C., FREDERICK, D. P., KON, S., GUMBMANN, M. R. and WAGNER, J. R. 1974. Conditions for the autolysis of alpha-galactosides and phytic acid in California small white beans. J. Food Sci. 39, 766–769.

BEN-GERA, I. and ZIMMERMAN, G. 1972. Changes in nitrogenous constituents of staple foods and feeds during storage. 1. Decrease in the chemical availability of lysine. J. Food Sci. Technol. 9(3), 113–118.

BILS, R. F. and HOWELL, R. W. 1963. Biochemical and cytological changes in developing soybean cotyledons. Crop Sci. 3, 304–308.

BOOKWALTER, G. N. 1976. Storage stability of corn-based foods used in food aid programs. Institute of Food Technologist Annual Meeting, Anaheim, CA, 1976, Abstract No. 336.

BOOKWALTER, G. N. 1977. Corn-based foods used in food aid programs: Stability characteristics—A Review. J. Food Sci. 42, 1421.

BOONVISUT, S. and WHITAKER, J. R. 1976. Effect of heat, amylase, and disulfide bond cleavage on the in vitro digestibility of soybean proteins. J. Agric. Food Chem. 24, 1130–1135.

BURGOS, A., CAVINESS, C. E., FLOYD, J. I. and STEPHENSON, E. L. 1973. Comparison of the amino acid content and availability of different soybean varieties in the broiler chick. Poult. Sci. 52, 1822–1827.

CALLOWAY, D. H., HICKEY, C. A. and MURPHY, E. L. 1971. The reduction in intestinal gas forming properties of legumes by traditional and experimental food processing methods. J. Food Sci. 36, 251–255.

CATSIMPOOLAS, N., CAMPBELL, T. B. and MEYER, E. W. 1968A. Immunochemical study of changes in reserve proteins of germinating soybean seed. Plant Physiol. 43, 799–805.

CATSIMPOOLAS, N., EKENSTAM, C., ROGERS, D. A. and MEYERS, E. W. 1968B. Protein subunits in dormant and germinating soybean seed. Biochim. Biophys. Acta 168, 122–131.

CHAPMAN, G. W. JR., ROBERTSON, J. A. and BURDICK, D. 1976. Chemical composition and lipoxygenase activity in soybeans as affected by genotype and environment. J. Am. Oil Chem. Soc. 53, 54–56.

CHEN, P. S. 1970. Soybeans for health, longevity and economy. 3rd edition, Provoker Press, St. Catherine, Ontario, Canada.

CHEN, L. H. and PAN, S. H. In press, A. Comparison of antinutritive factors in ungerminated and germinated seeds. Phytates. J. Food Sci.

CHEN, L. H. and PAN, S. H. In press, B. Comparison of antinutritional factors in ungerminated and germinated seeds. Hemagglutinins. J. Food Sci.

CHURELLA, H. R. and VIVIAN, V. 1976. The effect of phytic acid in soy infant formulas on the availability of minerals for the rat. Fed. Proc. 35. Abstract No. 2972.

CHURELLA, H. B., YAO, B. C. and THOMSON, W. A. B. 1976. Soybean trypsin inhibitor activity of soy infant formulas and the nutritional significance for the rat. J. Agric. Food Chem. *24*, 393–396.

COLLINS, J. L. and McCARTY, I. E. 1969. Mechanical harvesting and shelling of vegetable-type soybeans. Tenn. Home Sci. Prog. Rep. No. 69, 1–4, Knoxville, TN.

COLLINS, J. L., McCARTY, I. E. and SWINGLE, H. D. 1971. Shelling mature green beans with a roller-type sheller. Tenn. Farm. Home Sci. Prog. Rep. No. 78, 1–4, Knoxville, TN.

COLLINS, J. L. and SANDERS, G. G. 1973. Deep-fried snack food prepared from soybeans and onions. Food Technol. *25*(5), 46–50.

COLLINS, J. L. and SANDERS, G. G. 1976. Changes in trypsin inhibitor activity in some soybean varieties during maturation and germination. J. Food Sci. *41*, 168–172.

CRISTOFARO, E., MOTTU, F. and WUHRMANN, J. J. 1974. Involvement of the raffinose family of oligosaccharides in flatulence. *In* Sugars in Nutrition, H. L. Sipple and K. W. McNutt (Editors). Academic Press, New York.

DEODHAR, A. D., LAL, M. S., SHARMA, Y. K. and MEHTA, S. K. 1973. Chemical composition of vegetable type varieties of soybeans. Indian J. Nutr. Diet. *10*(3), 134–138.

DESIKACHAR, H. S. R. and DE, S. S. 1947. Role of inhibitors in soybeans. Science *106*, 421–422.

DOUGHERTY, R. H. and KNAPP, F. W. 1972. Vegetable-type soybeans as dry bean products. Proc. Fla. State Hort. Soc. *85*, 187–190.

EAST, J. W., NAKAYAMA, T. O. M. and PARKMAN, S. B. 1972. Changes in stachyose, raffinose, sucrose and monosaccharides during germination of soybeans. Crop Sci. *12*, 7–9.

EDJE, O. T. and BURRIS, J. S. 1970. Physiological and biochemical changes in deteriorating soybean seeds. Proc. Assoc. Offic. Seed Anal. *60*, 158–166.

ELIAS, L. G., CONDE, A., MUNZ, A. and BRESSANI, R. 1973. Effect of germination and maturation on the nutritive value of common beans (*Phaseolus vulgaris*). *In* Nutritional Aspects of Common Beans and Other Legume Seeds as Animal and Human Foods, W. G. Jaffe (Editor). Archivos Latin Americanos de Nutricion, Caracas, Venezuela, pp. 139–163.

EVERSON, G. J., STEENBOCK, A., CEDERQUIST, D. C. and PARSONS, H. T. 1944. The effect of germination, the stage of maturity, and the variety upon the nutritive value of soybean protein. J. Nutr. *27*, 225–229.

FOOD AND DRUG ADMINISTRATION. 1973. Food label information panel nutrition labeling. Federal Register *38*, 6950, March 14.

FORDHAM, J. R., WELLS, C. E. and CHEN, L. H. 1975. Sprouting of seeds and nutrient composition of seeds and sprouts. J. Food Sci. *40*, 552–556.

FOSTER, G. H. and HOLMAN, L. E. 1973. Grain breakage caused by commercial handling methods. Marketing Research Report No. 968, Agricultural Research Service, U.S. Department of Agriculture, Washington, D.C.

FRIEDLANDER, A. and NAVARRO, S. 1972. The role of phenolic acids in the browning, spontaneous heating and deterioration of stored soybeans. Experientia *28*, 761–763.

GOULD, D. H. and MacGREGOR, J. T. 1977. Biological effects of alkali-treated protein and lysinoalanine: An overview. *In* Protein Crosslinking: Nutritional and Medical Consequences, M. Freidman (Editor). Advances in Experimental Medicine and Biology, Plenum Press, New York.

GUPTA, A. K. and DEODHAR, A. D. 1975. Variation in trypsin inhibitor activity in soybean (Glycine max.). Indian J. Nutr. Diet. *12*, 81–84.

GUPTA, A. K., WAHIE, N. and DEODHAR, A. D. 1976. Protein quality and digestibility in vitro of vegetable and grain-type soybeans. Indian J. Nutr. Diet. *13*, 244–251.

HESSELTINE, C. W. and WANG, H. L. 1972. Fermented soybean food products. *In* Soybeans: Chemistry and Technology, Vol. 1, Proteins, A. K. Smith and S. J. Circle (Editors). Avi Publishing Company, Westport, CT, pp. 389–419.

HILL, L. D. (Editor). 1976. World Soybean Research. Proceedings of the World Soybean Research Conference, The Interstate Printers and Publishers, Inc., Danville, IL.

HONIG, D. H., WARNER, K. and RACKIS, J. J. 1976. Toasting and hexane:ethanol extraction of defatted soy flakes. Flavor of flours, concentrates and isolates. J. Food Sci. *41*, 642–646.

HORAN, F. E. 1976. Use of soy protein for food. *In* World Soybean Research, L. D. Hill (Editor). Proceedings of the World Soybean Research Conference, The Interstate Printers and Publishers, Inc., Danville, IL, pp. 775–788.

HYMOWITZ, T. and COLLINS, F. I. 1974. Variability of sugar content in seed of *Glycine max.* and *G. soya.* Agron. J. *66*, 239–241.

JAFFE, W. G. (Editor). 1973. Nutritional aspects of common beans and other legume seeds as animal and human foods. Proceedings of meeting Ribeirao Prito S.P. Brazil, November 1973. Publisher Archivos Latinoamericanos de Nutricion, Apartados 2049, Caracas, Venezuela.

JAYA, T. V., KRISHNAMURTHY, K. S. and VANKATARAMAN, L. V. 1976. Effect of germination and cooking on the protein efficiency ratio of some legumes. Nutr. Rep. Int. *12*(3), 175–184.

JOHNSON, D. W. 1976. Marketing and economic production—summing up. *In* World Soybean Research, L. D. Hill (Editor). Proceedings of the World Soybean Conference, The Interstate Printers and Publishers, Inc., Danville, IL, pp. 1014–1017.

KAKADE, M. L., HOFFA, D. E. and LIENER, I. E. 1973. Contribution of trypsin inhibitors to the deleterious effects of unheated soybeans fed to rats. J. Nutr. *103*, 1772–1778.

KAKADE, M. L., RACKIS, J. J., McGHEE, J. E. and PUSKI, G. 1974. Determination of trypsin inhibitor activity of soy products. A collaborative analysis of an improved procedure. Cereal Chem. *51*, 376–382.

KAKADE, M. L., SIMONS, N. R., LIENER, I. E. and LAMBERT, J. W. 1972. Biochemical and nutritional assessment of different varieties of soybeans. J. Agric. Food Chem. *20*, 87–90.

KALBRENER, J. E., ELDRIDGE, A. C., MOSER, H. A., HONIG, D. H., RACKIS, J. J. and WOLF, W. J. 1971. Sensory evaluation of commercial soy flours, concentrates and isolates. Cereal Chem. *48*, 595–600.

KAPOOR, U., KUSHWAH, H. S. and DATTA, I. C. 1975. Studies on gross chemical composition and amino acid content of soybean varieties. Indian J. Nutr. Diet. *12*(2), 47–52.

KAWAMURA, S. 1967. Quantitative paper chromatography of sugars of the cotyledon, hull, and hypocotyl of soybeans of selected varieties. Kagawa Univ. Fac. Technol. Bull. *15*, 117–131.

KIM, W. J., SMIT, C. J. B. and NAKAYAMA, T. O. M. 1973. The removal of oligosaccharides from soybeans. Lebensm.-Wiss. Technol. *6*, 201–204.

KNUCKLES, B. E., deFREMERY, D. and KOHLER, G. O. 1977. Coumesterol content of fractions obtained during wet processing of alfalfa. J. Agric. Food Chem. *24*, 1177–1180.

KROBER, O. A. 1956. Methionine content of soybeans as influenced by location and season. J. Agric. Food Chem. *4*, 254–257.

KROBER, O. A. and CARTTER, J. L. 1966. Relation of methionine content to protein levels in soybeans. Cereal Chem. *43*, 320–325.

KU, S., WEI, L. S., STEINBERG, M. P., NELSON, A. I. and HYMOWITZ, T. 1976. Extraction of oligosaccharides during cooking of whole soybeans. J. Food Sci. *41*, 361–364.

KWOLEK, W. F. and BOOKWALTER, G. N. 1971. Predicting storage stability from time-temperature data. Food Technol. *25*, 51–63.

KYLEN, A. M. and McCREADY, R. M. 1975. Nutrients in seeds and sprouts of alfalfa, lentils, mung beans and soybeans. J. Food Sci. *40*, 1008–1009.

LABUZA, T. P. 1972. Nutrient losses during drying and storage of dehydrated foods. Crit. Rev. Food Technol. *3*, 217–240.

LIENER, I. E. 1972. Nutritional value of food protein products. *In* Soybeans: Chemistry and Technology, Vol. 1, Proteins, A. K. Smith and A. K. Circle (Editors). Avi Publishing Co., Westport, CT, pp. 203–277.

LIENER, I. E. 1974. Phytohemagglutinins: Their nutritional significance. J. Agric. Food Chem. *22*, 17–22.

LIST, G. R., EVANS, C. D., WARNER, K., BEAL, R. E., KWOLEK, W. F., BLACK, L. T. and MOULTON, K. J. 1977. Quality of oil from damaged soybeans. J. Am. Oil Chem. Soc. *54*, 8–14.

McKINNEY, L. L., WEAKLEY, F. B., CAMPBELL, R. E. and COWAN, J. C. 1958. Changes in the composition of soybeans during sprouting. J. Am. Oil Chem. Soc. *35*, 364–366.

MECHAM, D. K., HANAMOTO, M. M., BEAN, M. M., FELLERS, D. A. and GUADAGNI, D. G. 1976. Soy-fortified wheat flour blends. 3. Storage stability of ingredients and incomplete blends. Cereal Chem. *53*, 405–412.

MILNER, M. 1975. How can science expand world protein resources? *In* Conference Papers, Soya Protein Conference and Exhibition 1975, London, United Kingdom, October, American Soybean Association, Hudson, IA.

MOSER, H. A., EVANS, C. D., CAMPBELL, R. E., SMITH, A. K. and COWAN, J. C. 1967. Sensory evaluation of soy flours. Cereal Sci. Today *12*(7), 296–299, 314.

MUELLER, D. C., KLEIN, B. and VAN DUYNE, F. O. 1974. Cooking with soybeans. Ill. Coop. Ext. Ser. Circ. 1092.

NATIONAL ACADEMY OF SCIENCE. 1973. Toxicants occurring naturally in foods. Food Nutrition Board, National Academy of Science, National Research Council, Washington, D.C.

NATIONAL SCIENCE FOUNDATION. 1976. Protein resources and technology. Status and Research Needs (Research recommendations and summary, Vol. 1, Washington, D.C.).

NATIONAL SOYBEAN RESEARCH COORDINATING COMMITTEE. 1977. A report on national soybean research needs, National Agricultural Research Policy Advisory Committee.

NELSON, A. I., STEINBERG, M. P. and WEI, L. S. 1976. Illinois Process for preparation of soy milk. J. Food Sci. *41*, 57–61.

PALMER, R., McINTOSH, A. and PUSZTAI, A. 1973. The nutritional evaluation of kidney beans (*Phaseolus vulgaris*). The effect of nutritional value of seed germination and changes in trypsin inhibitor content. J. Sci. Food Agric. *24*, 937–944.

PERRY, A. K., PETERS, C. R. and VAN DUYNE, F. O. 1976. Effect of variety and cooking method on cooking times, thiamine content and palatability of soybeans. J. Food Sci. *41*, 1330–1334.

PIXTON, S. W. and WARBURTON, S. 1971A. Moisture content/relative humidity equilibrium of some cereal grains at different temperatures. J. Stored Prod. Res. *6*, 283–293.

PIXTON, S. W. and WARBURTON, S. 1971B. Moisture content/relative humidity equilibrium at different temperatures of some oilseeds of economic importance. J. Stored Prod. Res. 7, 261–269.

PIXTON, S. W. and WARBURTON, S. 1975. The moisture content/equilibrium relative humidity relationship of soya meal. J. Stored Prod. Res. *11*, 249–251.

PRIVETT, O. S., DOUGHERTY, K. A., ERDAHL, W. L. and STOLYHWO, A. 1973. Studies on the lipid composition of developing soybeans. J. Am. Oil Chem. Soc. *50*, 516–520.

PROTEIN ADVISORY GROUP OF THE UNITED NATIONS. 1973. PAG Statement No. 22, PAG Bulletin, Vol. 3, No. 2, United Nations, N.Y.

RACKIS, J. J. 1974. Biological and physiological factors in soybeans. J. Am. Oil Chem. Soc. *51*, 161A–174A.

RACKIS, J. J. 1975. Oligosaccharides of food legumes: alpha-galactosidase activity and the flatus problem. *In* Physiological Effects of Food Carbohydrates, A. Jeanes and J. Hodge (Editors). American Chemical Society Symposium Series No. 15, American Chemical Society, Washington, D.C., pp. 207–222.

RACKIS, J. J. 1976. Flatulence problems associated with soy products. *In* World Soybean Research, L. D. Hill (Editor). Proceedings of the World Soybean Research Confrence, The Interstate Printers and Publishers, Inc., Danville, IL, pp. 892–903.

RACKIS, J. J. 1977. Enzymes in soybean processing and quality control. *In* Enzymes in the Beverage and Food Industry, R. L. Ory and A. J. St. Angelo (Editors). American Chemical Society Symposium Series No. 47, American Chemical Society, Washington, D.C., pp. 244–265.

RACKIS, J. J. and AKERS, H. 1976. Soybean Market Survey report, Morocco, Algeria, Libya, Kenya, Zambia and Nigeria, American Soybean Association, Hudson, IA.

RACKIS, J. J. and ANDERSON, R. L. 1977. Mineral availability in soy protein products. Food Prod. Dev. 11(10), 38, 40, 44.

RACKIS, J. J., ANDERSON, R. L., SASAME, H. A., SMITH, A. K. and VAN ETTEN, C. H. 1961. Amino acids in soybean hulls and oil meal fractions. J. Agric. Food Chem. *9*, 409–412.

RACKIS, J. J., HONIG, D. H., SESSA, D. J. and MOSER, H. A. 1972. Lipoxygenase and peroxidase activities of soybeans as related to the flavor profile during maturation. Cereal Chem. *49*, 586–597.

RACKIS, J. J., HONIG, D. H., SESSA, D. J. and STEGGERDA, F. R. 1970. Flavor and flatulence factors in soybean protein products. J. Agric. Food Chem. *18*, 977–982.

RACKIS, J. J., McGHEE, J. E. and BOOTH, A. N. 1975A. Biological threshold levels of soybean trypsin inhibitor by rat bioassay. Cereal Chem. *52*, 85–92.

RACKIS, J. J., McGHEE, J. E., HONIG, D. H. and BOOTH, A. N. 1975B. Processing soybeans into foods: Selected aspects of nutrition and flavor. J. Am. Oil Chem. Soc. *52*, 249A–253A.

RACKIS, J. J., McGHEE, J. E., LIENER, I. E., KAKADE, M. L. and PUSKI, G. 1974. Problems encountered in measuring trypsin inhibitor activity of soy flour. Report of a collaborative analysis. Cereal Sci. Today *19*, 513–516.

ROBINSON, W. B., BOURNE, M. C. and STEINKRAUS, K. H. 1971. Development of soy-based foods of high nutritive value for use in the Philippines. Natl. Technol. Inf. Serv., U.S. Department of Commerce, Washington, D.C., PB-213-758, January.

ROCKLAND, L. B., HAYES, R. J., METZLER, E. A. and BINDER, L. J. 1967. Process for producing quick-cooking legumes. U.S. Patent No. 3,318,708, May 9.

ROCKLAND, L. B. and METZLER, E. A. 1967. Quick-cooking lima and other dry beans. Food Technol. *21*(3A), 26A–30A.

ROEHM, J. N. and PRIVETT, O. S. 1970. Changes in the structure of soybean triglycerides during maturation. Lipids *5*, 353–358.

RUBEL, A., RINNE, R. W. and CANVIN, D. T. 1972. Protein oil and fatty acid in developing soybean seeds. Crop Sci. *12*, 739–741.

SESSA, D. J., WARNER, K. and RACKIS, J. J. 1976. Oxidized phosphatidylcholines from defatted soybean flakes taste bitter. J. Agric. Food Chem. *24*, 16–21.

SINCLAIR, P., VETTEL, R. S. and DAVIS, C. A. 1974. Soybeans in family meals. USDA Home and Garden Bull. No. 208.

SINGH, H. and PRIVETT, O. S. 1970. Studies on the glycolipids and phospholipids of immature soybeans. Lipids *5*, 692–697.

SMITH, A. K. and CIRCLE, S. J. (Editors). 1972. Soybeans: Chemistry and Technology, Vol. 1, Proteins. Avi Publishing Co., Westport, CT.

SMITH, A. K., RACKIS, J. J., ISNARDI, P., CARTTER, J. L. and KROBER, O. A. 1966. Nitrogen solubility index, isolated protein yield, and whey protein content of several soybean strains. Cereal Chem. *43*, 261–290.

SMITH, J. M. and VAN DUYNE, F. O. 1951. Other soybean products. *In* Soybeans and Soybean Products, Vol. II, K. S. Markley (Editor). John Wiley and Sons, N.Y., pp. 1055–1078.

SPATA, J. A., NELSON, A. I. and Singh, S. 1974. Developing a soyabean dal for India and other countries. World Crops, March/April.

STANDEL, B. R. 1963. Nutritional value of proteins of Oriental soybean foods. J. Nutr. *8*, 279–285.

STEGGERDA, F. R., RICHARDS, E. A. and RACKIS, J. J. 1966. Effects of various soybean products on flatulence production in the adult man. Proc. Soc. Exp. Biol. Med. *121*, 1235–1239.

THAPAR, V. K., BRIJ, P. and SINGH, R. 1974. Changes in some nonprotein nitrogenous compounds during germination of soybeans. Plant Biochem. J. *1*, 11–15.

TURNER, R. H. and LIENER, I. E. 1975. The selective removal of hemagglutinins in the nutritive value of soybeans. J. Agric. Food Chem. *23*, 484–487.

USDA. 1971. Vegetables in family meals. Home and Garden Bull. No. 105, U.S. Department of Agriculture, Washington, D.C.

USDA-FARMER COOPERATIVE SERVICE. 1976. Edible soy protein operational aspects of producing and marketing. FCS Res. Rep. 33, January.

VENKATARAMAN, L. V., JAYA, T. V. and KRISHNAMURTHY, K. S. 1976. Effect of germination on the biological value, digestibility coefficient and net protein utilization of some legume proteins. Nutr. Rep. Int. *13*(2), 197–212.

VOROB'EV, N. V. 1967. Dynamics of the formation of lipids during the development of soybean seeds. Biokhim. Fiziol. Maslich. Rast. *2*, 332–344.

WELCH, R. M. and CAMPEN, D. R. 1975. Iron availability to rats from soybeans. J. Nutr. *105*, 253–256.

WHYTE, K. C. 1973. The Complete Book of Sprouting. Troubador Press, San Francisco, CA.

YANNAI, S. and ZIMMERMAN, G. 1970. Influence of controlled storage of some staple foods on their protein nutritive value in lysine limited diets. 3. Protein nutritive value and antitryptic activity of soybean meal and peanut meal and protein nutritive value and free gossypol content of cottonseed meal. J. Food Sci. Technol. 7(4), 190–196.

ZIMMERMAN, G., BEN-GERA, I., WEISSMANN, S. and YANNAI, S. 1969. Storage under controlled conditions of dry staple foods (defatted milk powder, wheat, soybeans and defatted soybean meal) and its influence on their protein nutritive value. Dtsch. Ges. Chem. Apparatiwes. Monogr. *63*, 347–380.

CHAPTER 3

ROLE OF HORMONES IN RIPENING AND SENESCENCE

W. B. MC GLASSON

*Plant Physiology Unit, Division of Food Research
CSIRO, and School of Biological Sciences
Macquarie University, North Ryde, N.S.W. 2113*

ABSTRACT

The topic is discussed in the context that treatments arising from work on the role of hormones in ripening and senescence are an adjunct to refrigeration and controlled atmospheres for the preservation of fresh fruits and vegetables after harvest. It is also recognized that chemical treatments arising from such work must undergo costly development and safety testing before they are accepted.

The present status of knowledge on the five principal types of plant hormones (ethylene, auxins, gibberellins, cytokinins and inhibitors) will be reviewed and the deficiencies indicated. It is concluded from this review that it is still not possible to define the role of any of the principal hormones in ripening and senescence. The clearest case can be made for ethylene and findings of substantial practical value have resulted. The picture for the other plant hormones is unclear although senescence retarding effects have been obtained by treatments with auxins, gibberellins and cytokinins. Examination of the natural levels of the hormones in fruits highlights the considerable diversity among species and emphasizes the lack of close correlations between the levels of hormones and stage of development. An example of the physiological variation among ripening mutants in the tomato will be used to illustrate one of the more promising lines of research.

INTRODUCTION

It is universally accepted that fruits and vegetables, both fresh and processed, play an important role in human nutrition and general well-being. They provide a high proportion of human dietary intake of vitamins and minerals and in some societies they supply a large part of daily energy requirements. There is ample evidence that the regular availability of fresh fruits and vegetables is important for the general sociological welfare of communities. Maintenance of continuity of supplies of these commodities requires an integrated system beginning with the farmer and proceeding through transport, storage and distribution to the consumer. The postharvest horticulturist enters the picture from the time decisions have to be made about harvest maturity. Because population centers are frequently located long distances from the most suitable production areas or because many fruits and vegetables are highly seasonal the postharvest horticulturist is asked to devise methods for extending the period that fresh commodities can be kept in sound condition.

The most effective method of keeping fresh commodities sound is to refrigerate them. In more recent times the benefits of refrigeration for slowing biochemical activity have been augmented by systems for reducing oxygen and/or increasing carbon dioxide concentrations in the storage or transport environment. Following the discovery of plant growth regulators (first ethylene and auxin, later cytokinins, gibberellins and inhibitors) a large amount of work has been done on these substances with the aim of regulating ripening and senescence, and the subject has been extensively reviewed *e.g.* Leopold and Kriedemann 1975; Looney 1973; McGlasson *et al.* 1978; Wittwer 1971.

The discovery of IAA (indoleacetic acid, natural auxin) also stimulated interest in synthetic growth regulators, which led to the development of a large number of auxins, including the very active 2,4-dichlorophenoxyacetic acid (2,4-D), growth retardants such as succinic acid 2,2-dimethylhydrazide (daminozide), synthetic cytokinins, and the ethylene-releasing compound 2-chloroethylphosphonic acid (ethephon) (Wittwer 1971). The period of most rapid development of new compounds, new knowledge, and new technologies appears to have ended about 1970. In the last few years there have been few noteworthy plant growth regulators discovered in postharvest horticulture. It is suggested that one major reason for this is the increased concern about adding foreign chemicals to foods. The high costs of development and safety testing of new compounds is no doubt acting as a deterrent to the chemical industry. A further reason is that basic knowledge of the mode of action of the plant growth regulators and consequently their role in various plant processes is still lacking. Thus it is an appropriate time to evaluate the present situation and to highlight the necessity for devising fresh paradigms which will stimulate new research approaches.

THE PRESENT STATUS

Five principal types or classes of plant growth regulators or hormones have been identified: ethylene, auxins, gibberellins, cytokinins, and inhibitors (principally abscisic acid). Since these compounds are universally present in plant tissues it is reasonable to presume that they play some part in growth, ripening, and/or senescence of plant organs. Broadly ethylene and abscisic acid are regarded as senescence promoters, while auxins, gibberellins, and cytokinins are thought to retard senescence.

Although it is convenient to categorize the plant hormones by function, a newcomer to this subject is immediately confronted with the fact that the known plant hormones have widely overlapping roles. This has evoked the proposition that regulation of plant processes is likely to involve the simultaneous action of two or more types of hormones (Leopold and Kriedemann 1975), and this is a recurrent theme in most discussions on the role of hormones in ripening and senescence. In this section the present status of knowledge will be summarized for each class of hormone.

Ethylene

Ethylene is the easiest growth substance to study because it can be measured directly by gas chromatography without destroying the tissue. Perhaps because of the ease and precision of measurement, work on ethylene has been voluminous, and most contemporary publications on the subject of hormones and ripening include data on this gas. It has been clearly established that higher plant tissues evolve ethylene throughout growth and development (Pratt and Goeschl 1969).

Fruits may be divided into climacteric and nonclimacteric types. Climacteric fruits are those in which ripening is associated with a distinct increase in respiration and ethylene production. In nonclimacteric fruits ripening is protracted and the attainment of the ripe state is not associated with a marked increase in respiration or ethylene production (Fig. 3.1). A further distinguishing feature is that treatment of climacteric fruits with ethylene or propylene stimulates both respiration and autocatalytic ethylene production whereas the same treatment applied to nonclimacteric fruits stimulates respiration only (McMurchie et al. 1972).

Because ripening is a dramatic event in climacteric fruits, they have attracted most attention from physiologists. It is widely accepted that ethylene plays a crucial role in the initiation and development of ripening in climacteric fruits but there is much less agreement about its role in the non-climacteric fruits. In terms of the patterns of ethylene production during development and ripening at least two periods can be distinguished: first, the main growth period before the onset of ripening, and second, ripening itself. Based on the responses of bananas, lemons, and oranges to propylene treatment, McMurchie et al. (1972) proposed that there are two systems for ethylene biogenesis. System 1 is responsible for

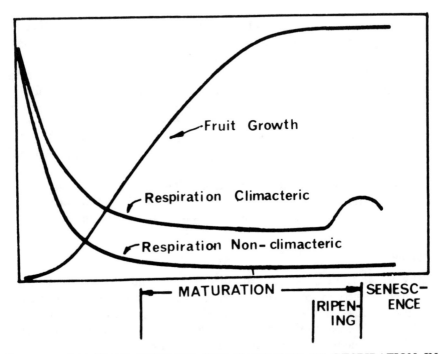

FIG. 3.1. DIAGRAM SHOWING THE PATTERNS OF RESPIRATION IN RELATION TO MATURATION, RIPENING AND SENESCENCE IN CLIMAC-TERIC AND NONCLIMACTERIC FRUITS

the low level of ethylene present in all fruit tissues, while System 2 produces the autocatalytic burst of ethylene production found in climacteric fruits such as the banana.

A continuing debate about the role of ethylene as an initiator of ripening concerns the timing of the rise in ethylene production in relation to the other changes associated with ripening. For ethylene to function as an initiator, either the sensitivity of the tissue to the pre-existing endogenous level of ethylene must change, or there must be an increase in the endogenous ethylene concentration, or both changes must occur. Measurements of ethylene concentrations have revealed fruits in which ethylene concentration clearly rises before the respiratory rise, others in which the rise coincides with the respiratory increase, and some in which ethylene concentration rises after the increase in respiration (McGlasson et al. 1978). However, the important considerations are not the exact timing of the rise in ethylene production, but rather what alters the sensitivity of fruit tissues to ethylene, how is the production of ethylene regulated, and what is its function during ripening.

There is extensive evidence that the sensitivity of fruits to applied ethylene changes during development. Extreme examples are the cantaloupe and tomato (Fig. 3.2). Continuous treatment of tomatoes with a saturating dose of ethylene approximately halves the time from harvest to the onset of ripening, whereas with cantaloupes a saturating dose completely obliterates differences in physiological age in fruits harvested from 30–90% of development. To demonstrate differences in physiological age in fruits such as the cantaloupe, much lower concentrations of ethylene must be applied, or treatments must be applied for short periods only. Another well-known example is the avocado. The commercial cultivars Fuerte and Hass do not ripen while attached to the tree and when harvested the fruits

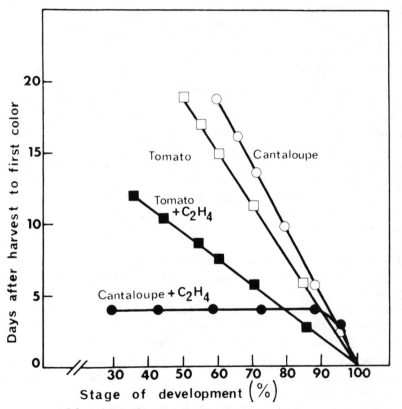

Adapted from Lyons and Pratt (1964) and McGlasson and Pratt (1964)

FIG. 3.2. RELATIONSHIP BETWEEN PHYSIOLOGICAL AGE AND RESPONSES OF DETACHED CANTALOUPE AND TOMATO FRUITS TO CONTINUOUS TREATMENT WITH ETHYLENE

Ethylene concentrations were 100 and 1000 μl/L respectively.

remain resistant to applied ethylene for 24 hr (Gazit and Blumenfeld 1970A). It has been suggested that either a ripening inhibitor is translocated from the branches and peduncle into the fruit or that these parts serve as a sink for a ripening hormone produced in the fruit (Tingwa and Young 1975). Attempts to isolate a ripening inhibitor have been unsuccessful. Recent research findings on exogenous treatments which affect the sensitivity of plant tissues to ethylene are those of Parups (1973) and Beyer (1976). Parups reported that a substituted benzothiadiazole could control ethylene-induced responses. Similarly Beyer reported that foliar treatment with silver ion specifically inhibited the action of applied ethylene in several plant tissues.

How ethylene production is regulated is unknown. It is generally accepted that methionine is the principal substrate for ethylene production and a cycle for the synthesis of methionine and its conversion to ethylene has been described (Yang 1975). Rhizobitoxine is a naturally occurring inhibitor of the conversion of methionine to ethylene. A high level of inhibition of ethylene production and a retardation of ripening has been obtained in apples vacuum infiltrated with rhizobitoxine solutions (Lieberman et al. 1975), but infiltration of tomato fruit and banana slices with the ethoxy analogue produced no significant effects (McGlasson unpublished). Rhizobitoxine and its synthetic analogues inhibit pyridoxal phosphate-dependent biochemical reactions (Owens et al. 1971), and therefore they can produce side-effects in treated tissues. To have practical value highly specific inhibitors of ethylene biosynthesis are needed. Other inhibitors of ethylene biogenesis have been reported. They include cobaltous ion (Lau and Yang 1976), benzylisothiocyanate in papaya (Patil and Tang 1974), and a protein from mung bean seedlings (Sakai and Imaseki 1973).

In climacteric fruits there is good evidence that ethylene functions as an integrator of the many biochemical processes which comprise ripening (Quazi and Freebairn 1970). Recent evidence indicates that although ethylene treatment hastens the destruction of chlorophyll (degreening), endogenous ethylene is not the primary trigger of color changes in oranges, a nonclimacteric fruit (Apelbaum et al. 1976).

In practical storage situations benefits can be obtained by maintaining ethylene at low levels in produce. One effective method is storage at low pressures (hypobaric storage) (Burg and Burg 1966). Useful results can be obtained more simply by destroying ethylene chemically (Scott and Gandanegara 1974; Scott and Wills 1974; Wild et al. 1976), or by ultraviolet irradiation of the storage atmosphere (Scott and Wills 1973; Scott et al. 1971). As Fig. 3.3 illustrates, a useful extension in the storage life of bananas can be achieved without refrigeration by storing the fruit in a modified atmosphere together with an ethylene absorbent.

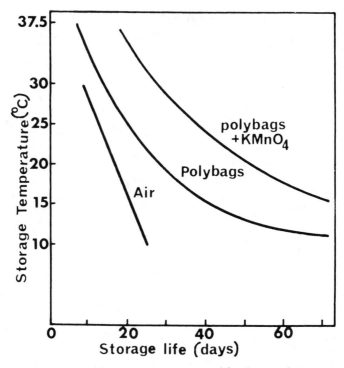

After Scott and Gandanegara (1974)

FIG. 3.3. INFLUENCE OF SEALED POLYETHYLENE BAGS WITH AND WITHOUT AN ETHYLENE ABSORBENT (KMnO₄) ON THE STORAGE LIFE OF GREEN BANANAS

Auxins

Proposals on the role of auxins in ripening and senescence have been based mainly on responses to treatment with auxins. Rather less is known about the natural levels because of the difficulty of assaying these hormones. It is widely thought that auxin concentrations are highest during the earliest stages of development and lowest during maturation, but no close correlations have been found between extractable auxin levels and rates of growth. Fig. 3.4–3.7 illustrate the magnitude of variation in the endogenous levels of auxins and the differences between species of fruits.

Interpretation of the results of application of auxins to plant tissues is complicated by at least two uncertainties. It is well known that relatively high concentrations of auxin, particularly the very active synthetic auxins such as 2,4-D, may stimulate ethylene production, and that the initial distribution of the applied auxin within the treated organ can markedly affect results (Frenkel and Dyck 1973; Vendrell 1969). Thus application of

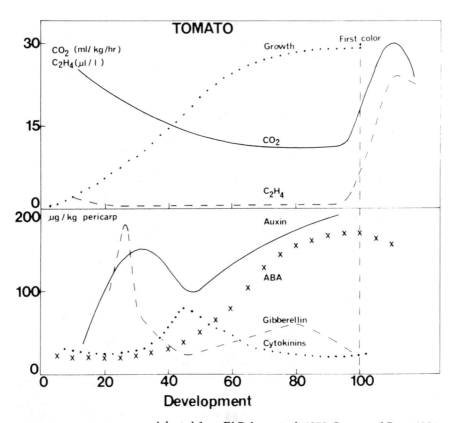

Adapted from El-Beltagy et al. 1976; Lyons and Pratt 1964;
McGlasson and Adato 1976; McGlasson et al. 1975

FIG. 3.4.   TOMATO—TRENDS IN FREE HORMONE LEVELS IN PERICARP
TISSUE DURING DEVELOPMENT AND RIPENING
Further work is required to establish trends in auxins and cytokinins.

auxin by dipping whole bananas and pears can advance ripening, but if
auxin is infused throughout the fruit tissue by vacuum techniques ripening can be delayed, although ethylene production is stimulated.

Largely as a result of these findings with bananas and pears, Frenkel
and his coworkers have proposed that ripening may be related to a deficiency in auxin (Frenkel 1975). To test this hypothesis the effects have
been examined of treatments which are thought to antagonize auxin
activity or reduce auxin content. The hypothesis and experimental approaches are attractive; some doubt exists however as to whether the
conclusions drawn from recent work have adequately distinguished the

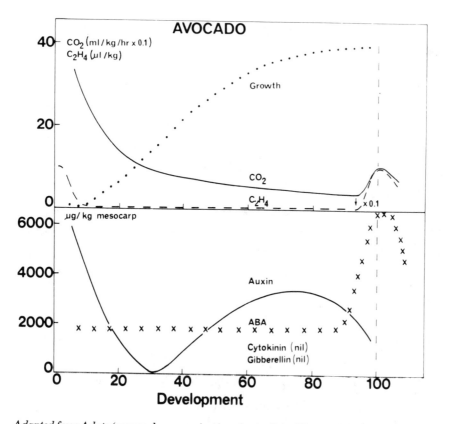

Adapted from Adato (personal communication, Agricultural Research Organization, Volcani Center, Israel); *Adato et al. 1976; Blumenfeld and Gazit 1972; Gazit and Blumenfeld 1970B, 1972; Kosiyachinda and Young 1975*

FIG. 3.5. AVOCADO—TRENDS IN FREE HORMONE LEVELS IN THE MESOCARP DURING DEVELOPMENT AND RIPENING

Although free cytokinins are shown as nil, Gazit and Blumenfeld (1970B) reported the presence of cytokinins which could be released by acid hydrolysis and that the level of this "bound" cytokinin decreased during development. Furthermore, an inhibitor (a substituted alkene) of cytokinin activity was found which increased during development (Bittner *et al.* 1971). The auxin curve indicates relative activity based on a bioassay. Actual concentrations are not given.

events preceding the onset of ripening and those which follow. From a practical view point findings which lead to acceptable treatments for slowing the rate of ripening (*i.e.* which increase shelf-life) could be valuable.

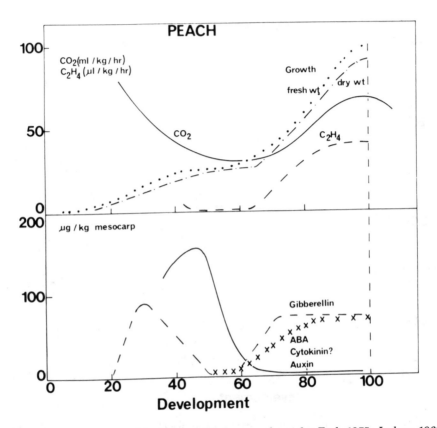

Adapted from Chalmers and van den Ende 1975; Jackson 1968;
Jerie and Chalmers 1976; Looney et al. 1974

FIG. 3.6. PEACH—TRENDS IN FREE HORMONE LEVELS IN THE
MESOCARP WITH EMPHASIS ON CHANGES DURING THE LAST HALF OF
DEVELOPMENT

Growth in the peach follows a double sigmoid pattern on the basis of both fresh and
dry weight with the latter lagging behind the changes in fresh weight. Ripening in
this fruit probably starts near the beginning of the last rapid growth stage.
Information is lacking about the levels of cytokinins and auxin during this stage.

Gibberellins

Knowledge of the levels and forms of gibberellins in plant tissue has
accumulated steadily since the development of sensitive and specific
bioassays in combination with gas liquid chromatography and mass spectrometry (Faull *et al.* 1974). Nevertheless, there is still a shortage of
information on gibberellins in mature tissues, and a role for gibberellins in

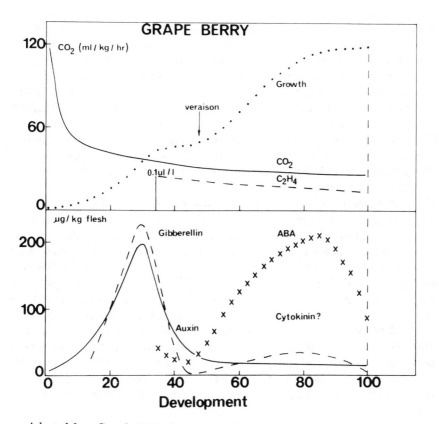

Adapted from Coombe 1960; Coombe and Hale 1973; Iwahori et al. 1968; Nitsch 1970

FIG. 3.7.   GRAPE BERRY—TRENDS IN FREE HORMONE LEVELS IN THE
FLESH WITH EMPHASIS ON THE LAST HALF OF DEVELOPMENT
Growth in the grape berry follows a double sigmoid pattern. Ripening in terms of
sugar accumulation and pigment changes in colored cultivars starts at the beginning of the last rapid growth stage. Information is lacking about the levels of
cytokinins during this stage.

ripening and senescence has been inferred mainly from their involvement
in juvenility and developmental processes in plants and the responses of
various tissues to treatments with gibberellins.

Interpretation of studies with applied chemicals is frequently con-
founded by the fact that such treatments may stimulate ethylene production as mentioned previously in the discussion on auxins. Treatment of
vegetative tissues and fruits with gibberellin has, in some cases, stimulated ethylene production while in others it had no effect or even reduced
ethylene production. Factors thought to influence this response are the

natural level of gibberellin (high in juvenile tissues but usually low in mature tissues) and the method of application (McGlasson et al. 1978).

The examples which most clearly imply a role for gibberellins in retarding senescence are the responses of citrus and apricot fruits to treatments with gibberellic acid. Pronounced retardation of the loss of chlorophyll and the increase in carotenoids which accompany ripening has been obtained by preharvest treatments with gibberellic acid (Abdel-Gawad and Romani 1974; Eilati et al. 1969). These pigment changes are associated with the transformation of chloroplasts to chromoplasts, and in oranges treated with gibberellic acid chromoplasts have reverted back to chloroplasts (Thompson et al. 1967).

Other observations which bear on the question of the role of gibberellins in regulating the onset of ripening are those obtained from studies with tomatoes. Treatment of preclimacteric fruit had little or no effect on the onset of the respiratory increase, and no effect on the onset of coloration or on the increase in softening, but retarded both coloration and softening once ripening had begun (Babbitt et al. 1973). Gibberellin-treated fruit were found to contain greatly reduced levels of polygalacturonase activity. Lack of a direct role for gibberellin in regulating the onset of ripening is suggested by work with the Nr mutant tomato. Although fruits of this mutant ripen later, soften more slowly, and develop greatly reduced levels of polygalacturonase activity, gibberellin levels are normal (Hobson 1967; McGlasson and Franklin unpublished). The lack of a close correlation between endogenous gibberellin levels and ripening is illustrated in Fig. 3.4–3.7.

Cytokinins

Leaf senescence and fruit ripening may be induced by detachment from the parent plant. Senescence in leaves can generally be delayed by treatment with cytokinins (Letham 1967) but results with fruits have been variable and their role in fruit ripening is unclear. The responses of fruits to treatments with cytokinins are similar to those obtained with gibberellins. Application of the synthetic cytokinin benzyladenine did not affect the time of onset of the respiratory climacteric in apples, apricots, and avocados (Abdel-Gawad and Romani 1974; Smock et al. 1962; Tingwa and Young 1975), but has been reported to delay the loss of chlorophyll and increase in carotenoids in some fruits (Abdel-Kader et al. 1966; Eilati et al. 1969). The effects of benzyladenine on rates of respiration and ethylene production have been variable but in those tissues where a clear retardation of senescence has been obtained the rate of respiration has been lowered, e.g. in strawberries (Dayawon and Shutak 1967) and broccoli (Dedolph et al. 1962).

The work on tomatoes by Varga and Bruinsma (1974) provides valuable insight into the role of cytokinins in fruit ripening. They increased the levels of extractable cytokinins in seeded fruits by reducing the ratio of foliage to fruits. The resulting high levels of endogenous cytokinins did not affect the onset of ripening (first color) but there was a reduction in the rate of coloration. Unfortunately, no data were given on other parameters of ripening in this study. An example of independent action of cytokinins on individual parameters of ripening was described by Wade and Brady (1971). They showed that in banana slices treated with a ripening concentration of ethylene the respiratory climacteric developed normally and the rate of starch hydrolysis was not affected by infiltrated kinetin although the loss of chlorophyll in the peel was delayed.

Information on changes in the levels of endogenous cytokinins, particularly during maturation and ripening, is sparse (Fig. 3.4–3.7). More data will be required before the role of endogenous cytokinins in ripening and senescence can be defined. On present knowledge the action of endogenous cytokinins may be indirect through effects on other hormones such as the gibberellins and ABA.

Abscisic Acid and Related Compounds

Abscisic acid (ABA) is the most recently discovered member of the five principal types of plant hormones. After ethylene it is the easiest hormone to assay. The standard method involves recovery from plant tissues by solvent extraction, then methylation followed by gas liquid chromatography using an electron capture detector. This detector is highly sensitive to the methyl ester of ABA and other compounds with electron capturing properties. ABA was first associated with bud dormancy, leaf and fruit abscission, and later with water stress and closure of leaf stomates. More recently it has been shown to accumulate in senescent tissues (Milborrow 1974).

Much work has been done on the determination of ABA levels and on its application to leaves and fruits, but its role in senescence and ripening is no clearer than for the other non-gaseous hormones. ABA treatments advance ripening in the climacteric fruits tomato, avocado, and banana (Adato and Gazit 1976; Bruinsma et al. 1975; Kader et al. 1973; Mizrahi et al. 1975), and in the nonclimacteric grape berry (Coombe and Hale 1973). At least in the tomato and the grape, the ABA treatment did not act through stimulated ethylene production.

A puzzling feature of the results with tomatoes and avocados is that a relatively high concentration of endogenous ABA is present in these fruits well before the onset of ripening. The addition of as little as $1/600$th of the endogenous level advances ripening in the tomato. A possible explanation for this apparent anomaly is that exogenous ABA enters intracellular compartments other than those in which endogenous ABA accumulates. This explanation is supported by studies with ^{14}C labelled ABA. About 80% of added ^{14}C-ABA is converted to a conjugated form, presumably a glycoside, within 24 hr (McGlasson and Franklin unpublished), whereas only about 12% of endogenous ABA is present in a conjugated form (McGlasson and Adato 1976).

Previous authors have proposed that ABA is involved in the initiation of increased ethylene production in climacteric fruits (Dilley 1969; Sacher 1973), and conversely that increased ABA levels are a consequence of a rise in ethylene production as found in the avocado (Adato et al. 1976). There is clear evidence that ethylene treatment can stimulate ABA accumulation in plant tissues (Adato and Gazit 1976; Goldschmidt et al. 1973; Mayak and Halevy 1972; Milborrow 1974). However, examination of some naturally ripened fruits shows that ABA either begins to accumulate when ethylene production is at a low level, e.g. tomato and grape (Fig. 3.4 and 3.7), or it accompanies the rise in ethylene production, e.g. avocado and peach (Fig. 3.5 and 3.6). Recent data lead to the conclusion that endogenous ABA is not involved directly in the regulation of ethylene synthesis, nor is its accumulation controlled by ethylene.

Although a direct involvement of ABA with ethylene production seems unlikely at least in fruit ripening, it may exert its effects by interacting with other hormones. Osborne et al. (1972) reported the discovery of a senescence factor which was separable from ABA in extracts of Eunonymus and bean leaves, and they proposed that ABA induces the production of a senescence factor which in turn controls ethylene production.

NEW RESEARCH APPROACHES

The foregoing summaries lead to the general conclusion that it is still not possible to define the role of any of the main types of hormones in ripening and senescence. Perhaps the clearest case has been established for ethylene. Certainly the practical value of maintaining ethylene at low levels in nonclimacteric plant organs and in fruits which are harvested and stored at a preclimacteric stage has been established. The senescence-retarding effects demonstrated in some tissues treated with auxins, gibberellic acid, and cytokinins suggest a natural role for these hormones. However, the situation for ABA, particularly in relation to fruit ripening, seems ambiguous.

The comparisons of endogenous levels of hormones in different species of fruits (Fig. 3.4–3.7) illustrate the diversity among species, the large changes which take place in a fruit during development and senescence, and the lack of close correlations between the levels of extractable hormones and the stage of development. An example of genetic diversity within a single species, the tomato, is illustrated in Fig. 3.8. Fruits of the

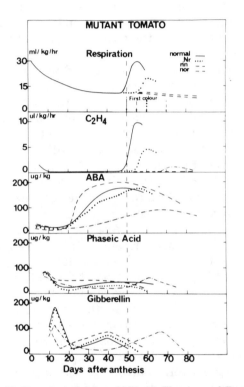

From McGlasson and Adato 1976; McGlasson and Franklin (unpublished)

FIG. 3.8. SCHEMATIC DIAGRAMS OF CHANGES IN RESPIRATION, ETHYLENE EVOLUTION AND IN THE CONCENTRATIONS OF ABA, PHASEIC ACID AND ACIDIC GIBBERELLINS DURING DEVELOPMENT AND SENESCENCE OF A NORMAL TOMATO CULTIVAR AND THE RIPENING MUTANTS *NR, RIN* AND *NOR*

Fruits of these mutant strains in a common genetic background (Rutgers) reach full size in about 50 days after anthesis. Red color in Rutgers coincides with the end of growth and follows by about two days the onset of the respiratory climacteric and the associated increase in ethylene production. Coloring (yellow-orange) in *Nr* begins at about 60 days and softening is delayed. The respiratory climacteric and the rise in ethylene production are also delayed and attenuated. *Rin* and *nor* are both non-climacteric. Coloring is markedly delayed in *rin* (53 days, yellow) and *nor* (70 days, yellow-orange). Coloring in *nor* is accompanied by a small increase in ethylene production.

mutants *Nr, rin* and *nor* fail to undergo many of the changes associated with normal ripening. Although fruits of these mutants grow normally and take the same time as normal fruits to reach full size, the time course and concentrations of ethylene, ABA, phaseic acid (a metabolite of ABA), and gibberellins vary widely between strains. These findings do not deny a role for these hormones in fruit ripening, but they do indicate that hormone levels and their metabolism are secondary to some other genetic control system, and highlight the shortcomings of studies on gross hormone levels *per se*.

It is proposed that research on the role of hormones on ripening and senescence should proceed as follows:

1. Studies on fresh plant organs or parts used for food should seek to define the stage of development when senescence or ripening begins. The assumption here is that these events are genetically programmed, and a better understanding will be gained of the control systems involved if it is known when the genetic program begins to operate. It has been long recognized that different tissues within an organ develop and differentiate at different rates. There are many reports on hormone levels in seed tissues in relation to early fruit development, and some which recognize differences between regions in the timing of the onset of ripening. It is necessary to extend these studies at both the tissue and cellular level.

2. Studies on the biosynthesis and metabolism of the hormones should continue. Such studies could reveal control points for either inhibiting synthesis of hormones such as ethylene or retarding the degradation of such hormones as auxins or gibberellins. It is to be hoped that naturally occurring regulatory compounds will be discovered which may be acceptable as food treatments.

3. To utilize genetic diversity within species, *e.g.* the tomato ripening mutants. Work with genetic variants can help to resolve the role of the known types of hormones at the various stages in development and senescence of plant organs and may lead to the discovery of new regulants and to improved commercial varieties.

BIBLIOGRAPHY

ABDEL-GAWAD, H. and ROMANI, R. J. 1974. Hormone-induced reversal of color change and related respiratory effects in ripening apricot fruits. Physiol. Plant. *32*, 161–165.

ABDEL-KADER, A. S., MORRIS, L. L. and MAXIE, E. C. 1966. Effect of growth-regulating substances on the ripening and shelf-life of tomatoes. HortScience *1*, 90–91.

ADATO, I. and GAZIT, S. 1976. Response of harvested avocado fruits to supply of indole-3-acetic acid, gibberellic acid, and abscisic acid. J. Agric. Food Chem. 24, 1165–1167.

ADATO, I., GAZIT, S. and BLUMENFELD, A. 1976. Relationship between changes in abscisic acid and ethylene production during ripening of avocado fruits. Aust. J. Plant Physiol. 3, 555–558.

APELBAUM, A., GOLDSCHMIDT, E. E. and BEN-YEHOSHUA, S. 1976. Involvement of endogenous ethylene in the induction of color change in Shamouti oranges. Plant Physiol. 57, 836–838.

BABBITT, J. K., POWERS, M. J. and PATTERSON, M. E. 1973. Effects of growth regulators on cellulase, polygalacturonase, respiration, color, and texture of ripening tomatoes. J. Amer. Soc. Hort. Sci. 98, 77–81.

BEYER, E. M. 1976. A potent inhibition of ethylene in plants. Plant Physiol. 58, 268–271.

BITTNER, S., GAZIT, S. and BLUMENFELD, D. A. 1971. Isolation and identification of a plant growth inhibitor from avocado. Phytochem. 10, 1417–1421.

BLUMENFELD, A. and GAZIT, S. 1972. Gibberellin activity in the developing avocado fruit. Physiol. Plant. 27, 116–120.

BRUINSMA, J., KNEGT, E.and VARGA, A. 1975. The role of growth-regulating substances in fruit ripening. In Facteurs et Régulation de la Maturation des Fruits. Centre National de la Recherche Scientifique, Paris.

BURG, S. P. and BURG, E. A. 1966. Fruit storage at subatmospheric pressures. Science 153, 314–315.

CHALMERS, D. J. and VAN DEN ENDE, B. 1975. A reappraisal of the growth and development of peach fruit. Aust. J. Plant Physiol. 2, 623–634.

COOMBE, B. G. 1960. Relationship of growth and development to changes in sugars, auxins and gibberellins in fruit of seeded and seedless varieties of Vitis vinifera. Plant Physiol. 35, 241–250.

COOMBE, B. G. and HALE, C. R. 1973. The hormone content of ripening grape berries and the effects of growth substance treatments. Plant Physiol. 51, 629–634.

DAYAWON, M. M. and SHUTAK, V. G. 1967. Influence of N^6-benzyladenine on the postharvest rate of respiration of strawberries. HortScience 2, 12.

DEDOLPH, R. R., WITTWER, S. H., TULI, V. and GILBART, D. 1962. Effect of N^6-benzylamino purine on respiration and storage behavior of broccoli (Brassica oleracea var. italica). Plant Physiol. 37, 509–512.

DILLEY, D. R. 1969. Hormonal control of fruit ripening. HortScience 4, 111–114.

EILATI, S. K., GOLDSCHMIDT, E. E. and MONSELISE, S. P. 1969. Hormonal control of color changes in orange peel. Experientia 25, 209–210.

EL-BELTAGY, A. S., PATRICK, J. P., HEWETT, E. W. and HALL, M. A. 1976. Endogenous growth regulator levels in tomato fruits during development. J. Hort. Sci. 51, 15–30.

FAULL, K. F., COOMBE, B. G. and PALEG, L. G. 1974. Extraction and characterization of gibberellins from Hordeum vulgare L. seedlings. Aust. J. Plant Physiol. 1, 183–198.

FRENKEL, C. 1975. Role of oxidative metabolism in the regulation of fruit ripening. In Facteurs et Régulation de la Maturation des Fruits. Centre National de la Recherche Scientifique, Paris.

FRENKEL, C. and DYCK, R. 1973. Auxin inhibition of ripening. Plant Physiol. 51, 6–9.

GAZIT, S. and BLUMENFELD, A. 1970A. Response of mature avocado fruits to ethylene treatments before and after harvest. J. Amer. Soc. Hort. Sci. *95*, 229–231.

GAZIT, S. and BLUMENFELD, A. 1970B. Cytokinin and inhibitor activities in avocado fruit mesocarp. Plant Physiol. *46*, 334–336.

GAZIT, S. and BLUMENFELD, A. 1972. Inhibitor and auxin activity in the avocado fruit. Physiol. Plant. *27*, 77–82.

GOLDSCHMIDT, E. E., GOREN, R., EVEN-CHEN, Z. and BITTNER, S. 1973. Increase in free and bound abscisic acid during natural and ethylene induced senescence in citrus fruit peel. Plant Physiol. *51*, 879–882.

HOBSON, G. E. 1967. The effect of alleles at the 'Never Ripe' locus on the ripening of tomato fruit. Phytochem. *6*, 1337–1341.

IWAHORI, S., WEAVER, R. J. and POOL, R. M. 1968. Gibberellin-like activity in berries of seeded and seedless Tokay grapes. Plant Physiol. *43*, 333–337.

JACKSON, D. I. 1968. Gibberellin and the growth of peach and apricot fruits. Aust. J. Biol. Sci. *21*, 209–215.

JERIE, P. H. and CHALMERS, D. J. 1976. Ethylene as a growth hormone in peach fruit. Aust. J. Plant Physiol. *3*, 429–434.

KADER, A. A., MORRIS, L. L. and LYONS, J. M. 1973. Effect of abscisic acid on ripening of tomato fruits. HortScience *8*, 264 (Abstract).

KOSIYACHINDA, S. and YOUNG, R. E. 1975. Ethylene production in relation to the initiation of respiratory climacteric in fruit. Plant Cell Physiol. *16*, 595–602.

LAU, O. and YANG, S. F. 1976. Inhibition of ethylene production by cobaltous ion. Plant Physiol. *58*, 114–117.

LEOPOLD, A. C. and KRIEDEMANN, P. E. 1975. Plant Growth and Development. McGraw-Hill Inc., New York.

LETHAM, D. S. 1967. Chemistry and physiology of kinetin-like compounds. Ann. Rev. Plant Physiol. *18*, 349–364.

LIEBERMAN, M., KUNISHI, A. T. and OWENS, L. D. 1975. Specific inhibitors of ethylene production as retardants of the ripening process in fruits. *In* Facteurs et Régulation de la Maturation des Fruits. Centre National de la Recherche Scientifique, Paris.

LOONEY, N. E. 1973. Control of fruit maturation and ripening with growth regulators. Acta Horticulturae. No. 34, 397–406.

LOONEY, N. E., McGLASSON, W. B. and COOMBE, B. G. 1974. Action of succinic acid-2,2-dimethylhydrazide and (2-chloroethyl) phosphonic acid. Aust. J. Plant Physiol. *1*, 77–86.

LYONS, J. M. and PRATT, H. K. 1964. Effect of maturity and ethylene treatment on respiration and ripening of tomato fruits. Proc. Amer. Soc. Hort. Sci. *84*, 491–500.

McGLASSON, W. B. and ADATO, I. 1976. Changes in the concentrations of abscisic acid in fruits of normal and *Nr, rin* and *nor* mutant tomatoes during growth, maturation and senescence. Aust. J. Plant Physiol. *3*, 809–817.

McGLASSON, W. B. and PRATT, H. K. 1964. Effects of ethylene on cantaloupe fruits harvested at various ages. Plant Physiol. *39*, 120–127.

McGLASSON, W. B., DOSTAL, H. C. and TIGCHELAAR, E. C. 1975. Comparison of propylene-induced responses of immature fruit of normal and *rin* mutant tomatoes. Plant Physiol. *55*, 218–222.

McGLASSON, W. B., WADE, N. L. and ADATO, I. 1978. Phytohormones and fruit ripening. *In* Plant Hormones and Related Compounds—A Comprehensive Treatise. Vol. 2. D. S. Letham, P. B. Goodwin and T. J. V. Higgins (Editors). ASP Biological and Medical Press B. V. Amsterdam. (In press).

McMURCHIE, E. J., McGLASSON, W. B. and EAKS, I. L. 1972. Treatment of fruit with propylene gives information about the biogenesis of ethylene. Nature *237*, 235–236.

MAYAK, S. and HALEVY, A. H. 1972. Interrelationships of ethylene and abscisic acid in the control of rose petal senescence. Plant Physiol. *50*, 341–346.

MILBORROW, B. V. 1974. The chemistry and physiology of abscisic acid. 1974. Annu. Rev. Plant Physiol. *25*, 259–307.

MIZRAHI, Y., DOSTAL, H. C., McGLASSON, W. B. and CHERRY, J. H. 1975. Effect of abscisic acid and benzyladenine on fruits of normal and *rin* mutant tomatoes. Plant Physiol. *56*, 544–546.

NITSCH, J. P. 1970. Hormonal factors in growth and development. *In* The Biochemistry of Fruits and their Products, Vol. 1, A. C. Hulme (Editor). Academic Press, London and New York.

OSBORNE, D. J., JACKSON, M. B. and MILBORROW, B. V. 1972. Physiological properties of abscision accelerator from senescent leaves. Nature, New Biol. *240*, 98–101.

OWENS, L. D., LIEBERMAN, M. and KUNISHI, A. 1971. Inhibition of ethylene production by rhizobitoxine. Plant Physiol. *48*, 1–4.

PARUPS, E. V. 1973. Control of ethylene-induced responses in plants by a substituted benzothiadiazole. Physiol. Plant. *29*, 365–370.

PATIL, S. S. and TANG, C. 1974. Inhibition of ethylene evolution in papaya pulp tissue by benzylisothiocyanate. Plant Physiol. *53*, 585–588.

PRATT, H. K. and GOESCHL, J. D. 1969. Physiological roles of ethylene in plants. Annu. Rev. Plant Physiol. *20*, 541–584.

QUAZI, M. H. and FREEBAIRN, H. T. 1970. The influence of ethylene, oxygen, and carbon dioxide on the ripening of bananas. Bot. Gaz. *131*, 5–14.

SACHER, J. A. 1973. Senescence and postharvest physiology. Annu. Rev. Plant Physiol. *24*, 197–224.

SAKAI, S. and IMASEKI, H. 1973. Properties of the proteinaceous inhibitor of ethylene synthesis: Action on ethylene production and indoleacetyl-aspartate formation. Plant Cell Physiol. *14*, 881–892.

SCOTT, K. J. and GANDANEGARA, S. 1974. Effect of temperature on the storage life of bananas held in polyethylene bags with ethylene absorbent. Tropic. Agric. Trin. *51*, 23–26.

SCOTT, K. J. and WILLS, R. B. H. 1973. Atmospheric pollutants destroyed in an ultraviolet scrubber. Lab. Prac. *22*, 103–106.

SCOTT, K. J. and WILLS, R. B. H. 1974. Reduction of brown heart in pears by absorption of ethylene from the storage atmosphere. Aust. J. Exp. Agric. Anim. Husb. *14*, 266–268.

SCOTT, K. J., WILLS, R. B. H. and PATTERSON, B. D. 1971. Removal by ultraviolet lamp of ethylene and other hydrocarbons produced by bananas. J. Sci. Food Agric. *22*, 496–497.

SMOCK, R. M., MARTIN, D. and PADFIELD, C. A. S. 1962. Effect of N^6-benzyladenine on the respiration and keeping quality of apples. Proc. Amer. Soc. Hort. Sci. *81*, 51–56.

THOMPSON, W. W., LEWIS, L. N. and COGGINS, C. W. 1967. The reversion of chromoplasts to chloroplasts in Valencia oranges. Cytologia *32*, 117–124.

TINGWA, P. O. and YOUNG, R. E. 1975. Studies on the inhibition of ripening in attached avocado (*Persea americana* Mill.) fruits. J. Amer. Soc. Hort. Sci. *100*, 447–449.

VARGA, A. and BRUINSMA, J. 1974. The growth and ripening of tomato fruit at different levels of endogenous cytokinins. J. Hort. Sci. *49*, 135–142.

VENDRELL, M. 1969. Reversion of senescence: effects of 2,4-dichloro-phenoxyacetic acid on respiration, ethylene production, and ripening of banana fruit slices. Aust. J. Biol. Sci. *22*, 601–610.

WADE, N. L. and BRADY, C. J. 1971. Effects of kinetin on respiration, ethylene production and ripening of banana fruit slices. Aust. J. Biol. Sci. *24*, 165–167.

WILD, B. L., McGLASSON, W. B. and LEE, T. H. 1976. Effect of reduced ethylene levels in storage atmospheres on lemon keeping quality. HortScience *11*, 114–115.

WITTWER, S. H. 1971. Growth regulants in agriculture. Outlook Agric. *6*, 205–217.

YANG, S. F. 1975. Ethylene biosynthesis in fruit tissues. *In* Facteurs et Régulation de la Maturation des Fruits. Centre National de la Recherche Scientifique, Paris.

CHAPTER 4

BIOCHEMICAL AND PHYSIOLOGICAL EFFECTS OF MODIFIED ATMOSPHERES AND THEIR ROLE IN QUALITY MAINTENANCE

W. G. BURTON

Formerly Food Research Institute
Norwich, England

ABSTRACT

Modified atmospheres influence deterioration resulting from biochemical reactions. Effective concentrations of the atmospheric components are those at the seat of reaction. These depend upon the dynamic equilibria established between ambient concentrations and production or consumption; involving the Michaelis constants of the enzymes; diffusion, both in solution and in the intercellular space; equilibration between the gas phase and dissolved gas; and diffusion through the outer integument. Individuals in apparently uniform samples of single cultivars vary, both in the rates of production and consumption and in permeability. Precise control of internal concentrations is scarcely possible. Thus safe lower levels of ambient oxygen are not those, giving $< 0.1\%$ internal O_2, which can markedly influence the activity of cytochrome-c oxidase, the main terminal oxidase of respiration; but those (often 2–3%, giving internal 1–2%) which eliminate low-oxygen-affinity oxidase activity. One physiologically important result is the reduction of ethylene production.

Some reactions are beneficially influenced by atmospheres which adversely affect others: 2–3% O_2 reduces apple scald but leads to off-flavors in raspberries and strawberries; high CO_2 delays loss of chlorophyll but can lead to discolorations such as brown-heart in apples, and texture faults in leaves. There are many facets of quality and many of the biochemical reactions involved in its loss are unknown.

Modified atmospheres can reduce rotting by pathogens, often by delaying ripening of fruit, ripe fruit being more susceptible to attack, rather than because of direct effects on the pathogens.

INTRODUCTION

We can classify the forms of deterioration to which harvested plant material is subject into three groups: microbiological attack, water loss and biochemical change. This excludes an obvious form of deterioration in which the plant material itself is not actively concerned, namely attack by animal pests, and after this mention I shall say no more about this.

The immediate effect of modified atmospheres is upon biochemical reactions in which the constituents of the atmosphere participate. There are secondary effects upon reactions which are dependent upon, or can be influenced by, these primary reactions. The reactions affected may be in the commodity we are storing or in pathogens which attack it, and thus a modified atmosphere has a potential scope for inhibiting or retarding such attack. In practice, however, atmospheres which inhibit pathogens can well be harmful to the host, or to its reaction to the pathogen, and this limits the scope for their use.

The obvious method of delaying continued metabolic change in storage—such as opening of inflorescences, lignification, break of dormancy and various forms of physiological deterioration which we can describe as senescence—is to reduce the temperature. In practice, this is almost invariably successful in giving a useful extension of storage life. Controlled atmosphere storage may then be used to give a further extension, and this is the typical form of its use. It may also be considered as an alternative to low temperature storage for those commodities which are susceptible to low temperature injury.

If we are thinking of the practical application of controlled atmosphere storage we are thus usually concerned with a commodity of which the useful storage life is terminated by metabolic change which is not sufficiently controlled by a low temperature. Very often we are thinking of commodities such as pome fruits, which are harvested immature or unripe, and which ripen in storage to a stage at which they are marketable. We wish to control the rate of ripening, and we wish to avoid its continuing to a stage of unacceptable over-ripeness, or senescence, during the desired commercial life in store.

The earliest experiments on controlled atmosphere storage were based, first on some evidence that, in broad terms, storage life was inversely correlated with the intensity of respiration; and, second, on the application of the Law of Mass Action to the simple concept of respiration being the oxidation of a sugar, through at least partly reversible steps, to give carbon dioxide and water. This concept would suggest that if the carbon dioxide in the atmosphere were augmented, or the oxygen decreased, respiration would be reduced and storage life extended.

Whatever over-simplifications may now be found in the theoretical basis, these early experiments were successful, and commercial use of modified atmospheres started in the 1920's and developed rapidly. There was then little knowledge of the biochemistry of plant metabolism. The individual reactions involved in the extension of storage life were not identified, and in fact in many cases are not identified to this day. In this paper I shall not be able in any way to give a detailed description of the biochemical effects of modified atmospheres. I shall describe a number of physiological effects, and I shall attempt to apply a little simple physics and logic to a consideration of the application of modified atmospheres.

The reactions which we hope to influence occur in the cells of the commodity, localized to a greater or less degree, for instance in the mitochondria. It is at the seat of the reaction that the modification of the atmosphere is effective, and in considering the effects of a modification we must know what the effective change is. It could be very different from the modification of the ambient atmosphere.

MODIFIED ATMOSPHERES

Exogenous and Endogenous Gases

Exogenous gases, such as oxygen and nitrogen—and I will restrict further consideration to oxygen—reach the centers of reaction by gaseous diffusion, in response to a concentration gradient, through the outer integument of the material and through its intercellular spaces. The final movement into the individual cell, and to the oxidase system which combines the oxygen, and thus activates the gradient, is in solution, again of course following a concentration gradient.

Endogenous gases, such as carbon dioxide and ethylene, are produced at the appropriate center of enzyme action in the cell, thus increasing the local concentration of dissolved gas. The resultant concentration gradient leads to movement in solution to the surface of the cell, where equilibration occurs with the gas in the adjacent intercellular space, thus raising its local partial pressure and activating gaseous diffusion through the tissue to the surface. Here the gas encounters the barrier of whatever forms the outer integument, and must diffuse through it in response to the gradient between the partial pressures in the intercellular space and the ambient atmosphere.

In any plant material, stored under constant conditions, a state of dynamic equilibrium is reached in which the concentrations of dissolved gases, both exogenous and endogenous, at the points of reaction, are those which lead to concentration gradients large enough to maintain the movement in of oxygen at the rate at which it is being combined at the equilibrium molecular concentration; and the movement out of the endogenous gases at the rate at which they are being produced.

In most comparatively massive organs, such as pome fruits, nearly all the movement of gases to and from the seat of reaction is by diffusion in the gas phase. Only distances which we can estimate to be a few, or at most tens of micrometers, are traversed in solution. Perhaps it will be as well to dispose of this spacially minor, though physiologically essential movement first. Movement into and in solution is very much slower than in the gas phase—by a factor of about 300,000. For a consideration of it I could not do better than refer you to Goddard (1947). In the present context it is sufficient to know that at the normal rate of oxygen uptake observed in the cells of an apple fruit, if the intercellular atmosphere contained 10% O_2, a high-oxygen-affinity enzyme such as cytochrome-c oxidase would maintain virtually anaerobic conditions at any point which depended upon diffusion in solution for a distance of as much as about 2 mm. It is therefore quite possible to visualize a commodity which is essentially anaerobic because a high proportion of its cells are not in contact with an interconnected system of intercellular spaces, even though the oxygen concentration in the spaces themselves is practically the same as that in the surrounding air.

Oxygen Status

Burton (1950) considered the question of oxygen status in detail for the potato tuber, which has only 1–2% of its volume as intercellular space, and showed that despite this small volume the cells were practically saturated with oxygen at the partial pressure in the space. We must conclude that an adequate proportion of the surface area of almost every individual cell in the tuber is in contact with an interconnected system of intercellular spaces. The intercellular space of the apple fruit is very much larger than that of the potato—Smith (1938A) reported it to be as much as 36% of the volume of the fruit in some Lord Derby apples. This does not necessarily mean that it is more efficient. It could be very large in some parts of the fruit but not in contact with every cell. The results of Brändle (1968) however, suggest that, at least outside the vascular region, the tissue is fairly uniformly aerated. His results might also suggest that the vascular region constitutes some barrier to diffusion in the gas phase, in that he found aeration to be less effective inside it than outside it, though again fairly uniform.

In general we are probably on fairly safe ground if we assume the intercellular space of most plant organs to be an efficient interconnected system throughout a large part of their volume, and the plant sap to be nearly saturated with oxygen at the intercellular partial pressure, and only slightly super-saturated with carbon dioxide.

If we turn now to diffusion in the gas phase in the intercellular space, it is clear from a great deal of work, from the classical experiments of Devaux (1890, 1891) onwards, that there is little resistance to diffusion through most of the space at the rates encountered in plant respiration. Even in the potato, with its very small space, the resistance is so slight that it is difficult to detect any consistent oxygen gradient from periphery to center, either by analysis (Burton 1950) or by oxygen electrode measurements (Brändle 1968). In some instances there may be local barriers to diffusion—for example in the vascular region of the apple fruit, but the evidence for such barriers is not consistent. As stated before, the results of Brändle would support such a concept, but on the other hand there are several results which do not (Smith 1947; Hardy 1949; Hulme 1951; Burg and Burg 1965).

Although the intercellular space may offer little resistance to gaseous diffusion, and may ramify to such an extent that the cell sap is practically equilibrated with the gases in it, this does not necessarily mean that the oxygen and carbon dioxide status of the tissue is little different from that of the ambient atmosphere. Between the intercellular atmosphere and the ambient atmosphere there is the barrier of whatever constitutes the outer integument. This, suberized or cuticularized, is largely impermeable to the movement of gases except where it is pierced by pores, such as lenticels or, in the case of leafy vegetables, stomata. The latter are very numerous—about 100–300 per mm^2 on at least one surface of the leaf—and when they are open the surface of the leaf presents little barrier to diffusion and its internal atmosphere must approximate very closely to that of its surroundings. Lenticels are very much less frequent—there is roughly 1 per cm^2 on the surface of the potato tuber. They are much larger than stomata, but by no means all the opening is effective in permitting gaseous diffusion because of the infilling cells—in fact as a rough measure we may take their permeability to be limited to that of the intercellular spaces immediately underlying them. As a result, an integument pierced by infrequent lenticels presents an appreciable barrier to diffusion. We may obtain a measure of this resistance from a knowledge of the concentrations of gas immediately inside and outside it, and the rate of movement through it. These are simple to determine with sufficient accuracy for our purpose, and the results of some such determinations of resistance are given in Table 4.1, not expressed in terms of the coefficient of resistance, 'k', which would involve also measurement of the length of the diffusion path, but in the purely practical terms of ml of gas per kg of commodity per 1% difference in concentration between inside and outside.

Given a knowledge of the rate of CO_2 output or O_2 uptake, we can readily calculate the composition of the internal atmosphere immediately under the skin for samples of fruit or potatoes with skin permeabilities similar to

TABLE 4.1

RATES OF DIFFUSION THROUGH THE INTEGUMENT, AND INTERNAL
ATMOSPHERES IMMEDIATELY UNDER THE INTEGUMENT, OF SOME
SAMPLES OF APPLE, PEAR AND POTATO

Commodity	Internal Atmosphere (%) Immediately Under Skin		Diffusion Through Skin, ml kg^{-1}h^{-1}per 1% gradient		Reference
	CO_2	O_2	CO_2	O_2	
Apple, freshly harvested, 12° C					
Cox's Orange Pippin	3.9	16.1	2.3	1.8[1]	Kidd and West (1949A, B)
King Edward VII	2.5	17.9	1.8	1.5[1]	Kidd and West (1949A, B)
Sturmer Pippin	1.8	18.9	3.7	3.3[1]	Kidd and West (1949A, B)
Pear, 5 days after harvest, 12° C					
Conference	1.8	18.9	2.4	2.1[1]	Kidd and West (1949A, B)
Potato, stored					
Arran Consul, 10° C	1.8	18.2	1.0	0.7[1]	Burton (1950)
Arran Consul, 25° C	7.0	13.0	0.8	0.7[1]	Burton (1950)
Majestic, 10° C	2.2	18.3	1.1	0.9	Burton (1974B)

[1]Derived from CO_2 output on the assumption that RQ = 1.

those illustrated. For example with an uptake and output of 4ml kg^{-1} h^{-1} we would expect that in air the peripheral intercellular space of the potatoes would contain about 4% CO_2 and 16% O_2, while that of the fruit would contain about 1–2% CO_2 and from 18 to nearly 20% O_2. Allowing for observed maximum gradients in the flesh, and for increased resistance in the vascular region as suggested by Brändle's determinations of oxygen status, we might expect the internal atmosphere in the whole of the apple flesh outside the vascular region to contain 2–4% CO_2 and 17–19% O_2; but inside that region to contain about 6–9% CO_2 and 12–15% O_2. In the potato we would expect 4–5% CO_2 and 15–16% O_2 throughout.

Burton (1974A) calculated probable internal atmospheres of apples in typical gas mixtures, allowing for changes in respiration, caused by the modified atmospheres, as found by Fidler and North (1967). For example, at 4° C, in 11% O_2 and 10% CO_2, he suggested equilibrium to be established at about 8–10% O_2 and 11–12% CO_2 in the intercellular space outside the vascular region, and about 5–6% O_2 and 13–14% CO_2 within it. In an atmosphere containing 3% O_2 and 5% CO_2 we might expect the intercellular space to contain 1–2% O_2 and 6–7% CO_2 in the outer zone, and less than 1% O_2 and 8–9% CO_2 in the inner zone. The basis for Burton's calculations is given in Table 4.2.

TABLE 4.2
CALCULATED INTERCELLULAR ATMOSPHERES OF APPLES IN AIR AND
MODIFIED ATMOSPHERES

	Air	11% O_2 10% CO_2	3% O_2 5% CO_2
1. Ambient atmosphere:	Air	11% O_2 10% CO_2	3% O_2 5% CO_2
2. Assumed rate of O_2 uptake, ml $kg^{-1}h^{-1}$:	4	2.4[1]	1.6[1]
3. Assumed rate of CO_2 output, ml $kg^{-1}h^{-1}$:	4	1.6[1]	1.3[1]
4. Partial pressure difference (atm) across the skin necessary to activate:			
assumed O_2 uptake[2]:	1.2–2.7%	0.7–1.6%	0.5–1.1%
assumed CO_2 output[2]:	1.1–2.2%	0.4–0.9%	0.4–0.7%
5. Approximate intercellular atmosphere below the skin (derived from 1 and 4 above)			
O_2:	18–20%	9–10%	c. 2%
CO_2:	1–2%	10–11%	c. 6%
6. Approximate gradient of both O_2 and CO_2 (atm):			
from below skin to outside vascular region[3]:	ca 1%	<1%	<1%
from outside vascular region to core[3]:	ca 5%	ca 3%	up to 2%

[1]Proportionate reduction in modified atmosphere based upon the results of Fidler and North (1967).
[2]Based upon the figures, derived from Kidd and West (1949A), given in Table 4.1.
[3]Based upon the determinations of oxygen status by Brändle (1968).

Changing Atmospheric Pressure

Apart from changing the concentrations of the gases in the atmosphere, modified atmospheres are attainable by changing the atmospheric pressure. Reducing this to one-tenth reduces the effective partial pressure of oxygen to the equivalent of 2% at normal atmospheric pressure. Allowing for reduction in the uptake of oxygen consequent upon the reduced partial pressure, and for correspondingly reduced carbon dioxide output, one might expect the partial pressure of oxygen in the intercellular space to be about 0.01 atm (about 1kPa) near the periphery, falling to something approaching zero near the center. The partial pressure of carbon dioxide might be expected to range from perhaps 0.02–0.03 atm in the inner zone to perhaps 0.01 atm at the periphery.

Commodities are not uniform of course, and I would not wish to suggest that my calculations give anything more than an indication of typical values. Kidd and West (1949A) found the intercellular CO_2, near the periphery of six individual Cox's Orange Pippin apples, to vary from 3.37–6.50% and the output of CO_2 from 6.2–8.0 $mlkg^{-1}h^{-1}$, the two variables being uncorrelated. The peripheral intercellular oxygen varied from

12.72–16.32%. If we placed these apples in 5% CO_2 and 3% O_2 we might expect the peripheral intercellular CO_2 to vary from 6.0–7.2% and the O_2 from 0–1.3%. I think you should bear this variability in mind when you are considering recommendations for optimum modified atmospheres specified to within, say, 0.1%. However accurately you think you can specify an ambient atmosphere, your commodity will not always oblige with equal precision.

Suppressing Enzyme Activity

If we are wishing to prevent the operation of an oxidase with a comparatively low affinity for oxygen, then the fact that we cannot achieve precision in our control of internal atmosphere does not matter. For example, let us suppose that we can improve storage life by preventing the operation of polyphenolase. Values given for the Michaelis constant of potato polyphenolase vary from $1 \times 10^{-4}M$ (Abukhama and Woolhouse 1966) to about $5 \times 10^{-4}M$ (Mapson and Burton 1962). Any reduction in oxygen concentration below normal ambient will reduce oxygen uptake by the enzyme, and we can virtually eliminate it by the time we have achieved an internal oxygen concentration of 1–2%, given, in apples, as well as in potatoes, by an ambient of 2–4%. Whatever the variability in the sample, we can reasonably guarantee that if we use a modified atmosphere containing 3% O_2, very little polyphenolase activity will be possible. It matters little whether we have achieved an internal concentration of 2.5% or of 0.5%, the result will be much the same.

The situation is very different if, to increase storage life, we need to reduce the main respiratory process with the linked diversion of high energy compounds into the metabolic pathways. The terminal oxidase is probably mainly cytochrome-c oxidase. The Michaelis constant for the combination of this with oxygen is variously reported in the literature at values ranging from $3 \times 10^{-6}M$ (Thimann et al. 1954) to $7 \times 10^{-8}M$ (Burton 1974B). In any case the affinity for oxygen is very high and we will not influence the basic respiration appreciably unless the oxygen concentration in the intercellular space is reduced to well below 1%—in fact with the probable lower value for the Michaelis constant, to about 0.1%. If we are to decrease the basal metabolism to any marked extent in the bulk of the tissue, then the center of a fruit or tuber will be anaerobic. In any case 0.1% O_2 is itself very close to anaerobic, and individual variation between fruit of even one cultivar is such that precision of that order may be thought to be unattainable.

It may therefore be suggested that the beneficial effects of storage in low oxygen result more from the suppression of the activity of comparatively low-oxygen-affinity enzymes such as polyphenolase, b-type cytochromes, ascorbic acid oxidase and glycolic acid oxidase, than from suppression of the basal metabolism mediated by cytochrome-c oxidase. These beneficial effects should be attainable in ambient oxygen concentrations of not less than about 2% oxygen. Further reduction in oxygen concentration, to a level influencing uptake by cytochrome-c oxidase, could have marked effects upon the metabolism of the stored material, possibly beneficial though not necessarily so, and with danger of anaerobiosis in part of the material.

Increasing CO_2 in Atmosphere

Turning now to the effects of increasing the carbon dioxide in the atmosphere, we might expect that the reactions immediately influenced would be reversible decarboxylations such as those involving pyruvate, citrate and α-ketoglutarate. The further effects, with the accumulation and re-channelling of respiratory intermediates, could be expected to be far-reaching, but I know of no detailed elucidation of the pathways by which storage characteristics, such as texture, color and flavor are influenced. Some effects are known, as for example the increase in citric and succinic acids in pears (Williams and Patterson 1964).

Ethylene

I have not hitherto mentioned ethylene, of great importance as a ripening hormone. Some of the effects of low oxygen must stem from the fact that the post-climacteric production of ethylene by ripening fruit is oxygen dependent, clearly at levels involving low-oxygen-affinity enzymes rather than cytochrome-c oxidase. Burg (1962) found production by apples to be affected as the ambient oxygen concentration was reduced below 8%, and to be halved by 2.5% O_2. Burton (1974A) discussed ethylene diffusion in the tissue and the probable physiologically active level.

RESULTS USING MODIFIED ATMOSPHERES

Let us now consider what results have been achieved by the use of modified atmospheres.

Low Oxygen Concentrations

First, low oxygen, by which I mean oxygen concentrations of 2–3%, sufficient to prevent reactions involving low-oxygen-affinity oxidases. These include prevention of discolorations at the cut surfaces of broccoli (Smith 1938B) and lettuce (Singh et al. 1972), typically caused by the oxidation of phenolic substances by phenolases; and of the loss of carotene (Baumann 1973) and sucrose (Hansen and Rumpf 1973) in stored carrots;

prevention of the formation of *iso*-coumarin, with resultant bitter flavor, in carrots (Hansen and Rumpf 1973); and of the incidence of scald in stored apples (Fidler and North 1961), which may result from oxidation products of α-farnesene (Huelin and Murray 1966). Low oxygen has been observed to delay the loss of chlorophyll in stored apples (Kidd and West 1934) and Savoy cabbage (Stoll 1973), but not in lettuce (Singh *et al*. 1972), and we must await a detailed biochemical elucidation of chlorophyll degradation to explain any discrepant results.

It may be mentioned in passing that any marked climacteric rise in oxygen uptake by stored fruit is a physical impossibility in low oxygen, as the partial pressure which activates the diffusion of oxygen into the fruit cannot exceed 2–3%, even with zero partial pressure in the peripheral intercellular space.

Modified atmospheres are not normally considered for the storage of potatoes, but low oxygen concentrations have marked stimulatory effects upon the break of bud dormancy in potato tubers, which occurs most readily in 2–5% O_2, and also upon subsequent sprout growth (optimal growth in *ca* 5% O_2).

There are deleterious as well as advantageous effects of low oxygen. For example, raspberries and strawberries develop off-flavors in 3% O_2.

There have recently been reports of the use of very low oxygen concentrations for apple storage: 1.25% has been specified, coupled with 0.75% CO_2, degrees of precision for which I cannot understand the biophysical basis. Such a low oxygen concentration could well reduce combination of O_2 with cytochrome-c oxidase in part of the tissue, but it is at a level at which sample variation could lead to anaerobiosis in some fruit.

Increased CO_2 Concentrations

Both texture and loss of chlorophyll are influenced by increased concentrations of carbon dioxide. Thus 10% CO_2 will retain the green color of apples, though it cannot be employed if the cultivar is susceptible to CO_2 injury; and up to 15% may improve the appearance of green vegetables and salad crops (Smith 1938B; McGill *et al*. 1966; Wang *et al*. 1971), though there may be undesirable effects on flavor.

Change in texture during the ripening of fruit results from reactions such as the enzymic breakdown of protopectin (Hulme and Rhodes 1971). These changes are retarded in apples to an increasing extent as the carbon dioxide concentration is raised up to 12%. The effects on green vegetables may not be desirable, Smith (1938B) stating that the leaves of broccoli stored in 5% O_2, 15% CO_2 were brittle.

Smith (1939) found the 'blowing' of broccoli to be prevented at 3.3° C by 10% CO_2, the effect being retained after three days subsequently in air at

10–15° C. He also (Smith 1965) found elongation of the stipes of stored button mushrooms to be depressed by increased CO_2, which also delayed the opening of the caps. These effects are in contrast to the influence on the sprout growth of potatoes, which at 10° C is stimulated by increasing the concentration of ambient CO_2 to 2–4%, giving a concentration in the sap, optimal for growth, of about $2 \times 10^{-3}M$ (Burton 1958).

Increased CO_2 has been investigated as a possible means of preventing the undesirable accumulation of reducing sugars in potatoes held at moderately low temperatures. 5% CO_2 has been found to prevent this at 5° C, although it leads to the accumulation of sucrose (Denny and Thornton 1941), suggesting an influence on the reactions leading to the formation of hexoses from sucrose, in contrast to the effect of low oxygen, which inactivates the sucrose synthesizing system (Nelson and Auchincloss 1933; Harkett 1971).

Rotting by pathogens can be much reduced by controlled atmosphere storage. 50% CO_2 will, for instance, completely suppress fungal rotting of blackcurrants at 4.4° C (see *e.g.* Smith 1957). Most commodities will not tolerate such high levels of CO_2, but useful control of apple rots is given by 8–10%, as in the original unscrubbed controlled atmosphere stores, and also by 2–3% O_2. These latter effects are not caused by direct control of the pathogen by the atmosphere—in the case of low oxygen, Shaw (1969) suggested that for such control a concentration as low as 0.5% might be necessary (see also Follstad 1966) and damage to the commodity would be likely. In considering the effects of controlled atmosphere storage on rotting it must be remembered that a rot is the result of an interaction between pathogen and host. The atmosphere affects the pathogen, but it also affects the host, and the net effect may be beneficial or otherwise. Reduced fungal rotting of apples in controlled atmosphere storage probably results in the main from its effect in delaying ripening and the resultant increased susceptibility of the fruit (Kidd *et al.* 1927), which means that 2–3% O_2 can be beneficial despite there being no direct effect on the pathogen. Shaw (1969) concluded that the beneficial effects of 5% and 20% CO_2 on *Botrytis* and *Rhizopus* rots of strawberries resulted from effects on the fruit, rendering them a substrate less favorable to the pathogens.

Commercial Applications

Commercial controlled atmosphere storage is concerned with retaining the marketability of commodities—primarily with their appearance, flavor and texture. Retention of nutritive value should, however, be mentioned, although only briefly. Fundamentally it is the most important function of storage, though it may mainly be achieved by the overall prevention of loss. Nutritive value may decline, as happens with the loss of vitamin-C from stored potatoes, but despite this the nutritive value of stored potatoes is still considerable, and retention of their edible weight in storage is a worthwhile objective.

The fruits and vegetables which are the main subjects of controlled atmosphere storage, particularly the pome fruits, are perhaps not of great nutritional significance, at least so far as is known, except as possible sources of dietary fiber. They provide variety and pleasant adjuncts to the diet. Controlled atmosphere storage enables them to be placed on the market over a greater period of time than would otherwise be the case. This contributes to the variety of the diet, but not necessarily to the nutrition of the nation. If it were devoted to preventing the loss of ascorbic acid in, for example, potatoes or other important antiscorbutic foods, the comment might be different, but evidence of effects of modified atmospheres on vitamin-C retention does not yet permit of practical conclusions (see *e.g.* Barker and Mapson 1952; McGill *et al.* 1966).

BIBLIOGRAPHY

ABUKHAMA, D. A. and WOOLHOUSE, H. W. 1966. Preparation and properties of o-diphenol: oxygen oxidoreductase from potato tubers. New Phytologist 65, 477–487.

BARKER, J. and MAPSON, L. W. 1952. The ascorbic acid content of potato tubers. 3. The influence of storage in nitrogen, air and pure oxygen. New Phytologist 51, 90–115.

BAUMANN, H. 1973. Preservation of carrot quality in various storage conditions. (Summary) Symposium, International Society for Horticultural Science, Freising-Weihenstephan, Sept. 3-7, 1973.

BRÄNDLE, R. 1968. Die Verteilung der Sauerstoffkonzentrationen in fleischigen Speicherorganen (Äpfel, Bananen und Kartoffelknollen). Bericht der Schweizerischen botanischen Gesellschaft 78, 330–364.

BURG, S. P. 1962. The physiology of ethylene formation. Ann. Rev. Plant Physiol. 13, 265–302.

BURG, S. P. and BURG, E. A. 1965. Gas exchange in fruits. Physiologia Plantarum 18, 870–884.

BURTON, W. G. 1950. Studies on the dormancy and sprouting of potatoes. I. The oxygen content of the potato tuber. New Phytologist 49, 121–134.

BURTON, W. G. 1958. The effect of the concentrations of carbon dioxide and oxygen in the storage atmosphere upon the sprouting of potatoes at 10° C. European Potato Journal 1(2), 47–57.

BURTON, W. G. 1974A. Some biophysical principles underlying the controlled atmosphere storage of plant material. Ann. Appl. Biol. 78, 149–168.

BURTON, W. G. 1974B. The oxygen uptake, in air and in 5% O_2, and the carbon dioxide output, of stored potato tubers. Potato Res. 17, 113–137.

DENNY, F. E. and THORNTON, N. C. 1941. Carbon dioxide prevents the rapid increase in the reducing sugar content of potato tubers stored at low temperatures. Contributions of the Boyce Thompson Institute for Plant Research 12, 79–84.

DEVAUX, H. 1890. Atmosphère interne des tubercules et racines tuberculeuses. Bulletin. Société botanique de France, *37*, 273–279.

DEVAUX, H. 1891. Étude expérimentale sur l'aération des tissus massifs. Annales des sciences naturelles (Botanique), 7 ser., *14*, 297–395.

FIDLER, J. C. and NORTH, C. J. 1961. Effect of various pre-treatments, and of concentration of oxygen in storage, on scald. Bulletin of the International Institute of Refrigeration, Annexe 1961-1, 175–178.

FIDLER, J. C. and NORTH, C. J. 1967. The effect of conditions of storage on the respiration of apples. I. The effects of temperature and concentrations of carbon dioxide and oxygen on the production of carbon dioxide and uptake of oxygen. J. Hort. Sci. *42*, 189–206.

FOLLSTAD, M. N. 1966. Mycelial growth rate and sporulation of *Alternaria tenuis, Botrytis cinerea, Cladosporium herbarum* and *Rhizopus stolonifer* in low oxygen atmospheres. Phytopathology *56*, 1098–1099.

GODDARD, D. R. 1947. The respiration of cells and tissues. *In* The Physical Chemistry of Cells and Tissues, R. Höber (Editor). Churchill, London.

HANSEN, H. and RUMPF, G. 1973. Storage of carrots: The influence of the storage atmosphere on taste, wastage and contents of sucrose, fructose and glucose. (Summary) International Society for Horticultural Science, Symposium, Freising-Weihenstephan, Sept. 3–7, 1973.

HARDY, J. K. 1949. Diffusion of gases in fruit: The solubility of carbon dioxide and other constants for Cox's Orange Pippin apples. Report of the Food Investigation Board, London, for 1939, 105–109.

HARKETT, P. J. 1971. The effect of oxygen concentration on the sugar content of potato tubers stored at low temperature. Potato Res. *14*, 305–311.

HUELIN, F. E. and MURRAY, K. E. 1966. α-Farnesene in the natural coating of apples. Nature (London) *210*, 1260–1261.

HULME, A. C. 1951. Apparatus for the measurement of gaseous conditions inside an apple fruit. J. Exp. Botany *2*, 65–85.

HULME, A. C. and RHODES, M. J. C. 1971. Pome Fruits. *In* The Biochemistry of Fruits and their Products, Vol. 2, 333–373, A. C. Hulme (Editor). Academic Press, London.

KIDD, F. and WEST, C. 1934. Injurious effects of pure oxygen upon apples and pears at low temperatures. Report of the Food Investigation Board, London, for 1933, 74–77.

KIDD, F. and WEST, C. 1949A. Resistance of the skin of the apple fruit to gaseous exchange. Report of the Food Investigation Board, London, for 1939, 59–64.

KIDD, F. and WEST, C. 1949B. Carbon dioxide injury in relation to the maturity of apples and pears. Report of the Food Investigation Board, London, for 1939, 64–68.

KIDD, F., WEST, C. and KIDD, M. N. 1927. Gas storage of fruit. Special Report of the Food Investigation Board, Department of Scientific and Industrial Research, London, 30.

MAPSON, L. W. and BURTON, W. G. 1962. The terminal oxidases of the potato tuber. Biochem. J. *82*, 19–25.

McGILL, J. N., NELSON, A. I. and STEINBERG, M. P. 1966. Effects of modified storage atmosphere on ascorbic acid and other quality characteristics of spinach. J. Food Sci. *31*, 510–517.

NELSON, J. M. and AUCHINCLOSS, R. 1933. The effects of glucose and fructose on the sucrose content of potato slices. J. Am. Chem. Soc. *55*, 3769–3772.

SHAW, G. W. 1969. The effects of controlled atmosphere storage on the quality and shelf life of fresh strawberries with special reference to *Botrytis cinerea* and *Rhizopus nigricans*. Ph.D. thesis, University of Maryland.

SINGH, B., YANG, C. C. and SALUNKHE, D. K. 1972. Controlled atmosphere storage of lettuce 1. Effects on quality and the respiration rate of lettuce heads. J. Food Sci. *37*, 48–51; 2. Effects on biochemical composition of the leaves. J. Food Sci. *37*, 52–55.

SMITH, W. H. 1938A. Anatomy of the apple fruit. Report of the Food Investigation Board, London, for 1937, 127–133.

SMITH, W. H. 1938B. The storage of broccoli. Report of the Food Investigation Board, London, for 1937, 185–187.

SMITH, W. H. 1939. The gas storage of broccoli. Report of the Food Investigation Board, London, for 1938, 202–208.

SMITH, W. H. 1947. A new method for the determination of the composition of the internal atmosphere of fleshy plant organs. Ann. Botany (N.S.) *11*, 363–368.

SMITH, W. H. 1957. Accumulation of ethyl alcohol and acetaldehyde in black currents kep in high concentrations of carbon dioxide. Nature (London) *179*, 876–877.

SMITH, W. H. 1965. Storage of mushrooms. Report of the Ditton Laboratory for 1964–1965, 25.

STOLL, K. 1973. Tabellen zur Lagerung von Früchten und Gemüsen in kontrollierter Atmosphäre. Mitteilungen der Eidgenössischen Forschungsanstalt für Obst-, Wein- und Gartenbau, Wädenswil, Flugschrift 78.

THIMANN, K. V., YOCUM, C. S. and HACKETT, D. P. 1954. Terminal oxidases and growth in plant tissues. 3. Terminal oxidation in potato tuber tissue. Archives of Biochem. Biophys. *53*, 239–257.

WANG, S. S., HAARD, N. F. and DIMARCO, G. R. 1971. Chlorophyll degradation during controlled atmosphere storage of asparagus. J. Food Sci. *36*, 657–661.

WILLIAMS, M. W. and PATTERSON, M. E. 1964. Nonvolatile organic acids and core breakdown of Bartlett pears. J. Agric. Food Chem. *12*, 80–83.

Response of Edible Plant Tissues to Stress Conditions

CHAPTER 5

STRESS METABOLITES IN POSTHARVEST FRUITS AND VEGETABLES—ROLE OF ETHYLENE

NORMAN F. HAARD

Memorial University of Newfoundland
St. John's, Newfoundland

and

MILDRED CODY

New York University
New York, New York

ABSTRACT

Plant tissues invariably accumulate an array of unusual metabolites when subjected to physiological stimuli or trauma. The physiological function, chemical nature and potential hazard to human health of these compounds is discussed. Ethylene gas can act alone or in combination with trauma to elicit formation of certain stress metabolites. The significance of ethylene involvement in the elicitor response is discussed in relationship to the occurrence of the stress metabolites in marketable produce. Evidence is provided to support the hypothesis that ethylene affects terpene stress metabolite formation by activation of alternate chain respiration.

INTRODUCTION

A phenomenon in plants comparable to induced immunity in animals was first reported by Müller and Börger in 1940. These early observations led to the "Phytoalexin theory" of disease resistance in plants which holds that a chemical compound, a phytoalexin (from GK. "to ward off"), is produced by plant cells as a result of metabolic interaction between host and parasite. Earlier studies suggested that phytoalexins were not present in the plant before infection and that they acted to specifically inhibit parasitic action of nonpathogenic microorganisms of the host. While the phytoalexin concept does not entirely explain plant disease resistance, there is considerable evidence that chemical substances produced by the plant as a result of parasitic attack act as antimicrobial agents. The phytoalexin concept is summarized in several reviews (Grzelinska 1976; Kuć 1976; Ingham 1972; Uritani 1967; Tibor 1975).

Recent studies have shown that phytoalexins may occur in trace quantities in apparently healthy tissue and that accumulation is not dependent on infection by a specific microorganism. It is now clear that phytoalexins accumulate in response to a wide array of biological, physical and chemical stresses or stimuli. Hence, apparently innocuous events such as cut injury, exposure to ultraviolet light or ethylene gas, and temperature adversity will sometimes elicit plant tissue to accumulate these unusual metabolites. Examples of these nonspecific elicitors are shown in Table 5.1. As a result, the term "stress metabolite" has been coined to include compounds formed in response to injury, physiological stimuli and infectious agents. The term "stress metabolite," unlike "phytoalexin," is not intended to imply or convey an implicit functionality of the compound.

A second important development is the finding that stress metabolites, known to occur in marketable fruits and vegetables, are sometimes toxic to mammals as well as microorganisms. For example, sweet potato roots may accumulate several furanoterpenoid compounds which are toxic to mammals (Table 5.2). Consumption of diseased sweet potato roots by livestock has resulted in animal death due to 4-ipomeanol poisoning (Boyd *et al.* 1973; Wilson *et al.* 1970, 1971). Surveys in various areas have demonstrated that these terpenes are invariably present in slightly blemished sweet potatoes (Table 5.3; Wilson *et al.* 1971; Catalano *et al.* 1976; Martin *et al.* 1976; Coxon *et al.* 1975). Ipomeamarone is largely converted to an unknown oxidation product during normal cooking, although 4-ipomeanol is stable under normal cooking operations (Cody and Haard 1976). The alkaloid α-solanine which accumulates in massive quantities in certain cultivars of Irish potato tubers, has been cited as the cause of several human deaths; this compound is also stable to normal cooking operations (Jadhav and Salunkhe 1975). Epidemiological data and animal feeding studies led Renwick (1972) to hypothesize that consumption of blighted potatoes during early stages of pregnancy led to the development of severe birth defects, anecephaly and spina bifida. This hypothesis has not been supported by other studies (Poswillo *et al.* 1972; Chaube *et al.* 1973; Swinyard and Chaube 1973), although it should be pointed out that the terpene stress metabolite (rishitin) implicated can accumulate in potato tuber slices as a result of nonspecific stresses (UV light) in approximately 12 hr (Cheema and Haard 1978). Other stress metabolites of known or suspected toxicity include pisatin found in pea pods (Oku *et al.* 1976) and furanocoumarin (*e.g.* 8-methoxyporalen) from celery (Scheel *et al.* 1963). At this time relatively little is known about the toxicology of various other stress metabolites, although their chemical structures are such as to make them suspect. Since the tendency of a plant tissue to accumulate these compounds is closely related to its disease resistance characteristics, one is

TABLE 5.1

EXAMPLES OF NONSPECIFIC ELICITORS OF STRESS METABOLITES

Elicitor	Tissue	Stress Metabolite
cut injury	Irish potato tuber	β-solamarine
ethylene	carrot root	3-methyl 6-methoxyl-3-hydroxy-3,4-dihydro-isocoumarin
mercuric chloride	sweet potato root	ipomeamarone
chilling injury	sweet potato root	4-ipomeanol
DNA intercalating agents	pea pod	pisatin
ultra violet light	Irish potato tuber	rishitin
visible light	Irish potato tuber	solanine

TABLE 5.2
TOXICITY OF FURANOTERPENOID STRESS METABOLITES ISOLATED FROM SWEET POTATO ROOT

Compound	LD_{50} (mg/kg Mouse) Oral
Ipomeamarone	230[2]
4-ipomeanol	38[1]
1-ipomeanol	79[1]
ipomeanine	26[1]
1,4-ipomeadiol	104[1]

[1]Boyd et al. (1975).
[2]Tiara and Fukagawa (1958).

left with the frightening suggestion that plant breeders have engineered new cultivars with superior tolerance to biological stress and that in so doing they have created fruits and vegetables with a greater propensity for the accumulation of stress metabolites that may be toxic to mammals.

From these considerations it is clear that more information is needed regarding our understanding of these stress metabolites. Specifically, we need additional information on: (1) the toxicology of stress compounds and their degradation products, particularly with regard to long term consumption of low doses; (2) cultivar and varietal differences in the tendency to accumulate stress metabolites; (3) influence of stresses associated with

TABLE 5.3
IPOMEAMARONE CONTENT OF SWEET POTATO ROOTS

Sample No.	Condition of Root	Ipomeamarone Content μg/g Fresh Weight	mg/Root	Presence[1] of 4-Ipomeanol	Ipomeanine
1	Partially rotten, soft	774	85	+	?
2	Partially rotten, soft	351	39	+ +	?
3	Partially rotten, soft	871	261	+	?
4	Partially rotten, sprouting	123	11	ND	ND
5	Good	36	4	ND	+
6	Good, sprouting	7	1	+	?
7	Fair, sprouting	61	11	ND	ND
8	Good	ND	ND	ND	+
9	Good	ND	ND	ND	+
10	Internal black speckles	119	17	+ +	?
11	Good	11	1	+ +	?
12	Good	1	0.1	+	?
13	Partially rotten, soft	209	18	+ +	?
14	Good	35	6	ND	ND
15	Good	328	76	ND	+
16	Good	68	13	ND	ND
17	Rotten, soft	883	167	+	?
18	Good	22	7	ND	ND
19	Good	20	4	ND	+
20	Good	1	0.4	ND	ND
21	Good	3	0.7	ND	ND
22	Partially rotten, internal black speckles	844	159	+	?
23	Poor	411	129	ND	+

[1]ND = not detected; ? = possible presence of ipomeanine obscured due to presence of 4-ipomeanol; + = probably present in small amount; + + = presence clearly demonstrated (Coxon *et al*. 1975; Fd. Cosm. Tox. *13*, 87).

harvest, handling, storage and processing of fruits and vegetables on accumulations of stress metabolites and (4) influence of processing and cooking operations on the degradation of stress metabolites.

Chemical Nature of Stress Metabolites

Examples of stress metabolites currently known to accumulate in edible plant tissues are summarized in Table 5.4 and Fig. 5.1. These compounds fall into various chemical groupings—*e.g.* isoprene compounds (glycoalkaloids, terpenes), phenylpropanoid compounds (phenolics, coumarins) and isoflavanoid derivatives (phaseollins). More than one chemical class of stress compounds often accumulate in a given tissue (*e.g.* sweet potato and potato). A stimulus or trauma which elicits accumulation of one group of metabolites does not necessarily promote the formation of other chemical types in a given tissue. For example, ethylene gas stimulates accumulation of phenolic substances in intact sweet potato roots; terpenes accumu-

TABLE 5.4

EXAMPLES OF STRESS METABOLITES ISOLATED FROM FRUITS AND VEGETABLES

Phytoalexin	Source, Common Name	Specific Elicitor(s)	Nonspecific Elicitors
Pisatin	Pisum Sativum, L., Garden Pea	M. Fructicola et al.	UV irradiation, DNA intercalating compounds, antibiotics
Phaseollin	Phaseolus Vulgaris, L., Green bean	Colleotrichum et al. lindemuthanum	Monilicolin A, heavy metal ions, metabolic inhibitors, transriptional, trans. inhibitors
Rishitin	Solanum Tuberosum L., Irish Potato	Phytophthora infestans	HgOAc, AgNO₃
Chlorogenic acid	Solanum Tuberosum L., Irish Potato	Phytophthora infestans	ethylene
α-Solanine	Solanum Tuberosum L., Irish Potato		UV radiation, cut injury
β-Solamarine	Solanum Tuberosum L., Irish Potato	Synchytrium endobioticum	cut injury (DARK)
6-Methoxymellein	Davcas carota, carrot	Ceratocystis fimbriata	chilling, HgCl₂, C₂H₄
Ipomeamarone	Ipomea Batatas, Sweet potato	Ceratocystis fimbriata	chilling, HgCl₂, C₂H₄ PdCl₂, CuCl₂, SDS, amino acids, etc.
9-Oxonerolidol	Solanum melongena, Eggplant	Monilia fructicola et al.	?

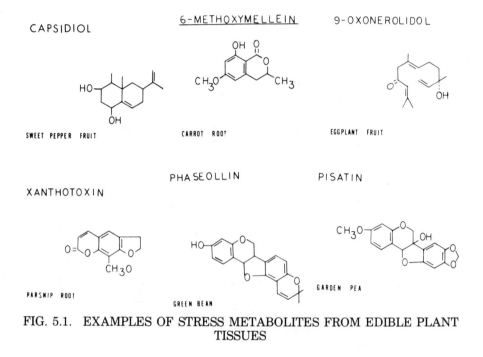

FIG. 5.1.   EXAMPLES OF STRESS METABOLITES FROM EDIBLE PLANT
TISSUES

late in response to storage at chilling temperature or subatmospheric pressure; cut-injury together with light irradiation promotes coumarin accumulation in the same tissues; while terpenes, coumarins and phenolics accumulate in black rot infected roots. Similarly, the glycoalkaloids, phenolics and terpenes of potato tuber appear to be under separate control (Currier and Kuć 1975). The situation is further complicated by the fact that each chemical grouping may consist of several closely related compounds. For example, while ipomeamarone is most often the principle terpene found in stressed sweet potato roots, in excess of nine additional furanoterpenes may also accumulate. Moreover, the principle stress metabolite which accumulates in response to a given stress is known to vary with variety (Price *et al.* 1976) and the physiological state of the tissue (Cheema and Haard 1978). For example, mercuric acetate readily elicits accumulation of rishitin in potato tubers out of cold storage but is ineffective on freshly harvested or conditioned tubers (Fig. 5.2).

We should also point out that various other substances also accumulate in plant tissues as a result of wounding. These include lignin, waxy deposits like suberin, ascorbic acid, phospholipids and melanin-like sub-

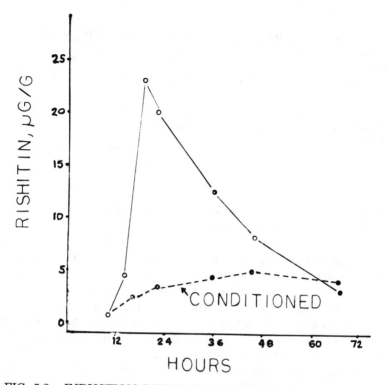

FIG. 5.2.   INDUCTION OF RISHITIN IN IRISH POTATO TUBER
DISCS BY MERCURIC ACETATE (5 mM)
Potatoes were stored at 4° C for several months prior to the experiment
(o — o) and subsequently conditioned for 7 days (● - - - ●).

stances resulting from enzymatic browning (all examples from wounded
potato tuber). In view of the diverse array of metabolic pathways which
may be induced by wounding, the variable nature of cultivars from the
same species and the influence of physiological status, any study of the
biochemical control of stress compounds is an extremely complex matter.

A Priori Evidence for the Role of Ethylene in Stress Metabolite Formation

The plant hormone ethylene has been implicated in a wide array of
physiological and biochemical events. Some examples are provided in
Table 5.5; the subject has been thoroughly reviewed by Abeles (1973).
Various lines of evidence point to a role for ethylene in stress metabolite
formation. These include the following: (1) ethylene is invariably formed
in diseased and traumatized plant tissue; (2) ethylene affects the disease

TABLE 5.5
SOME BIOLOGICAL PROPERTIES OF ETHYLENE

Induction	Inhibition
Fruit ripening	Mitosis
Abscission	DNA synthesis
Cell expansion	Cell expansion
Epinasty	Auxin transport
Enzyme synthesis	Stress metabolites
Stress metabolites	

resistance properties of plant tissue; (3) exogenous ethylene can elicit formation of certain stress metabolites.

Stress Ethylene.—A large body of literature now exists that shows ethylene production by plant tissue increases rapidly following trauma or physiological stimuli. Some examples are summarized in Table 5.6. Abeles has referred to such ethylene as stress ethylene. Evidence indicates that ethylene biogeneses by healthy and diseased tissues are mediated by different biosynthetic pathways (Sakai *et al.* 1970; Hislop *et al.* 1973), although it appears that ethylene induced by toxic chemicals is like ethylene formed in healthy tissue in that both originate from methionine (Abeles and Abeles 1972). There are several examples of host-pathogen systems which emanate unusually high levels of ethylene gas. In most instances ethylene in diseased tissue is of host origin (Williamson 1950), although in some cases pathogenic organisms themselves are capable of producing ethylene (Freebairn and Buddenhagen 1964; Ilag and Curtis 1968).

TABLE 5.6
SOME STRESS CONDITIONS KNOWN TO STIMULATE ETHYLENE
BIOGENESIS

Pathogenesis	(Sakai *et al.* 1970)
α-Radiation	(Maxie *et al.* 1965)
Chloride ion	(Rasmussen *et al.* 1969)
Toxic chemicals	(Abeles and Abeles 1972)
Low temperatures	(Cooper *et al.* 1969)
Waterlogging	(Kramer 1951)
Cut injury	(Imaseki *et al.* 1968)

Evidence points to a coincidental relationship between ethylene production and stress metabolite formation in sweet potato root (Imaseki et al. 1968; Kim et al. 1974). Simple mechanical wounding results in traces of ethylene formation and low levels of terpenes. Treatment of slices with chemicals, such as mercuric chloride, certain amino acids, fungal and weevil extracts, and sodium dodecylsulfate, leads to greater ethylene formation and proportionally greater quantities of terpenes. Infection of roots by C. fimbriata results in massive accumulation of terpenes and unusually elevated levels of ethylene. A similar relationship between endogenous ethylene formation and stress metabolite formation (coumarin derivatives) has been observed in carrot discs (Chalutz et al. 1969).

Ethylene and Disease Resistance.—In view of the phytoalexin concept the involvement of ethylene in disease resistance is of possible relevance to our interest in stress metabolite accumulation. Ethylene appears to effect the disease resistance of plants in several different ways. Under certain circumstances ethylene appears to confer resistance on otherwise suseptible varieties (Stahmann et al. 1966). Other work indicates ethylene can induce symptoms of disease (Dimond and Waggoner 1953; Jackson 1956; Smith et al. 1964). There is also evidence that exogenous ethylene can break natural resistance (De Munk and De Rooy 1971; Daley et al. 1970). Talboys (1972) has suggested a multiple role for ethylene in which the rate and type of response is concentration and/or rate dependent.

Phytoalexin Formation.—Certain stress metabolites accumulate in cut-injured tissue subjected to an ethylene atmosphere. The antifungal compound 3-methyl-6-methoxy-3-hydroxy-3,4-dihydroisocoumarin and related coumarins accumulate in carrot discs subjected to exogenous ethylene (Carlton et al. 1961; Chalutz et al. 1969). These compounds do not appear to be especially toxic to mammals although they impart bitterness to the tissue (Chalutz et al. 1969). The toxic metabolite pisatin accumulates in pea pods as a consequence of simple ethylene treatment (Chalutz and Stahmann 1969). Similarly, sweet potato roots incubated in an ethylene atmosphere accumulate various phenolic substances (Buesher et al. 1975). It would appear that direct induction of stress metabolites by ethylene relates to phenylpropanoid compound metabolism since the enzymes phenylalanine ammonia lyase and cinnamic acid-4-hydroxylase are induced by this hormone. Terpenes do not appear to be induced by ethylene gas; however, as will be shown by our data, ethylene can dramatically effect the induction of terpenes when other stresses are imposed on the tissue.

Exogenous Ethylene and Black Rot Infection of Sweet Potatoes.—Sweet potato root tissue infected with *C. fimbriata* and incubated in a continuous air stream accumulated detectable levels of ipomeamarone after 48 hours (Fig. 5.3). Ipomeamarone levels were optimal after approximately 120 hours and subsequently decreased. The decline in ipomeamarone was coincidental with the increase in a zone on thin layer plates judged to be 4-ipomeanol (Haard and Weiss 1976). Application of exogenous ethylene to the infected root tissue repressed the accumulation of ipomeamarone, as shown in Fig. 5.3, and accelerated the accumulation of presumptive 4-ipomeanol, as judged by thin-layer chromatography. Pretreatment of the tissue with benzylisothiocyanate, an inhibitor of ethylene biogenesis in Avocado fruit, appeared to reverse the influence of exogenous ethylene. These data led us to suggest that endogenously formed ethylene acts to facilitate the conversion of the 15 carbon terpene to the 9 carbon 4-ipomeanol. In our experience 4-ipomeanol is a common component of extracts obtained from marketable sweet potato root tissue. This finding is significant since the 9 carbon furanoterpenoids are considerably more toxic to mammals than are the 15 carbon compounds (Wilson *et al.* 1971; Table 5.2).

Endogenous Ethylene and Black Rot Infection.—Earlier studies showed that infection of roots with *C. fimbriata* resulted in a high rate of ethylene evolution lasting several days (Chalutz *et al.* 1969; Imaseki *et al.* 1968). We have incubated black rot infected roots in a hypobaric environment so as to increase the diffusivity of ethylene from the tissue and to prevent its accumulation in the storage atmosphere. The result of this experiment, shown in Fig. 5.4, was a promotion of ipomeamarone accumulation compared to air controls or to isobaric oxygen tension (10% O_2, 1 atm). Although the effect of the hypobaric environment was not evident during the early stages of infection, it is quite dramatic in samples incubated for a long term. These data are consistent with the repressive influence of exogenous ethylene and the influence of benzylisothiocyanate shown in Fig. 5.3. Thin-layer chromatography revealed that hypobaric stored tissue did not accumulate substantial levels of ipomeamaronol. This effect was also evident when roots were pretreated with exogenous ethylene and subsequently incubated under hypobaric conditions (Fig. 5.5). Pretreatment with ethylene was sufficient to repress ipomeamarone accumulation and accelerate ipomeamaronol accumulation while total terpenes decreased slightly. These data also support the view that ethylene acts to accelerate the formation of the hydroxylated terpene at the expense of ipomeamarone.

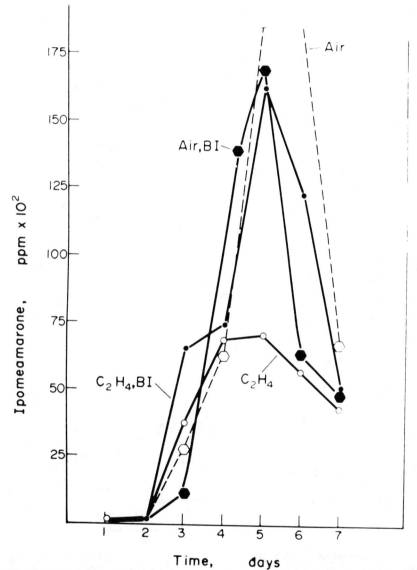

FIG. 5.3   IPOMEAMARONE CONTENT OF SWEET POTATO ROOT PLUGS
AFTER INFECTION WITH *C. fimbriata*

Roots were incubated in air (◯), 100 ppm ethylene in air (o), treated with benzylisothiocyanate and incubated in air (◉), or treated with isobenzylisothiocyanate and incubated in air containing 100 ppm ethylene (•). Mechanically wounded plugs did not accumulate appreciable ethylene during the course of this experiment.

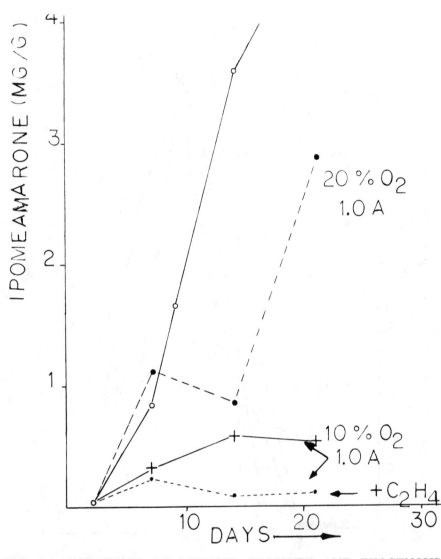

FIG. 5.4. INFLUENCE OF REDUCED PRESSURE AND EXOGENOUS ETHYLENE ON IPOMEAMARONE ACCUMULATION BY *C. fimbriata* IN-FECTED SWEET POTATO ROOTS
Ethylene concentration was 100 ppm in air.

Terpene Accumulation by Cut-Injured Sweet Potato Roots

Long term incubation of cut-injured roots results in furanoterpenoid accumulation and there is a coincidental relationship between endogenous ethylene and terpene accumulation (Table 5.7). Moreover, application of exogenous ethylene to cut-injured roots prior to storage at subatmospheric conditions also promotes the accumulation of terpenes (Fig. 5.6). A similar promotive effect of ethylene on terpenes is seen when cut-injured roots are exposed to exogenous ethylene in a continuous gas stream at 1 atm pressure (Table 5.8). Total furanoterpene accumulation in uninoculated sweet potatoes was generally greater at $10°$ C than at $22°$ C (Tables 5.7, 5.8). At reduced pressures, ipomeamarone and ipomeamaronol accumulations were greater at $22°$ C than at $10°$ C, but dehydroipomeamarone accumulations were much greater at $10°$ C than at $22°$ C. At $10°$ C over half of the total furanoterpene accumulation was as dehydroipomeamarone. Dehydroipomeamarone is a direct precursor of ipomeamarone; the conversion of dehydroipomeamarone to ipomeamarone might be slower at $10°$ C than at $22°$ C, resulting in the observed greater dehydroipomeamarone accumulation and lower ipomeamarone accumulation at $10°$ C. At atmospheric pressure accumulations of ipomeamarone, ipomeamaronol and dehydroipomeamarone were greater at $10°$ C than at $22°$ C. Ethylene accumulations were greater at $10°$ C than at $22°$ C in incubation chambers maintained at atmospheric pressure and at hypobaric pressure.

Cut-injured sweet potatoes incubated in hypobaric chambers at 0.1 atm with 100% O_2 bled into the chamber accumulated greater amounts of furanoterpenes than sweet potatoes incubated at 1 atm pressure with 10% O_2 or 20% O_2 or sweet potatoes incubated at 0.2 atm with 100% O_2 bled into the chamber. These data show that endogenously formed ethylene or exogenously applied ethylene can potentiate accumulation of terpenes in cut-injured sweet potato roots. Intact roots did not accumulate terpenes unless severe chilling condition persisted. The results are accordingly relevant to marketing conditions of storage and handling which may result in low level accumulations of these stress metabolites (see Table 5.3).

Terpene Accumulation Resulting from Heavy Metal Ions.—Some time ago, Uritani et al. (1960) showed that mercuric chloride (0.1%) induced formation of terpenes in sweet potato root slices. We have observed that ethylene gas acts, as in the case of cut-injured roots incubated for a long term, to synergistically promote mercuric chloride induced accumulation of ipomeamarone (Fig. 5.7). A similar potentiation of ipomeamarone accumulation was also observed when slices were treated with other heavy metal ions ($PtCl_4$, $PdCl_2$, $CuCl_2$). All terpenes observed on thin-layer plates increased along with ipomeamarone, although most other metabolites did not show the rapid decline evidenced by ipomeamarone in Fig. 5.7. The rapid kinetics of ipomeamarone accumulation by mercuric ion induction make the system more workable than the previously described sys-

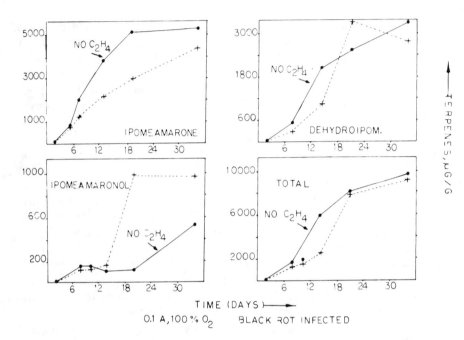

FIG. 5.5.   INFLUENCE OF ETHYLENE PRETREATMENT (100 ppm IN AIR, 1 HR) ON TERPENE ACCUMULATION IN *C. fimbriata* INFECTED SWEET POTATO ROOTS AT REDUCED PRESSURE

tems. During the time course of the experiments described in Fig. 5.7 control discs, treated or not treated with ethylene, did not accumulate appreciable levels of terpenes.

Mechanism of Ethylene Action

The mechanism of action of this hormone in plant tissue is by no means understood. We have pursued two lines of investigation regarding the role

TABLE 5.7
STRESS METABOLITE (FURANOTERPENOID) AND ETHYLENE
FORMATION IN CUT INJURED SWEET POTATO ROOTS

Storage Conditions	Total Furanoterpenoids[1] μg/g	C_2H_4 (ppm)[1]
Air, 22° C	0.0	0.0
100% O₂ (0.2 atm), 22° C	2.2	0.0
10% O₂, 10° C	60.5	1.3
Air, 10° C	92.6	1.7
100% O₂, (0.1 atm), 22° C	163.4	2.6

[1]Total furanoterpenoids and ethylene values are average values at 21–28 days incubation.

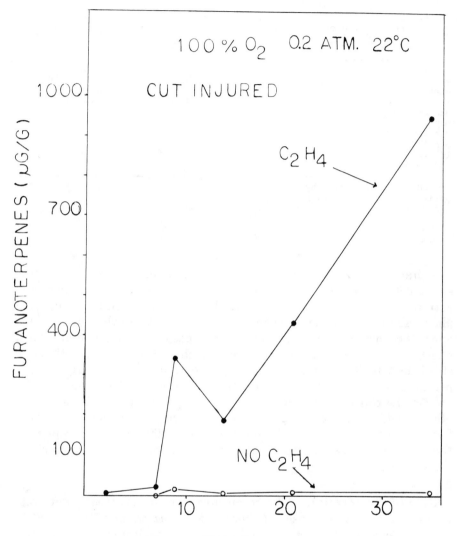

$100\% O_2$ 0.2 ATM. 22°C

CUT INJURED

C_2H_4

NO C_2H_4

FIG. 5.6.   INFLUENCE OF ETHYLENE PRETREATMENT (100 ppm IN AIR, 1 HR) ON TOTAL TERPENE ACCUMULATION IN CUT-INJURED SWEET POTATO ROOTS DURING INCUBATION AT REDUCED PRESSURE

of ethylene in stressed sweet potato root slices. These are the direct induction of key enzymes in the biosynthesis of stress metabolites and the activation of cyanide insensitive respiration.

TABLE 5.8
INFLUENCE OF EXOGENOUS ETHYLENE ON FURANOTERPENOID
ACCUMULATION IN CUT INJURED SWEET POTATO ROOTS

Storage Atmosphere	Exogenous C_2H_4 (100 ppm)	Total Furanoterpenoids[1] ($\mu g/g$)
Air, 22° C	−	0.0
Air, 22° C	+	106.5
Air, 10° C	−	92.6[2]
Air, 10° C	+	136.2
10% O_2, 10° C	−	60.5[3]
10% O_2, 10° C	+	98.2

[1]Total furanoterpenoid and ethylene values are average of 21–28 days incubation.
[2]Endogenous ethylene in continuous air stream was 1.7 ppm.
[3]Endogenous ethylene in continuous air stream was 1.3 ppm.

Enhanced Enzyme Activity.—It is well established that treatment of sweet potato root slices with exogenous ethylene results in increased activity of various enzymes (Table 5.9). The enhancement of phenylalanine ammonia lyase and cinnamic acid hydroxylase by ethylene appears to correlate with the ability of this gas to induce accumulation of phenylpropanoid compounds. The relationship of peroxidase activity to stress metabolite accumulation is not clear. It has been argued that peroxidases utilize peroxides liberated by various oxidases and thereby protect the cell from damage (Pegg 1976). Polyphenol oxidases, on the other hand, are thought to protect the cell from pathogens by catalyzing formation of antimicrobial quinones (Kosuge 1969). At the present time there is no evidence linking ethylene as an effector of enzymes to terpene biosynthesis.

Role of Respiration.—The respiratory activity of sweet potato root tissue increases in response to *C. fimbriata* (Akazawa and Uritani 1955). Greksack *et al.* (1972) have concluded that this respiratory stimulation is closely related to an increase in mitochondrial activity and not mitochondrial numbers. Alternate chain (cyanide insensitive) respiration is active in stressed sweet potato tissue (Solomos and Laties 1974). Recently, Solomos and Laties (1974, 1976) have suggested that ethylene acts, like cytochrome oxidase inhibitors, to activate alternate chain respiration in plant tissue (Fig. 5.8). We have obtained evidence that the potentiation of terpenes in sweet potato slices by ethylene relates to activation of alternate chain respiration (Haard 1977). Inhibitors of cytochrome oxidase (Fig. 5.9, 5.10) act to potentiate ipomeamarone accumulation by mercuric

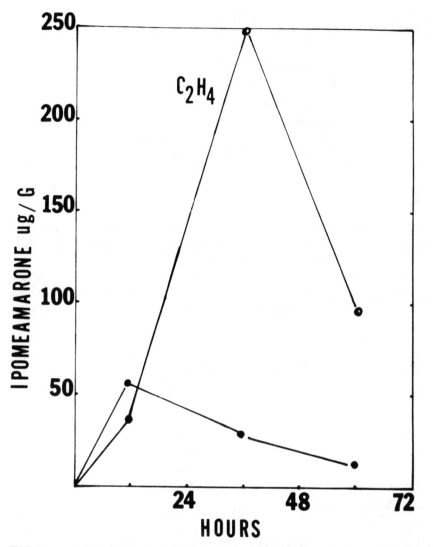

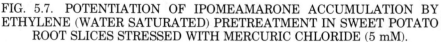

FIG. 5.7. POTENTIATION OF IPOMEAMARONE ACCUMULATION BY
ETHYLENE (WATER SATURATED) PRETREATMENT IN SWEET POTATO
ROOT SLICES STRESSED WITH MERCURIC CHLORIDE (5 mM).

chloride while having no measurable effect on terpenes in cut-injured
tissue. Alternatively, inhibitors of alternate chain respiration appear to
block mercuric chloride induction of terpenes (Table 5.10). Mercuric
cyanide shows comparable inducer properties as mercuric chloride, indicating that the potentiating effect of these agents is not simply due to

TABLE 5.9
ENZYMES ACTIVATED OR INDUCED IN STRESSED SWEET POTATO
ROOTS

Polyphenol oxidase
Soluble peroxidase
Bound peroxidase
Succinic dehydrogenase
Cytochrome oxidase
Phenylalanine ammonia lyase
Cinnamic acid 4-hydroxylase
3-Hydroxy-3-methyl glutamyl COA reductase
Pyrophosphomevalonate decarboxylase

metal liganding. Moreover, aging of sweet potato root slices for 24 hr, results in activation of alternate chain respiration (Solomos and Laties 1974) and stimulation of mercuric chloride induced ipomeamarone accumulation (Table 5.11). Preliminary studies indicate that ethylene acts similarly in potentiating mercuric acetate promotion of rishitin in white potato slices (Cheema and Haard, unpublished).

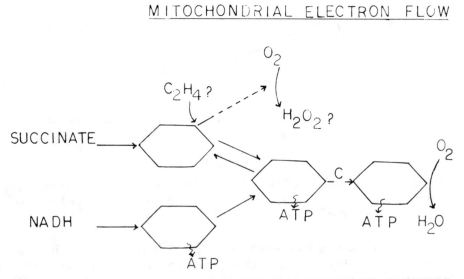

FIG. 5.8. SCHEME FOR MITOCHONDRIAL ELECTRON CARRIERS SHOWING PROPOSED SITE OF ETHYLENE ON ALTERNATE CHAIN RESPIRATION

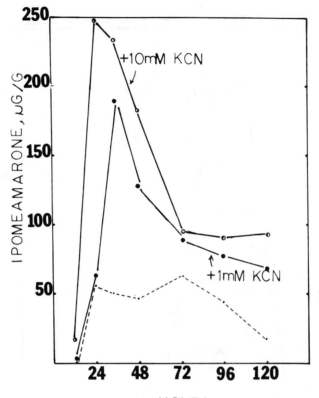

FIG. 5.9. POTENTIATION OF IPOMEAMARONE ACCUMULATION BY CYANIDE IN SWEET POTATO ROOT SLICES STRESSED WITH MERCURIC CHLORIDE (5 mM)

CONCLUSIONS

The contrary behavior of ethylene on terpene accumulation in stressed sweet potato root slices is comparable to the confused status of ethylene in relation to disease resistance. While ethylene itself can elicit accumulation of certain stress metabolites, it is only an effector of terpene accumulation in tissues otherwise stressed by fungal infection, heavy metal ions or cut-injury and temperature adversity. It is proposed that ethylene acts to potentiate terpene accumulation caused by nonspecific elicitors by virtue of its ability to switch on alternate chain respiration. Evocation of alternate chain respiration may: (1) function to decontrol respiration with a

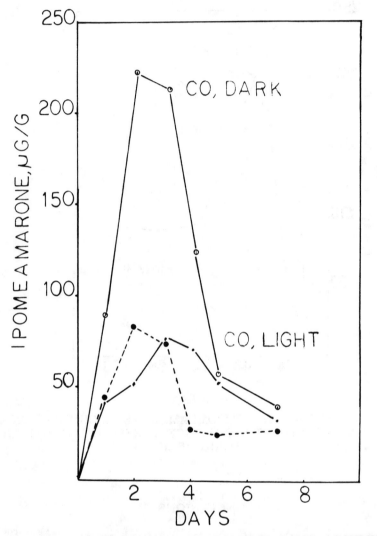

FIG. 5.10. POTENTIATION OF IPOMEAMARONE ACCUMULA-
TION BY CARBON MONOXIDE (DARK) IN SWEET POTATO ROOT
SLICES STRESSED WITH MERCURIC CHLORIDE (5 mM)

resulting mobilization of carbon flow; (2) permit respiration to continue
when intracellular ATP and NADH concentrations limit electron transfer
via cytochrome oxidase; (3) permit simultaneous electron transfer and
accumulation of reducing power; (4) generate heat and thereby differen-
tially affect biosynthetic reactions having different temperature coeffi-

TABLE 5.10
INFLUENCE OF RESPIRATORY INHIBITORS ON $HgCl_2$-INDUCTION OF IPOMEAMARONE[1]

Respiration Inhibitor (10 mM)	Ipomeamarone (μg/g fwb) Time After Treatment (hr)	
	24	48
None	17	101
KCN	63	235
KSCN	26	27
8-HQS	20	24
KCN + KSCN	36	49
KCN + 8 − HQS	27	34

[1]Tissue discs were immersed in $HgCl_2$ (5 mM) or $HgCl_2$ plus inhibitor or combinations of inhibitors as indicated and incubated at 20° C, 85% relative humidity. 8-HQS (8-hydroxy quinoline sulfate) and KSCN are inhibitors of alternate chain respiration. Data for ipomeamarone are representative of amounts of other furanoterpenoids as judged by thin-layer chromatography.

TABLE 5.11
INFLUENCE OF TISSUE SLICE AGING ON IPOMEAMARONE FORMATION[1]

Treatment	Preincubation Time (hr)	Ipomeamarone (μg/g)		
			Hr	
		24	48	78
$HgCl_2$	0	17	101	7
$HgCl_2$	24	344	337	100
$HgCl_2$ + KCN	0	65	229	50
$HgCl_2$ + KCN	24	165	125	182

[1]Tissue discs were immersed in $HgCl_2$ (5 mM) or $HgCl_2$ (5 mM) containing KCN (10 mM) immediately after cutting or after 24 hr aging and subsequently incubated at 20° C, 85% relative humidity. Tissue aged 24 hr and subsequently treated with water did not accumulate terpenes.

cients or causing volatilization of effector molecules; or (5) alter the redox potential of the cell and thereby affect an event which delimits terpene accumulation. Fig. 5.11 schematically depicts the proposed dual nature of the elicitor response for isoprene compounds. The apparent repressive effect of ethylene on ipomeamarone formation in black rot infected roots may relate to creating conditions favorable for ipomeamarone metabolism and/or for the synthesis of other metabolites such as chlorogenic acid and umbelliforone.

ACKNOWLEDGEMENTS

The work described in this paper was supported by a grant from the National Institute of Health, PHS 1226001086A1.

STRESS

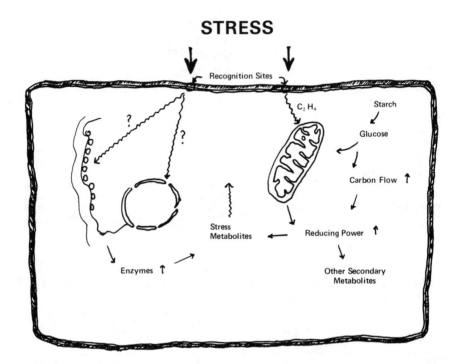

FIG. 5.11.   SCHEMATIC DIAGRAM OF A CELL SHOWING PROPOSED DUAL
NATURE OF ELICITOR RESPONSE

BIBLIOGRAPHY

ABELES, F. 1973. Ethylene in Plant Biology. Academic Press, N.Y.

ABELES, A. L. and ABELES, F. B. 1972. Biochemical pathway of stress induced ethylene. Plant Physiol. *50*, 496–498.

AKAZAWA, T. and URITANI, I. 1955. Respiratory increases and phosphorus and nitrogen metabolism in sweet potatoes infected with Black Rot. Nature *176*, 1071–1072.

BOYD, M. R., BURKA, L. T., HARRIS, T. M. and WILSON, B. J. 1973. Lung toxic furanterpenoids produced by sweet potatoes. Bochim. Biophys. Acta *337*, 184–195.

BOYD, M. R., BURKA, L. T. and WILSON, B. J. 1975. Distribution, excretion and binding of radioactivity in the rat after intraperotonial administration of the lung toxic furan, 4-ipomeanol. Toxicology & App. Pharmacol. *32*, 147–157.

BUESCHER, R. W., SISTRUNK, W. A. and BRADY, P. L. 1975. Effects of ethylene on metabolic and quality attributes of sweet potato. J. Food Sci. *40*, 1018–1020.

CARLTON, B. C., PETERSON, C. E. and TOLBERT, N. E. 1961. Effects of ethylene and oxygen on production of a bitter compound by carrot roots. Plant Physiol. *36*, 550–552.

CATALANO, E. A., HASLING, V. C., DUPUY, H. P. and CONSTANTIN, R. J.

1976. Ipomeamarone in blemished and diseased sweet potatoes. J. Agric. Food Chem. *25*, 94–96.

CHALUTZ, E., DE VAY, J. E. and MAXIE, E. C. 1969. Ethylene induced isocoumarin formation in carrot root tissue. Plant Physiol. *44*, 235–241.

CHALUTZ, E. and STAHMANN, M. A. 1969. Induction of pisatin by ethylene. Phytopathology *59*, 1972–1973.

CHAUBE, S., SWINYARD, C. A. and DAINES, R. H. 1973. Failure to induce malformations in fetal rats by feeding blighted potatoes to their mothers. Lancet *1*, 329–330.

CHEEMA, A. and HAARD, N. F. 1978. Induction of rishitin and lubinin in potato tuber discs by nonspecific elicitors and the influence of storage conditions. Physiol. Plant Path. (In press)

CODY, M. and HAARD, N. F. 1976. Influence of cooking on toxic stress metabolites in sweet potato root. J. Food Sci. *41*, 469.

COOPER, W. C., RASMUSSEN, G. K. and HUTCHENSON, D. J. 1969. Promotion of abscission of orange fruits by cycloheximide as related to site of treatment. Bioscience *19*, 443–444.

COXON, D. T., CURTIS, R. F. and HOWARD, B. 1975. Ipomeamarone, a toxic furanoterpenoid in sweet potatoes (*Ipomea batatas*) in the United Kingdom. Fd. Cosmet. Toxicol. *13*, 87–90.

CURRIER, W. W. and KUC, J. 1975. Effect of temperature on rishitin and steroid glycoalkaloid accumulation in potato tuber. Phytopathology *65*, 1194–1197.

DALEY, J. M., SEEVERS, P. M. and LUDDEN, P. 1970. Studies on wheat stem root resistance controlled in the 5-6 Locus. 3. Ethylene and disease reaction. Phytopathology *60*, 1648–1652.

DE MUNK, W. J. and DE ROOY, M. 1971. The influence of ethylene on the development of 5° C precooled "Apeldoom" tulips during forcing. Hort. Sci. *6*, 40–41.

DIMOND, A. E. and WAGGONER, P. E. 1953. The course of epinastic symptoms in fusarium wilt of tomatoes. Phytopathology *43*, 663–669.

FREEBAIRN, H. T. and BUDDENHAGEN, I. W. 1964. Ethylene production by *Pseudomonas solanacearum*. Nature *202*, 313–314.

GREKSACK, M., ASAHI, T. and URITANI, I. 1972. Increase in mitochondrial activity in diseased sweet potato root tissue. Plant & Cell Physiol. *13*, 1117–1121.

GRZELINSKA, A. 1976. Fitoaleksyny. Post. Biochem. *22*, 53–64.

HAARD, N. F. 1977. Potentiation of ipomeamarone by cyanide insensitive respiration in cut-injured sweet potato slices. Z. Pflanzenphysiol. Bd. *81*, 364–368.

HAARD, N. F. and WEISS, P. 1976. Influence of exogenous ethylene on ipomeamarone accumulation in black rot infected sweet potato roots. Phytochemistry *15*, 261–262.

HISLOP, E. C., ARCHER, S. A. and HOOD, G. V. 1973. Ethylene production by healthy and *Sclerotinia fructigena* infected apple peel. Phytochemistry *12*, 1281–1286.

ILAG, L. and CURTIS, R. W. 1968. Production of ethylene by fungi. Science *159*, 1357–1358.

IMASEKI, H., URITANI, I. and STAHMANN, M. A. 1968. Production of ethylene by injured sweet potato root tissue. Plant & Cell Physiol. *9*, 757–768.

INGHAM, J. L. 1972. Phytoalexins and other natural products as factors in plant disease resistance. Bot. Rev. *38*, 343–424.

JACKSON, W. T. 1956. The relative importance of factors causing injury to shoots of flooded tomato plants. Amer. J. Botany *43*, 637–639.

JADHAV, S. J. and SALUNKHE, D. K. 1975. Formation and control of chlorophyll and glycoalkaloids in tubers of S. tuberosum. Adv. Food Res. 21, 307–354.

KIM, W. K., OGUNI, I. and URITANI, I. 1974. Phytoalexin induction in sweet potato roots by amino acids. Agr. Biol. Chem. 38, 2567–2568.

KOSUGE, T. 1969. The role of phenolics in host response to infection. Ann. Rev. Phytopathol. 7, 195–222.

KRAMER, P. J. 1951. Cause of injury to plants resulting from flooding of the soil. Plant Physiol. 26, 722–736.

KUĆ, J. 1976. Phytoalexins. In Physiological Plant Pathology, R. Heitefuss and P. H. Williams (Editors). Springer Verlag, N.Y.

MARTIN, W. J., HASTING, V. C. and CATALANO, E. A. 1976. Ipomeamarone content in diseased and nondiseased tissues of sweet potato infected with different pathogens. Phytopathology 66, 678–679.

MAXIE, E. C., EAKS, I. C., SOMNER, N. F., RAE, H. C. and EL BATAL, S. 1965. Effect of gamma radiation on rate of ethylene and carbon dioxide evolution by lemon fruit. Plant Physiol. 40, 407–409.

MÜLLER, K. and BÖRGER, H. 1940. Arb. Biol. Rerchsanstalt. Lander. Forstu. Berlin 23, 189–231.

OKU, S. O., SHIRAISHI, T., UTSUMI, K. and SENO, S. 1976. Toxicology of a phytoalexin, pisatin, to mammalian cells. Proc. Japan Acad. 52, 33–36.

PEGG, G. F. 1976. The involvement of ethylene in plant pathogenesis. In Physiological Plant Pathology, R. Heitefuss and P. H. Williams (Editors). Encyclopedia of Plant Physiology, New Series Vol. 4, Springer-Verlag, N.Y.

PHAN, C. T. and SARKAR, S. K. 1975. Rev. Can. Biol. 23–32.

POSWILLO, D. E., SOPHER, D. and MITCHELL, S. 1972. Experimental induction of fetal malformation with blighted potato. Nature 239, 462–464.

PRICE, K. R., HOWARD, B. and COXON, D. T. 1976. Stress metabolite production in potato tubers infected by P. infestans, F. avenaceum and P. exigua. Physiol. Plant Pathology 9, 189–197.

RASMUSSEN, G. F., FURR, J. R. and COOPER, W. C. 1969. Ethylene production by citrus leaves grown in artificially salinized plots. J. Amer. Soc. Hort. Sci. 94, 640–642.

RENWICK, J. H. 1972. Hypothesis: Anecephaly and spina bifida are usually prevented by a specific but unidentified substance present in potato tubers. Br. J. Prev. Soc. Med. 26, 67–88.

SAKAI, S., IMASEKI, S. and URITANI, I. 1970. Biosynthesis of ethylene in sweet potato root tissue. Plant & Cell Physiol. 11, 737–745.

SCHEEL, L. D., PERNON, V. B., LARKIN, R. L. and KUPEL, R. E. 1963. The isolation and characterization of two phototoxic furanocoumarins from diseased celery. Biochemistry 2, 1127.

SMITH, W. H., MEIGH, D. F. and PARKER, J. C. 1964. Effect of damage and fungal infection on the production of ethylene by carnations. Nature 204, 92–93.

SMITH, K. A. and RUSSELL, R. S. 1969. Occurrence of ethylene and its significance in anaerobic soil. Nature 222, 769–770.

SOLOMOS, T. and LATIES, G. 1974. Similarities between the action of ethylene and cyanide in initiating the climacteric and ripening of avocados. Plant Physiol. 54, 506–511.

SOLOMOS, T. and LATIES, G. 1976. Induction of ethylene by cyanide resistant respiration. Biochem. Biophys. Res. Comm. 70, 663.

STAHMANN, M. A., CLARE, B. E. and WOODBURY, W. 1966. Increased disease resistance and enzyme activity induced by ethylene and ethylene production by black rot infected sweet potato tissue. Plant Physiol. 41, 1505–1512.

SWINYARD, C. A. and CHAUBE, S. 1973. Are potatoes teratogenic for experimental animals. Teratology *8*, 349–358.

TALBOYS, P. W. 1972. Resistance to vascular wilt fungi. Proc. Roy. Soc. (London) Ser. B. *181*, 319–332.

TIARA, T. and FUKAGAWA, Y. 1958. Nippen Nagei Kagaku Kaishi *32*, 513.

TIBOR, E. 1975. Fitoalexinek. Növenytermeles *24*, 359–370.

URITANI, I., URITANI, M. and YAMADA, H. 1960. Similar metabolic alterations induced in sweet potato by poisonous chemicals as by *C. fimbriata*. Phytopathology *50*, 30–34.

URITANI, I. 1967. Abnormal substances produced in fungus contaminated foodstuffs. J. Assoc. Off. Anal. Chem. *50*, 105–114.

WILLIAMSON, C. E. 1950. Ethylene: A metabolic product of disease and injury plants. Phytopathology *40*, 205–208.

WILSON, B. J., YANG, D. T. C. and BOYD, M. R. 1970. Toxicology of mold damaged sweet potatoes. Nature *227*, 521.

WILSON, B. J., BOYD, M. R., HARRIS, T. M. and YANG, D. T. C. 1971. A lung oedema factor from mold damaged sweet potatoes. Nature *231*, 52.

CHAPTER 6

TEMPERATURE STRESS IN EDIBLE PLANT TISSUES AFTER HARVEST

IKUZO URITANI

Laboratory of Biochemistry
Faculty of Agriculture
Nagoya University, Nagoya, Japan

ABSTRACT

Plants live healthily under the circumstances to which their ancestors adapted themselves, including plant tissues after harvest. The circumstances involve various kinds of factors, among which temperature should be important. It is normal to store or transport high-moisture tissues in the cold condition (0–$10°$ C) to repress the metabolism and maintain the freshness, flavor and nutrition just after the harvest. Some plant tissues are subjected to curing preceding cold storage to remove excess water in the tissues and to cure the outer wounded parts. Some plant tissues are dormant after harvest, often maintained by cold storage. Even after awaking from dormancy, the cold condition controls the germination. Fruits are normally harvested at the immature stage, and the maturation proceeds after harvest, but is repressed during cold storage. In tropical and subtropical plants, the tissues after harvest suffer from chilling injury during cold storage, causing cytological and metabolic alterations, which are reversible for a short period, but soon irreversible. When cold-stored tissues at this stage are transferred to a normal temperature room, some secondary metabolites are accumulated, and the tissues are easily attacked by saprophytic microorganisms, and cannot complete ripening. Chilling injury involves the irreversible disorganization of cellular membranes, composed of proteins and complex lipids. This disorder is based on the lipophilic nature of the proteins and the phase transition of the lipids. When some fruits such as tomato are stored above $30°$ C, they suffer from heat injury, inducing the disturbance of the normal ripeness.

In general, every species of plants grows and lives well under the circumstances to which the ancestor adapted itself. This is basically the case also in plant tissues after harvest.

The circumstances surrounding plants involve various kinds of factors.

Among them, temperature is an important one. Several times the earth underwent glacial periods, even after higher plants appeared on the earth. At every time of the glacial attack, plant species growing in groups were greatly changed, according to the fossil record. This implies that the growth of every species of plant is dependent upon temperature.

The materials in this paper belong to high-moisture tissues after harvest such as vegetables and fruits. Those plants are cultivated both in the temperate zone and in the tropical or subtropical one.

CHILLING STORAGE AND COLD CHAIN

It is normal to store or transport high-moisture tissues in the cold condition. The optimal temperature for storage and transportation is dependent upon the materials, but normally is in the range of 0–15° C. The chilling treatment is to repress the metabolism and maintain the freshness, flavor and nutrition as it was just after harvest. This is because the temperature coefficient (Q_{10}) of metabolic activities such as respiration is generally large in the range of 0–10° C compared to the Q_{10} in the range of 20–30° C, as shown in Fig. 6.1 (James 1955). Q_{10} in the low temperature

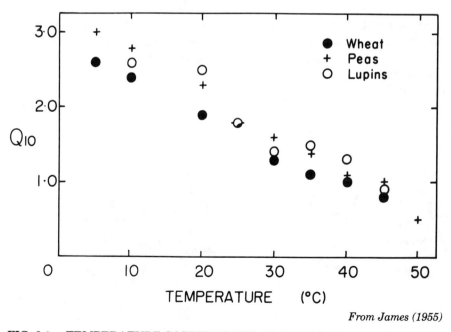

From James (1955)

FIG. 6.1.   TEMPERATURE COEFFICIENTS OF WHEAT, PEAS AND LUPINS

range seems to be larger in tropical or subtropical plants than in temperate zone plants (Chachin and Ogata 1977). However, we have to select carefully the optimal temperature for storage or transportation, because of the occurrence of chilling injury in high-moisture tissues which will be discussed in detail later.

CURING

Some plant tissues are subjected to curing preceding chilling storage. This is to remove excess water in the tissues and to cure the outer wounded parts, forming lignin layer or cork layer (Fig. 6.2) (Kahl 1973). In the case of sweet potato, the roots are left for several days at 30–33° C with about 60–80% relative humidity. Curing is performed also in the case of white potato, although in this case, room temperature is usually used. Curing is also useful for the storage of yam (*Dioscorea rotundata* Poir) (Passam *et al*. 1976).

DORMANCY

Some plant tissues are dormant after harvest. These include white potato tuber, onion bulb, *Dioscorea batatas* tuber, chestnut fruit, etc. Dormancy is often maintained by chilling storage and broken after some period. Even after awaking from dormancy, the cold condition controls the germination.

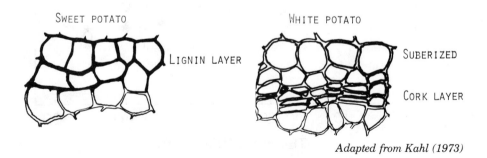

Adapted from Kahl (1973)

FIG. 6.2.   CURING AND WOUND HEALING

METABOLIC ALTERATIONS DURING STORAGE

Even if plant tissues after harvest are stored at their own optimal low temperatures, the contents of the metabolites are changed, affecting taste, flavor, and nutritive value. In general, reducing and nonreducing sugars are increased, along with a decrease in starch. Fluctuation of the content of organic acids is dependent upon plant species. For example, in banana, lemon and apple, organic acids increase a little, remain almost constant, and decrease, respectively (Biale and Young 1962). With different varieties or at different stages of maturation, fluctuations are observed even with the same species.

MATURATION AND AGEING

Fruits are normally harvested at the immature stage, and maturation proceeds after harvest at room temperature. When the fruits are stored in the cold, maturation is repressed. In the case of temperate zone plants, their fruits ripen again soon after transferring them to room temperature. However, when tropical or subtropical zone plants are stored at temperatures below certain critical points, normal ripening is disturbed because of chilling injury (Biale and Young 1962).

The ripening of fruits and the ageing of vegetables are repressed even more by the combination of low temperature treatment with atmospheric control. The storage is called CA storage. Here the CO_2 concentration is high, but O_2 concentration is low compared to air, in addition to the appropriate low temperature (Tamura 1977). Metabolic processes such as respiration are severely repressed, maintaining the freshness of vegetables and fruits.

CHILLING INJURY

As mentioned above, in the case of tropical and subtropical plants, the tissues after harvest may suffer chilling injury when stored below some critical point. Such chilling storage causes cytological and metabolic alterations which are reversible for a short period, but soon become irreversible (Lyons 1973; Shichi and Uritani 1956; Ogata 1977).

When chilling-sensitive tissues are stored in the range of about 10–5° C, some secondary metabolites such as polyphenols are accumulated (Lieberman et al. 1958); they are also sometimes attacked by some saprophytic microorganisms (Shichi and Uritani 1956).

When tissues that have been injured irreversibly by chilling are transferred to a nonchilling temperature, some secondary metabolites as mentioned above are more easily accumulated; they become easily infected by some saprophytic microorganisms such as *Rhizopus stolonifer,* and often in the case of fruits cannot ripen.

So far as is known, chilling injury involves the irreversible disorganization of cellular membranes, such as mitochondrial membrane, composed of proteins and complex lipids. This disorder is based on the lipophilic nature of the proteins (Yamaki and Uritani 1973B) and the phase transition of the lipids (Lyons and Raison 1970; Raison and Lyons 1970, 1971).

We have been studying the biochemical mechanism of chilling injury, using roots of sweet potato, *Ipomoea batatas* Lam. Understanding the basic mechanisms involved should aid in developing improved methods of storage of chilling-sensitive plant tissues, and in developing the most suitable temperature of storage of the particular species or variety of plant tissue.

We have many kinds of vegetables and fruits whose origins were in tropical or subtropical regions. Chilling-sensitive vegetables and fruits include sweet potato, yam (*Dioscorea*), cassava, eggplant, pepper, bean, corn, okra (*Abelmoschus esculentus*), pumpkin, cucumber, tomato, avocado, various species of citrus fruits, banana, melon, pineapple, papaya, mango, etc. (Raison and Lyons 1970; Ogata 1977; Murata 1977; Kozukue and Ogata 1972; Abe *et al.* 1974).

I would like to mention a possible mechanism of chilling injury, emphasizing studies on sweet potato roots. We used two cultivars, Okinawa 100 and Norin 1, very chilling-sensitive and less chilling-sensitive cultivars, respectively. Sometimes, white potato was used as a temperate zone plant.

MECHANISM OF CHILLING INJURY

Decreases in Respiration and Mitochondrial Activity

Decrease in Respiration.—Each sweet potato root was perpendicularly cut into two halves. One group was stored at 0° C (called chilled group), and the other at 10° C or 20° C (called control group). At appropriate periods, tissue was taken from each of the paired halves (chilled and control) from the same root. Then, thin discs (0.5 mm thick) were prepared from the inner part of each tissue, and the respiratory rate was measured at 25° C by the Warburg manometric method. The rate was decreased rapidly beginning with the 10th to 12th day in the case of the chilled group, but was almost constant in the case of the control group (Fig. 6.3) (Shichi

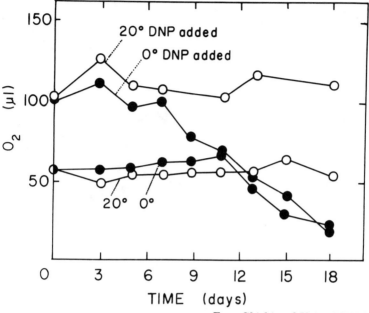

From Shichi and Uritani (1956)

FIG. 6.3. RESPIRATORY CHANGE RELATED TO THE DURATION OF STORAGE AT 0° C AND 20° C

Curves were made on the basis of O_2 uptake for 90 min at 25° C by Warburg manometric method. Final concentration of 2,4-dinitrophenol DNP was 5×10^{-5} M.

and Uritani 1956). Addition of 2,4-dinitrophenol as an uncoupler in a concentration of 5×10^{-5}M raised the respiratory rate about twofold. The rate of the chilled group decreased at about the same time as the group without 2,4-dinitrophenol, but that of the control group maintained the twofold increase through the whole period from 0–18 days. This suggests that not only respiratory activity but also oxidative phosphorylation activity are inhibited by chilling treatment (Shichi and Uritani 1956).

This accorded with the experimental results on respiration and oxidative phosphorylation of mitochondria prepared from both groups (Fig. 6.4) (Minamikawa *et al.* 1961; Lieberman *et al.* 1958).

Decrease in Respiratory Control Ratio in Mitochondria.—Intact mitochondria were prepared from both groups, the amount of oxygen uptake in state 3 and state 4 were measured by the oxygen electrode method, and the values of the respiratory control (RC) ratio were calcu-

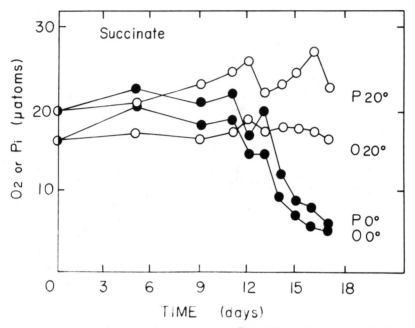

From Miinamikawa et al. (1961)

FIG. 6.4. RESPIRATORY (O) AND PHOSPHORYLATIVE (P) ACTIVITIES OF MITOCHONDRIA FROM SWEET POTATO ROOT TISSUES RELATED TO THE DURATION OF STORAGE AT 0° C AND 20° C
Curves were made by Warburg manometric method, when succinate was used as substrate.

lated. RC ratio decreased from the 10th to 12th day in the case of the chilled group, first by an increase in oxygen uptake in state 4 and then by a decrease in oxygen uptake in state 3 (Ohashi and Uritani 1972). Such sequential changes were clearly shown using cv. Norin 1 rather than cv. Okinawa 100, since the former cultivar was gradually injured owing to its less-sensitive nature. These data imply that oxidative phosphorylation in mitochondria is first inhibited, and then respiratory activity repressed by chilling treatment, as previously shown by Lieberman *et al.* (1958).

Decrease in the Activity of the Respiratory Enzyme System in Mitochondria.—A decrease in oxygen uptake in state 3 of mitochondria was observed when succinate was used as substrate in the case of the chilled group, which suffered slight chilling injury (Uritani *et al.* 1971). Addition of cytochrome c restored oxygen uptake approximately to the level of the control group. In the case where the chilled group suffered severe damage, only partial recovery was observed with cytochrome c.

Oxygen uptake in state 3 in the case of the chilled group was less inhibited by cyanide than that in the case of the control group. Addition of cytochrome c to mitochondria from the chilled group restored the cyanide-sensitive oxygen uptake to the level of the control group.

When malate was used as substrate, no effect of addition of cytochrome c or of NADH was observed. Except for malate dehydrogenase activity, there were no differences between the chilled group and the control group in respiratory enzyme activities. The activity of malate dehydrogenase was less in the chilled group than in the control group (Uritani et al. 1971).

Changes in Mitochondria and Mitochondrial Membrane

Change in the Sucrose Density Gradient Centrifugation Pattern of Mitochondrial Fraction.—Crude mitochondrial fractions were prepared from both the chilled group and the control group (stored for 8–14 days) and ultracentrifuged on sucrose density gradient. The results indicated that the mitochondria from the chilled group were located at a heavier density position than the control group mitochondria, when cytochrome c oxidase activity was used as the indicator of mitochondria (Fig. 6.5) (Yamaki and Uritani 1972A). On the other hand, when mitochondrial fractions from white potato were used, there was no difference between the chilled group and the control group.

Change in the Ratio of Phospholipid-P to Acid-Insoluble N.— When assayed for phospholipid-P and acid-insoluble N in pure mitochondrial preparations, the ratio of phospholipid-P content to acid-insoluble N content was lower in the chilled group mitochondria than in the control group mitochondria (Yamaki and Uritani 1972C). This suggests the release of phospholipid from the mitochondrial membrane in the chilled group.

Release of Various Species of Phospholipid from Mitochondrial Membrane.—Mitochondrial membrane was prepared by sonication from mitochondria purified by sucrose density gradient centrifugation. Then, time course analyses of the contents of various species of phospholipid were performed in both groups. The main species of phospholipid in sweet potato mitochondria were phosphatidylethanolamine (PE), phosphatidylcholine (PC), lysophosphatidylcholine (LPC), and an unknown phospholipid (called S4); their compositions were 40, 30, 10 and 20%, respectively. PE began to decrease after the fourth day of chilling storage and the decrease

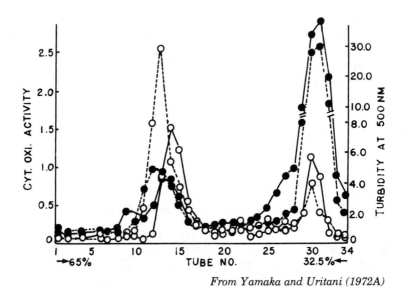

From Yamaka and Uritani (1972A)

FIG. 6.5. SUCROSE DENSITY GRADIENT CENTRIFUGATION
PATTERN OF CRUDE MITOCHONDRIAL FRACTIONS FROM 14
DAY-CHILLED GROUP AND CONTROL GROUP ---○---,–○–
Cytochrome c oxidase activity of mitochondrial fractions from sweet
potato tissues stored at 0° C (chilled group) and 10° C (control group),
respectively ---●---, –●–. Turbidity at 500 nm, chilled group
and control group.

continued up to the 14th day during which 40% of the total PE was lost.
Decreases in PC, LPC and S4 were indicated after 8 days of chilling
storage. The amounts of these species of phospholipid did not seem to be
changed further by prolonged chilling storage (Fig. 6.6) (Yamaki and
Uritani 1972B). However, such decreases were not observed in the case of
white potato. The release of the phospholipid accords with the above shown
data on the decrease in the ratio of phospholipid-P to acid-insoluble N.

Electron Microscopy.—Electron microscopy showed that the same
positions of both outer and inner membranes were partially broken in
mitochondria from the chilled group (14 day-chilling storage). This indicated that phospholipid was released from both membranes during chilling storage (Fig. 6.7) (Yamaki and Uritani 1972C).

Changes in Other Cellular Membranes

Change in Microsomal Membrane.—Microsomal fractions were prepared from both the chilled and the control group (20 day-chilling storage)

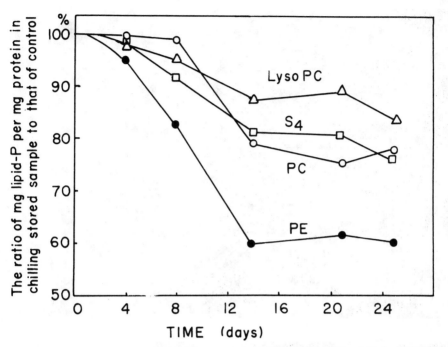

From Yamaki and Uritani (1972B)

FIG. 6.6. CHANGES IN LIPID COMPONENTS IN MITOCHONDRIAL MEMBRANES FROM SWEET POTATO TISSUES IN RESPONSE TO CHILL-ING STORAGE

and subjected to sucrose density gradient centrifugation. Investigations on the activity of antimycin A insensitive NADH-cytochrome-c reductase, a marker enzyme of microsomes, indicated that microsomal membrane was located at a heavier density position in the sucrose gradient in the chilled group than in the control group. Furthermore, the ratios of phospholipid-P content to acid-insoluble N content, and of NADH-cytochrome c reductase activity to acid-insoluble N content were less in microsomes from the chilled group. The results suggested that phospholipid and some membrane-bound enzymes were released from microsomes during chilling storage (Uritani *et al.* 1974).

Change in Vacuolar Membrane.—Electron microscopy showed that the vacuolar membrane in intact tissue of the chilled group (14-day chilling storage) was extensively broken, although no breakage occurred in those from the control group (Fig. 6.8) (Yamaki and Uritani 1973A).

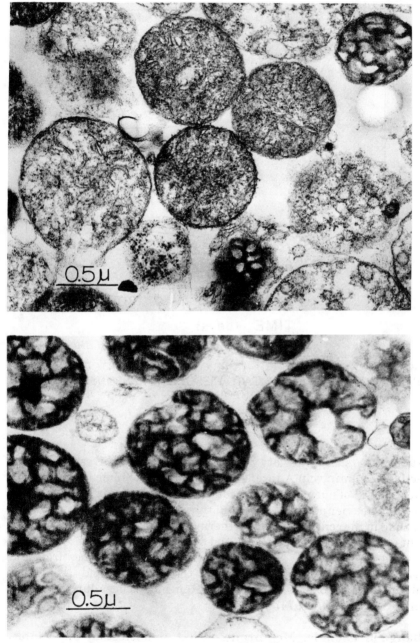

From Yamaki and Uritani (1972C)

FIG. 6.7.   ELECTRON MICROSCOPY OF MITOCHONDRIA FROM 14
DAY-CHILLED GROUP (A) AND CONTROL GROUP (B)

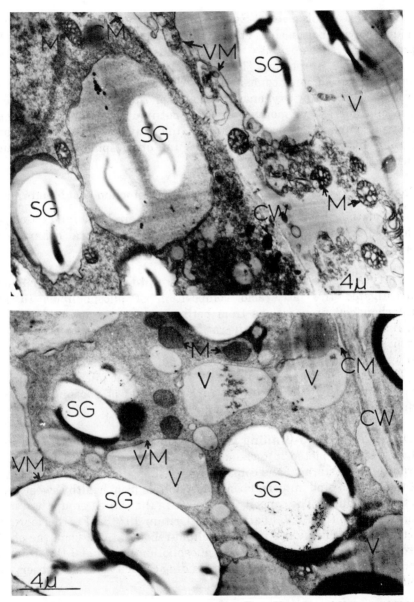

From Yamaki and Uritani (1973A)

FIG. 6.8.   ELECTRON MICROSCOPY OF 14 DAY-CHILLED GROUP (A)
AND CONTROL GROUP (B)
Sweet potato tissues were fixed by 5% glutaraldehyde for 3 hr, then by 1%
osmic acid for 3 hr: M, mitochondria; CW, cell wall; CM, cytoplasmic
membrane; VM, vacuolar membrane; ST, starch granule; and V, vacuolar
membrane.

Change in Polyphenol-Synthetic Membrane.—In response to cut injury, polyphenols such as chlorogenic acid are produced. Preceding this production, the enzymes involved, such as *trans*-cinnamic acid 4-hydroxylase, are formed (Tanaka *et al*. 1974). The latter enzyme is a multi-enzyme complex composed of NADPH-cytochrome c reductase and cytochrome P450 (an oxygenase), involves phospholipid, and is localized in a specific microsomal membrane. When cut-injured tissue containing this enzyme complex was incubated at 0° C, the activity was soon inactivated, perhaps by the release of some enzyme unit or phospholipid in the multi-enzyme complex (Uritani 1976).

Changes in Protein-Lipid Interaction in the Mitochondrial Membrane

All of the above results suggest that protein-lipid interaction was changed during chilling storage. Thus, some biochemical behavior pertaining to protein-lipid interaction was investigated.

Fatty Acid Composition of Phospholipid in Mitochondria.—We investigated the fatty acid composition of phospholipids in pure mitochondrial preparations from both sweet potato and white potato, against the assumption that the ratio of unsaturated fatty acid content to saturated fatty acid content was larger in sweet potato mitochondria than in white potato mitochondria (Uritani and Yamaki 1969).

This implied that we should think of the specific micro-localization in mitochondrial membrane of the various species of phospholipid which must be different in fatty acid composition.

Change in Lipid-Binding Capacity of Lipid-Depleted Mitochondria.—Lipid-depleted mitochondria (LDM) were prepared from chilled group (14-day storage) and control group (0-day storage) and the capacity of the LDM to bind mitochondrial phospholipid from healthy sweet potato was investigated. The capacity was lower in the chilled group than in the control group (Fig. 6.9) (Yamaki and Uritani 1973B). Furthermore, the activity of succinoxidase in phospholipid-rebound LDM from the chilled group was 50% less than in the phospholipid-rebound LDM from the control group. Such a change in the binding capacity was also observed in 2-day chilling-stored tissue.

Change in Arrhenius Plot of Succinoxidase Activity of Mitochondria.—The group of Lyons and Raison (1970), Lyons (1973), and Raison and Lyons (1970, 1971) indicated that the Arrhenius plot of activities of mitochondria prepared from tropical zone plants was discontinuous. We

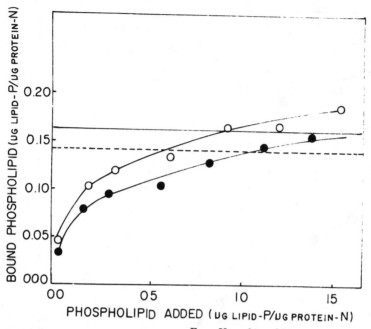

From Yamaki and Uritani (1973B)

FIG. 6.9. REBINDING OF MITOCHONDRIAL PHOS-
PHOLIPID FROM HEALTHY SWEET POTATO TO LIPID-
DEPLETED MITOCHONDRIA (LDM) FROM 14 DAY-CHILLED
GROUP (●) OR CONTROL GROUP (o)

Phospholipid rebound LDM from chilled group and control group
(healthy tissue without incubation), - - - - - and ———— is the
amount of μg of phospholipid-P per μg protein N in untreated
mitochondria from chilled group and control (healthy) group,
respectively.

repeated their experiments. The Arrhenius plot of succinoxidase activity
of both intact mitochondria and sonicated mitochondria from the control
group (0-day storage) showed two transition temperatures for activation
energy at 8–10° C and 16–18° C. Such figures were also observed in the
case of mitochondria from the chilled group (21-day chilling storage), but
the activation energy was lower than that of control group (Fig. 6.10)
(Yamaki and Uritani 1974).

A similar anomalous pattern was also observed with microsomal-bound
3-hydroxy-3-methylglutaryl coenzyme A reductase, induced in sweet
potato tissue for phytoalexin production in response to infection by
Ceratocystis fimbriata, a pathogenic fungus of sweet potato (Fig. 6.11)

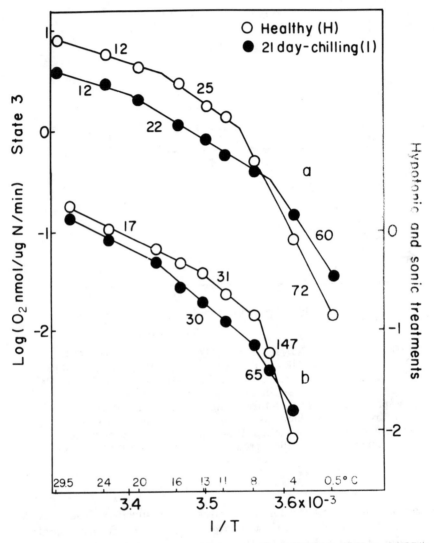

From Yamaki and Uritani (1974)

FIG. 6.10. ARRHENIUS PLOT OF SUCCINOXIDASE ACTIVITY OF MITOCHONDRIA FROM 21 DAY-CHILLED GROUP OR CONTROL (HEALTHY) GROUP

The number beside each line shows the activation energy in kcal/mole calculated from the slope of the line. a and b: the state 3 activity of succinoxidase of intact mitochondria and of mitochondrial fragments disrupted by hypotonic (0.02M K-phosphate buffer) and sonic treatments, respectively.

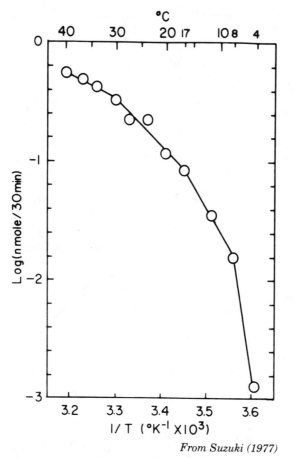

From Suzuki (1977)

FIG. 6.11.   ARRHENIUS PLOT OF THE ACTIVITY OF
MICROSOMAL-BOUND 3-HYDROXY-3-METHYLGLU-
TARYL COENZYME A REDUCTASE FROM *C. fimbriata*
INFECTED SWEET POTATO TISSUE

(Suzuki 1977), but not with *trans*-cinnamic acid 4-hydroxylase, a
membrane-bound enzyme which was synthesized in response to cut injury
accompanied by polyphenol production as mentioned above (Fig. 6.12)
(Tanaka 1977).

Some soluble enzymes such as phenylalanine ammonia-lyase (PAL) and
acid invertase are also synthesized in sweet potato tissue in response to cut
injury. The activity of purified PAL also showed a transition in the Ar-

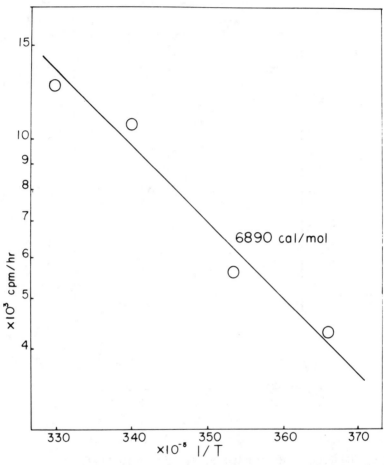

From Tanaka (1977)

FIG. 6.12. ARRHENIUS PLOT OF THE ACTIVITY OF
MEMBRANE-BOUND *trans*-CINNAMIC ACID 4-HYDROXYLASE
FROM CUT-INJURED INCUBATED SWEET POTATO TISSUE

rhenius plot around 17° C (Fig. 6.13) (Tanaka and Uritani 1977). PAL had
two Km values dependent on substrate concentration. The logarithms of
the Km values also showed transition temperatures in the range of 10–
20° C, when plotted against the reciprocal of absolute temperature (Fig.
6.14) (Tanaka and Uritani 1977). The physiological significance of these
transition temperatures for activity and Km values of PAL is not clear at
present.

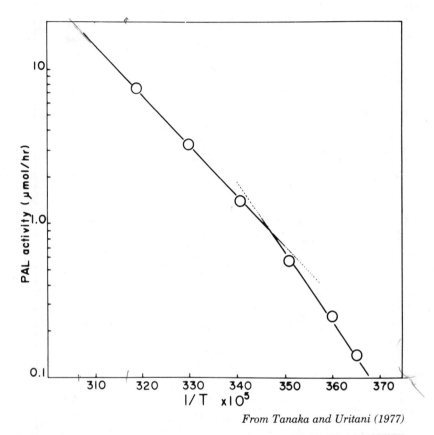

From Tanaka and Uritani (1977)

FIG.   6.13.   ARRHENIUS   PLOT   FOR   PHENYLALANINE
AMMONIA-LYASE IN TRIS-HCl BUFFER, pH 8.5

Effect of Temperature on Lipid-Binding Capacity of Lipid-Depleted Mitochondria (LDM).—In the binding of phospholipid to LDM from control group (0-day storage), the number of binding sites (n) and the dissociation constant (k) were obtained according to a linear form of the Langmuir-adsorption isotherm. The values changed conspicuously at a temperature between 10 and 15° C (Fig. 6.15) (Yamaki and Uritani 1974). That is, n and k values above the temperature were considerably higher than those below the temperature, in spite of the addition of species of phospholipid micelles whose phase transition temperatures were different from each other (Fig. 6.16) (Yamaki and Uritani 1974). Furthermore, the addition of different phospholipid micelles did not much alter the critical temperature. In similar binding experiments with chilled group (14-day

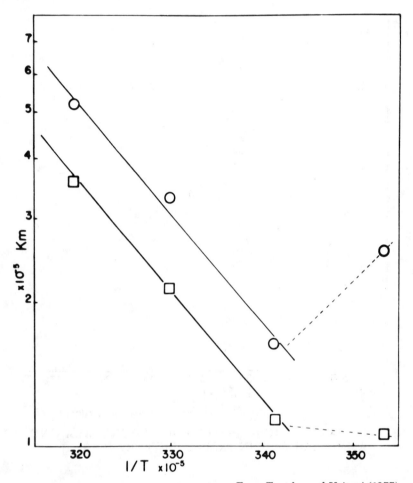

From Tanaka and Uritani (1977)

FIG. 6.14.   DEPENDENCE OF THE Km VALUES OF PAL ON
TEMPERATURE
The logarithms of Km values were plotted against the reciprocal of absolute temperature. ○ and □ are the concentrations above or below 30μM,
respectively.

chilling storage), the data on the relation between LDM-bound phospholipid content and phospholipid concentration in the solution did not
follow the hyperbolic curve, but showed a biphasic curve between 0–15° C.
Furthermore, n values obtained did not change much between the higher
and lower ranges above or below 10–15° C (Yamaki and Uritani 1974).

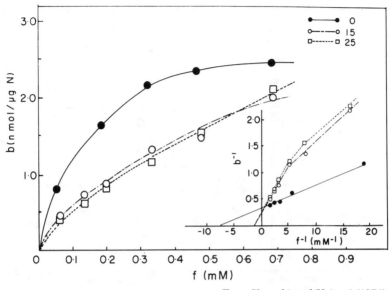

From Yamaki and Uritani (1974)

FIG. 6.15. TEMPERATURE DEPENDENCY OF REBINDING
OF MITOCHONDRIAL PHOSPHOLIPID MICELLES FROM
HEALTHY SWEET POTATO TO LIPID-DEPLETED
MITOCHONDRIA FROM HEALTHY SWEET POTATO
The horizontal axis indicates the concentration of added phospholipid micelles (f, mM) or the reciprocal of the concentration (f^{-1}, mM^{-1}). The vertical axis indicates the amount of rebound phospholipid (b, nmole/μg protein N) or the recipcocal of the amount (b^{-1}). $\bullet$, $\circ$ and $\square$ = 0° C, 15° C and 25° C, respectively.

Tentative Mechanism of Chilling Injury

All of the above data imply that chilling injury is due to an irreversible disorganization of the cellular membranes and is a kind of "membrane disease." Further, we suggest that the membrane protein of tropical plant tissues interacts with phospholipid by hydrophobic bonds. These bonds are broken by chilling treatment, because of the hydrophobic nature of the protein resulting in cold denaturation as well as of the phase transition of the phospholipid. In addition, the hydrophobic nature of the protein moiety may affect the temperature at which the phase transition of the phospholipid occurs, thus playing an important role in the initiation and irreversible nature of chilling injury.

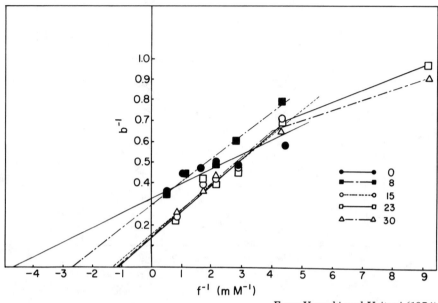

From Yamaki and Uritani (1974)

FIG. 6.16. TEMPERATURE DEPENDENCY OF REBINDING OF MITO-
CHONDRIAL PHOSPHOLIPID MICELLES FROM RAT LIVER TO LIPID-
DEPLETED MITOCHONDRIA FROM HEALTHY SWEET POTATO

The horizontal axis and vertical axis, same as in legend to Fig. 6.15. ● ,
■ , ○ , □ and △ = 0° C, 8° C, 15° C, 23° C and 30° C, respectively.

Possible Storage Methods of Not-Causing Chilling Injury and of Inducing Recovery from Chilling Injury

It is important to indicate the suitable temperature for safe storage in
each species or variety of tropical or subtropical plant tissues. For this
purpose, it may be useful to analyze the transition temperature of the
Arrhenius plot for some membrane-bound enzyme. The 1 or 2° C temperature above the transition point may be suitable for storage.

Furthermore, combination of temperature with other environmental
factors such as gas composition in the atmosphere might lower the transition temperature that might enable tropical plant tissues to be stored
without deterioration at some low temperature, where otherwise chilling
injury would occur.

HEAT INJURY

When some fruits such as tomato are stored above 30° C, they suffer from heat injury, inducing a disturbance of the normal ripening process. For example, when tomato fruit was harvested at the mature green stage and stored in air at 33° C, the respiratory rate was not enhanced, neither ethylene, lycopene nor polygalacturonase was produced, and ripening was stopped (Ogura *et al*. 1975A, B; 1976). When transferred to room temperature, ripening proceeded, but the rate was lower than in the case of tomato which was stored at room temperature from the initial time. Initial storage at 33° C may aid in improving the long life of fresh tomatoes. On the other hand, when tomatoes were stored at 4° C and then transferred to room temperature they did not ripen and showed chilling injury.

TEMPERATURE STRESS IN WOUNDED TISSUES

Plant tissues are often mechanically damaged during transportation in the cold chain system. That induces wound respiration, and the temperature of the atmosphere is raised (Greksak *et al*. 1972; Kahl 1973). Thus, wounding stimulates the production of some secondary metabolites, lowering the quality of vegetables and fruits, although they are in the cold chain system.

TEMPERATURE STRESS IN PATHOGENIC INFECTION

Pathogenic infection is determined by the specificity of host-parasite interaction. However, the speed of either infection or resistant action is dependent on temperature. Low temperatures often enhance infection, because the resistant action of plant tissues in response to infection is often severely lowered compared to the growth and development of the pathogens (Akazawa and Uritani 1961).

FREEZING

Tissues of plants who origins are in the temperature zone sometimes are somewhat resistant to mild freezing (0–5° C), showing cold hardiness (Mazur 1969). However, freezing often induces freezing injury, leading to the production of some secondary metabolites, as demonstrated by the formation of 6-methoxymellein in frozen carrot roots (Kuć 1976). Usually, it is not desirable to expose the tissues to temperatures below 0° C.

CONCLUSION

We have many kinds of temperature stress in edible plant tissues after harvest, and these provide many problems for storage. It would be desirable to understand the scientific bases of the various phenomena which are known empirically about the relation of temperature stress and the storage of plant tissues after harvest. Such understanding may be helpful in defining optimal storage conditions for postharvest vegetables and fruits to maintain good quality and prevent loss, thus contributing to the solution of the world food problem.

BIBLIOGRAPHY

ABE, K., IWATA, T. and OGATA, K. 1974. Chilling injury in eggplant fruits. 1. General aspects of the injury and microscopic observation of pitting development. J. Jap. Soc. Hort. Sci. 42, 402–407. (Japanese)

AKAZAWA, T. and URITANI, I. 1961. Influence of environmental temperatures on metabolic alterations related to disease resistance in sweet potato roots infected by black rot. Phytopathology 51, 668–674.

BIALE, J. B. and YOUNG, R. E. 1962. The biochemistry of fruit maturation. Endeavour 21, 164–174.

CHACHIN, K. and OGATA, K. 1977. Postharvest physiology of fruits and vegetables. In Storage of Fruits and Vegetables. K. Ogata (Editor). Kenpakusha Co., Tokyo. (Japanese)

GREKSAK, M., ASAHI, T. and URITANI, I. 1972. Increase in mitochondrial activity in diseased sweet potato root tissue. Plant & Cell Physiol. 13, 1117–1121.

JAMES, W. O. 1955. The rate of respiration. In Plant Respiration. Oxford University Press, London.

KAHL G. 1973. Genetic and metabolic regulation in differentiating plant storage tissue cells. Bot. Rev. 39, 274–299.

KOZUKUE, N. and OGATA, K. 1972. Physiological and chemical studies of chilling injury in pepper fruits. J. Food Sci. 37, 708–711.

KUĆ, J. A. 1976. Phytoalexins. In Physiological Plant Pathology. R. Heitefuss and P. H. Williams (Editors). Springer-Verlag, Berlin, Heidelberg.

LIEBERMAN, M., CRAFT, C. C., AUDIA, W. V. and WILCOX, M. S. 1958. Biochemical studies of chilling injury in sweet potatoes. Plant Physiol. 33, 307–311.

LYONS, J. M. 1973. Chilling injury in plants. Ann. Rev. Plant Physiol. 24, 445–466.

LYONS, I. M. and RAISON, J. K. 1970. Oxidative activity of mitochondria isolated from plant tissues sensitive and resistant to chilling injury. Plant Physiol. 45, 385–389.

MAZUR, P. 1969. Freezing injury in plants. Ann. Rev. Plant Physiol. 20, 419–448.

MINAMIKAWA, T., AKAZAWA, T. and URITANI, I. 1961. Mechanism of cold injury in sweet potatoes. 2. Biochemical mechanism of cold injury with special reference to mitochondrial activities. Plant & Cell Physiol. 2, 301–309.

MURATA, T. 1977. Physiological injury. *In* Storage of Fruits and Vegetables. K. Ogata (Editor). Kenpakusha Co., Tokyo. (Japanese)

OGATA, K. 1977. Chilling storage. *In* Storage of Fruits and Vegetables. K. Ogata (Editor). Kenpakusha Co., Tokyo. (Japanese)

OGURA, N., NAKAGAWA, H., TAKEHANA, H. 1975A. Effect of high temperature-short term storage of mature green tomato fruits on changes of their chemical composition after ripening at room temperature. J. Agr. Chem. Soc. Japan. 49, 189–196. (Japanese)

OGURA, N., NAKAGAWA, H. and TAKEHANA, H. 1975B. Effect of storage temperature of tomato fruits on changes of their polygalacturonase and pectinesterase activities accompanied with ripening. J. Agr. Chem. Soc. Japan. 49, 271–274. (Japanese)

OGURA, N., HAYASHI, R., OGISHIMA, T., ABE, Y., NAKAGAWA, H. and TAKEHANA, H. 1976. Ethylene production by tomato fruits at various temperatures and effect of ethylene treatment of the fruits. J. Agr. Chem. Soc. Japan. 50, 519–523. (Japanese)

OHASHI, H. and URITANI, I. 1972. The mechanism of chilling injury in sweet potato. 9. The relation of chilling to changes in mitochondrial respiratory activities. Plant & Cell Physiol. 13, 1065–1073.

PASSAM, H. C., READ, S. J. and RICKARD, J. E. 1976. Wound repair in yam tubers: Physiological processes during repair. New Phytol. 77, 325–331.

RAISON, J. K. and LYONS, J. M. 1970. The influence of mitochondrial concentration and storage on the respiratory control of isolated plant mitochondria. Plant Physiol. 45, 382–385.

RAISON, J. K. and LYONS, J. M. 1971. Hibernation: alteration of mitochondrial membranes as a requisite for metabolism at low temperature. Proc. Nat. Acad. Sci. U.S.A. 68, 2092–2094.

SHICHI, H. and URITANI, I. 1956. Alterations of metabolism in plants at various temperatures. Part 1. Mechanism of cold damage of sweet potato. Bull. Agr. Chem. Soc. Japan. 20 suppl., 284–288.

SUZUKI, H. 1977. A biochemical study of metabolic regulation of isoprenoid biosynthesis in diseased sweet potato root tissue. D. A. thesis, Nagoya University. (Japanese)

TAMURA, T. 1977. CA storage. *In* Storage of Fruits and Vegetables. K. Ogata (Editor). Kenpakusha Co., Tokyo. (Japanese)

TANAKA, Y. 1977. Personal communication. Lab. of Biochem., Fac. of Agric., Nagoya University.

TANAKA, Y., KOJIMA, M. and URITANI, I. 1974. Properties, development and cellular-localization of cinnamic acid 4-hydroxylase in cut-injured sweet potato. Plant & Cell Physiol. 15, 843–854.

TANAKA, Y. and URITANI, I. 1977. Purification and properties of phenylalanine ammonia-lyase in cut-injured sweet potato. J. Biochem. 81, 963–970.

URITANI, I. 1976. Chemical mechanism of chilling injury in plants. Chemical Regulation of Plants (Journal) 11, 47–55. (Japanese)

URITANI, I., HYODO, H. and KUWANO, M. 1971. Mechanism of cold injury in sweet potatoes. 4. Biochemical mechanism of cold injury with special reference to mitochondrial activities. Agr. Biol. Chem. 35, 1248–1253.

URITANI, I., YAMAKI, S., TAKADA, K., SHIBATA, A. and TANAKA, Y. 1974.

Change in microsomal membrane in chilling-stored sweet potatoes. J. Jap. Biochem. Soc. *46*(8), 726. (Japanese)

URITANI, I. and YAMAKI, S. 1969. Mechanism of chilling injury in sweet potatoes. 3. Biochemical mechanism of chilling injury with special reference to mitochondrial lipid components. Agr. Biol. Chem. *33*, 480–487.

YAMAKI, S. and URITANI, I. 1972A. Mechanism of chilling injury in sweet potatoes. 5. Biochemical mechanism of chilling injury with special reference to mitochondrial lipid components. Agr. Biol. Chem. *36*, 47–55.

YAMAKI, S. and URITANI, I. 1972B. The mechanism of chilling injury in sweet potato. 6. Changes of lipid components in the mitochondrial membrane during chilling storage. Plant & Cell Physiol. *13*, 67–79.

YAMAKI, S. and URITANI, I. 1972C. Mechanism of chilling injury in sweet potato. 7. Changes in mitochondrial structure during chilling storage. Plant & Cell Physiol. *13*, 795–805.

YAMAKI, S. and URITANI, I. 1973A. Morphological changes in chilling injured sweet potato root. Agr. Biol. Chem. *37*, 183–186.

YAMAKI, S. and URITANI, I. 1973B. Mechanism of chilling injury in sweet potato. 10. Change in lipid-protein interaction in mitochondria from cold-stored tissue. Plant Physiol. *51*, 883–888.

YAMAKI, S. and URITANI, I. 1974. Mechanism of chilling injury in sweet potato. 12. Temperature dependency of succinoxidase activity and lipid-protein interaction in mitochondria from healthy or chilling-stored tissue. Plant & Cell Physiol. *15*, 669–680.

CHAPTER 7

PATHOLOGICAL DISEASES OF FRESH FRUITS AND VEGETABLES

JOSEPH W. ECKERT

University of California
Riverside, CA

ABSTRACT

Postharvest diseases of fruits and vegetables may be traced to: (1) infection or infestation of the product during the period of its development in the field and (2) infection through mechanical injuries during the harvesting operation and through physiological injuries caused by an unfavorable storage environment. The subsequent progress of an infection depends upon the enzymatic capabilities of the pathogen and the physiological status of the host tissues, in terms of available moisture, vulnerability to attack by the macerating enzymes of the pathogen, and the defense mechanisms which can be developed before the infection process is complete. Resistance to disease development usually decreases after the product reaches full maturity and with time after harvest, but these events can be regulated to some extent by the postharvest environment and by chemical treatments.

Certain compounds which directly inhibit the growth of pathogenic fungi have been highly successful in controlling a number of postharvest diseases. The strategy of chemical control has concentrated on treatments that protect the product from infection or that terminate the growth of the pathogen in incipient infections. Fungicide treatments may also suppress the symptoms and side effects of the disease. Antimicrobial agents rarely, if ever, show true therapeutic or curative action against established infections on harvested produce. Certain plant growth regulators may retard the development of pathogens by delaying the onset of senescence of the host tissues. Several compounds with an imidazole nucleus, i.e., thiabendazole, benomyl, and imazalil have provided excellent control of numerous postharvest diseases, but some pathogenic fungi and all bacteria are tolerant of these compounds. Furthermore, strains of fungi that are tolerant of thiabendazole and benomyl have emerged when these treatments have been utilized intensively.

The search continues for compounds which will control decays of several fruits and vegetables that are caused by the fungi Alternaria *and Geotrichum and the bacterium* Erwinia. *Broad spectrum antimicrobial agents which do not favor the selection of tolerant strains of pathogens would have a distinct advantage over those fungicides that have a single biochemical target which have been developed over the past few years.*

INTRODUCTION

Fresh fruits and vegetables are often produced in areas distant from population centers and frequently they mature at a time of the year when consumer demand is weak or when the market is glutted with the product. These circumstances may necessitate a period of several weeks or months for storage and shipment before the product reaches the consumer. Substantial decay losses may occur during this period if the product is not treated with an effective inhibitor of microbial growth and (or) stored in an environment unfavorable to disease development. Postharvest losses of 25–50% of a crop are commonplace in some tropical countries where refrigeration facilities are not available and appropriate chemical treatments are not utilized (Coursey and Booth 1972; Eckert 1975). Decay losses in technologically advanced areas are more difficult to assess because they vary greatly with season, farm, and postharvest handling practices. Some idea of the magnitude of the problem can be obtained from published surveys (U.S. Dept. Agr. 1965) and from decay records of untreated produce in large scale experimental shipments (Eckert 1977).

Aside from the obvious loss of one or more units of an edible product, postharvest deterioration may have several less evident consequences: (1) Partial or total loss of consumer packages with only one or several diseased units. (2) Reduced postharvest life of the product due to accelerated ripening or senescence triggered by ethylene released from a few diseased fruit in a package or storage room (Peacock 1973; Wild *et al.* 1976). (3) Possible contamination of the edible product with a mycotoxin elaborated by the disease-inducing microorganism, e.g., patulin produced by *Penicillium expansum* in diseased apples and stone fruits (Buchanan *et al.* 1974; Sommer *et al.* 1974) or furanoterpenoid metabolites in sweet potatoes infected with *Ceratocystis fimbriata* (Boyd and Wilson 1972). (4) Softening of processed fruits by heat-tolerant macerating enzymes secreted by Rhizopus stolonifer in incipient infections on apricots and peaches (Harper *et al.* 1972).

Postharvest losses are highly leveraged in terms of both time and money. Despite months of the best agronomic practices in the production of the crop, deterioration in a few days or weeks after harvest may wipe out a large investment in harvesting, packaging, storage and transportation, the combined costs of which may be several-fold greater than the total value of the product in the field. Postharvest deterioration is a serious problem not only for the producer and the distributor of fresh fruits and vegetables, but also adversely influences the availability and cost of these commodities to the consumer. The staggering postharvest losses of staple root crops and fruits in developing countries of the tropics clearly indicate that the solution to the world food shortage is not wholly agronomic in nature (Coursey and Booth 1972). In technologically advanced societies, losses of fresh fruits and vegetables in the home have an obvious impact upon the family food bill. More indirectly, postharvest diseases are often the major consideration in reaching decisions that influence the consumer price: requirements and duration of storage, mode of transportation, possible utilization of economizing practices such as mechanical harvesting, bulk handling, and consumer packaging.

Many fundamental aspects of the biology and control of postharvest diseases have been discussed in an earlier symposium on this subject (Eckert 1975). The present paper will examine in greater depth the nature of the host-pathogen interaction involved in postharvest diseases and will evaluate the impact of environmental factors upon the magnitude of losses in storage and transport. This analysis will reveal possibilities for manipulation of the postharvest environment to maximize host resistance to infection and to discourage the development of pathogenic microorganisms. Finally, the successes, problems and novel approaches to the use of chemical treatments to control postharvest diseases will be updated since the last symposium.

NATURE AND CAUSE OF POSTHARVEST DISEASES

Ripening fruits and vegetables are susceptible to attack by a variety of pathogenic microorganisms which were unable to parasitize them during the period of their development on the plant. A mature living plant product should not be viewed, however, simply as a nutritive medium capable of supporting the growth of the wide variety of saprophytic microorganisms. While it is true that extracts of many plant products support the growth of common fungi and bacteria, the living fruit or vegetable is quite resistant to attack by most microoganisms. The extent of microbial deterioration is determined in each individual case by the physiological capabilities of the microorganisms and the properties of the specific plant product.

The brown rots, exemplified by *Monilinia fructicola* on stone fruits, *Gloeosporium* spp. on apples, and *Diplodia* stem-end rot of citrus fruits (Fig. 7.1) may become a serious problem when a substantial portion of the crop is infected at the time of harvest, but usually the diseased fruit remain

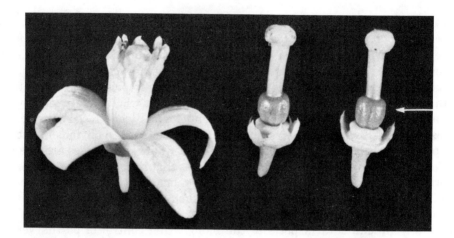

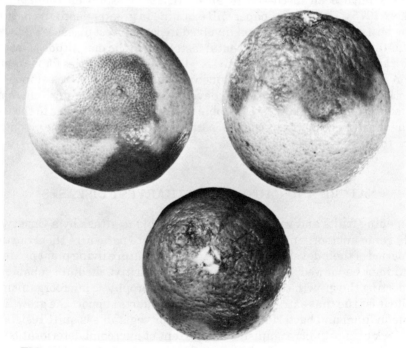

FIG. 7.1.   FLOWERS AND FRUIT OF THE VALENCIA ORANGE
Upper photo, right, shows young fruitlet (arrow) resting upon the floral disc and sepals. Petals of flower (left) have been removed in photos of young fruitlets. Fruit is in natural orientation on tree, revealing the accessibility of disc and inner surface of sepals to airborne spores. Lower photo (by A. A. McCornack) shows development of stem end rot by *Diplodia natalensis* initiated in the button (sepal + disc) of the fruit.

firm and the pathogen does not spread readily during storage and shipment. On the other hand, the "soft rots" of fruits and vegetables caused by *Rhizopus, Geotrichum, Sclerotinia* and *Erwinia* are a serious group of postharvest diseases which progress rapidly under optimum conditions and the extracellular enzymes of the pathogens may macerate the fleshy tissues of the host into a watery mass within a few days. In addition, the soft-rot pathogens may spread by contact to adjacent units in the same container, creating pockets of decayed produce.

The development of disease may be divided into two stages—infection and symptom expression. These two events may be discrete and separated by several months (latent infections) or they may take place in a continuous sequence if the environment is favorable for development of the pathogen.

Latent Infections

The phenomenon of latent infection in plant diseases is due to a transient resistance of the host to the extensive development of the pathogenic microorganism (Verhoeff 1974). Certain postharvest diseases may arise from infections on the flower parts or young fruit, in which case the period of latency can be as long as several months. In other diseases, short-term latent infections may be initiated in the field several weeks or less before the crop matures.

Conidia of *Botrytis cinerea* are abundant in the atmosphere of strawberry and raspberry plantations during flowering in the spring. These spores may germinate in a water drop on the petals of the flower. The pathogen in the necrotic petal then moves from the diseased flower part into the receptacle (strawberry) and therein forms a latent infection. This infection becomes the site of disease development after the berries are harvested several months later (Jarvis 1962). Similarly, spores of *Colletotrichum gloeosporioides* germinate in water on the surface of avocado, banana, citrus, mango and papaya fruits at any time during the development of the fruit on the plant (Baker 1938; Brown 1975; Stanghellini and Aragaki 1966). The germ tube of the fungus terminates in several hours in a durable structure known as an appressorium. These appressoria function as a quiescent stage on citrus, avocado and papaya fruits, giving rise to infection hyphae when the fruits begin to ripen after harvest. On bananas and mangoes, however, the newly formed appressoria immediately give rise to infection hyphae which pierce the cuticle of the fruit and form latent infections in the epidermal cell walls.

The stem-end rots of citrus fruits caused by *Diplodia, Phomopsis,* and *Alternaria* arise from quiescent infections in the stem button (calyx + disc) of the fruit (Fig. 7.1). These infections are initiated at any stage in the development of the fruit when environmental conditions are favorable for spore dispersal and germination. Quiescent infections of these fungi in the button of the fruit do not become active until this organ becomes senescent and begins to separate from the fruit after harvest (Brown and Wilson 1968).

Gray mold *(Botrytis cinerea)* is the most serious disease of grapes during storage at low temperature; it originates from late-season infections in the vineyard. In Australia, quiescent infections of *Monilinia fructicola* established in peaches and apricots before harvest are considered to be a significant factor in the selection of a fungicide to control this disease after harvest (Kable 1971). Brown rot of citrus fruits which develops in storage after harvest arises from incipient infections initiated by zoospores of *Phytophthora* spp. which splash onto low-hanging fruit several days before harvest. Fruit infected by this fungus earlier in the season usually fall from the tree or show obvious symptoms which leads to rejection at the time of harvest.

Infections Through Lenticels

The lenticels are "open" during certain stages in the development of apples and potatoes, permitting the entry of pathogenic microorganisms which are unable to penetrate the uninjured surface of their host. Lenticel rotting of apples, a major storage disease of fruit grown in Europe and other humid areas, arises from latent infections of *Gloeosporium* spp. which develop in lenticels of fruit during periods of relatively high temperature and humidity late in the summer (Edney 1958, 1976). The lenticels of potato tubers are a common site of initiation of bacterial soft rot after harvest. The lenticels of most tubers are contaminated with cells of the bacterium *Erwinia carotovora* at the time of harvest, but the bacteria remain quiescent in the lenticels until the tubers are subjected to the conditions which increase their susceptibility to decay, such as mechanical pressure, free water, and low oxygen tension in the tubers (Lund 1971; Lund and Wyatt 1972; Lund and Kelman 1977; Pérembelon and Lowe 1975). The surface of potato tubers may also be contaminated with strains of *Erwinia carotovora* which may be moved into the lenticels during the washing operation after harvest.

Infection During and After Harvest

In contrast to the fungus pathogens discussed in the preceding section, most microorganisms responsible for postharvest diseases are unable to penetrate the surface barriers of the host. Therefore, mechanical and physiological injuries arising during and after harvest are the usual point of entry for these "wound invading" pathogens, which as a group cause the most devastating postharvest diseases.

The injury created by severing the product from the plant is a common site of invasion by wound pathogens. Examples of diseases which arise in this fashion are crown rot of banana, pedicel rot of pineapple, and stem-end rots of mango, papaya, avocado and green pepper. A certain degree of random mechanical injury is inevitable in the course of harvesting, processing and packaging fruits and vegetables, even when these operations are carried out with reasonable care. Propagules of pathogenic fungi and bacteria which are abundant in the atmosphere and on the surface of maturing fruits and vegetables in the field, enter through these "breaks" in the surface of the hosts. For example, it is virtually impossible to avoid surface-skinning some varieties of Irish potatoes and sweet potatoes during harvesting and handling due to the immature cork layer and the active cambium at this stage of tuber development. The nature of the injury is also important. Deep injuries do not dry out rapidly and are difficult to impregnate with a fungicide. Various tissues of the host may differ in resistance to attack by pathogens and clean cuts mobilize defense mechanisms of the host more rapidly than scrapes or bruises. Specific examples of these phenomena will be discussed with other facets of host resistance in the following section.

The anticipated increase in injury resulting from mechanical harvesting compared with hand harvesting has been documented in many trials. The control of postharvest diseases incited by wound pathogens in mechanically harvested produce is one of the great challenges for postharvest pathology in future years. Bruises and excessive pressure may activate latent infections without necessarily rupturing the surface of the host. Blue mold decay *(Penicillium expansum)* is frequently initiated in the lenticel region of apples (Wright and Smith 1954) and bacterial soft rot *(Erwinia carotovora)* in the lenticels of potatoes by excessive pressure which crushes cells around the lenticels. Bruising the surface of bananas likewise stimulates the development of latent infections of *Colletotrichum musae.* Finger-stem-rot results from twisting the individual bananas on the hand and a lenticular rot develops at points where the ridges of the fingers have rubbed against the fiberboard shipping container (Stover 1972).

Physiological injuries caused by cold, heat, oxygen deficiency, and other environmental agents also predispose fruit to attack by wound pathogens. It is not surprising that conditions such as ice damage and cold-induced pitting which visibly damage the surface of the fruit cause an increase in decay by wound-invading pathogens. Environmental stresses may also increase the susceptibility of the host without producing obvious symptoms of injury. These effects of the environment will be dealt with in a later section.

FACTORS INFLUENCING DISEASE SEVERITY

The capability of a microorganism to initiate a postharvest disease, as well as the final outcome, depends upon a number of factors that can conveniently be associated with (1) the microorganism, (2) the host, or (3) the environment.

Microorganism

For development of a postharvest disease, the environment in the host tissues in terms of temperature, pH, nutrients and water potential must be favorable for growth of the microorganism. Second, the microorganism must elaborate enzymes which macerate the host tissue and also cause a release of nutrients suitable for growth of the microorganism from the macerated host cells. Fungi such as *Rhizopus stolonifer, Geotrichum candidum* and *Ceratocystis paradoxa* grow very slowly, if at all, below 10° C, whereas *Botrytis, Cladosporium* and *Penicillium* are capable of pathogenic growth even at 1° C. Spores of fungi which initiate latent infections in the field must be capable of growth and infection in drops of pure water or dilute nutrients which diffuse from the host surface. Spores of *Botrytis* may germinate in pure water, but are unable to infect cabbage leaves or strawberry fruit unless supplied with exogenous nutrients (Jarvis 1962; Yoder and Whalen 1975A). *Penicillium digitatum* and *Rhizopus stolonifer* require complex nutrients both for germination and infection of citrus fruits and carrots, respectively (Menke *et al.* 1964; Pelser and Eckert 1977). In the case of "wound pathogens," the required nutrients for development of the microorganisms are supplied by the broken cells of the injured area. Pathogens which invade through lenticels of the host are supplied with nutrients which leak from the cells surrounding the lenticel, especially after injury, anaerobic conditions or tissue senescence.

The water potential deficit is a significant factor in the colonization of low moisture grains by microorganisms (Bothast 1978). Some fresh products such as potatoes are more susceptible to bacterial soft rot when they are turgid (Pérembelon and Lowe 1975), while slightly wilted carrots and cabbage are most susceptible to invasion by *Botrytis* (Goodliffe and Heale 1977; Yoder and Whalen 1975A, B). These effects most likely reflect the physiological status of the host rather than a direct effect upon the pathogen, since most microorganisms involved in postharvest diseases grow well at water potential deficits larger than those likely to be encountered in the tissues of palatable fruits and vegetables (Burton 1973; Pérembelon and Lowe 1975).

Plant cell walls are held together by intercellular materials composed mainly of pectic polysaccharides. Development of a postharvest disease depends upon the capability of the pathogen to secrete enzymes that depolymerize these insoluble pectic polymers, leading to loss of tissue coherence and separation of the individual cells, a process referred to as tissue "maceration." The cells of the macerated tissues increase in permeability and die, allowing outward diffusion of host metabolites which may be used as substrates for growth by the pathogen.

Pathogens involved in postharvest diseases produce endo polygalacturonases, endo pectin lyases, and endo pectate lyases (Bateman and Millar 1966). Although the random hydrolysis of α-1, 4 glycosidic linkages in polygalacturonans is sufficient to cause tissue maceration and cell death, cellulases and hemicellulases may also be involved in pathogenesis. Recently, a proteinaceous factor has been isolated which stimulates tissue maceration by fungal enzymes, but the factor is essentially devoid of activity alone or in *in vitro* assays with pure enzymes (Ishii 1977).

The endo polygalacturonases are enzymes which hydrolyze the α-1, 4 galacturonide linkages of pectic acid with different degrees of randomness to produce a series of oligo galacturonides. These enzymes have been associated with tissue maceration by several microorganisms that cause postharvest diseases, e.g., *Penicillium, Rhizopus, Geotrichum, Sclerotinia* and the bacterium *Erwinia carotovora,* as well as many nonpathogenic microorganisms growing in culture (Rombouts and Pilnik 1972). Maceration of citrus peel by *Geotrichum* (Tóibín 1974), sweet potatoes by *Rhizopus* (Spalding 1969), and persimmons by *Gloeosporium* (Tani 1967) can be accounted for entirely by the action of extracellular endo polygalacturonases produced by the pathogens.

The endo pectin lyases split the α-1, 4 galacturonide linkages of pectin to different degrees of randomness by a transelimination mechanism so that a series of C_4–C_5 unsaturated methylated oligo galacturonides are produced. The pectin lyases which have been characterized are all of fungal origin. Pectin lyases are probably the principal cause of maceration of citrus fruit peel attacked by *Penicillium digitatum* and *Penicillium italicum* (Bush and Codner 1970).

The third group of enzymes associated with postharvest diseases, the endo pectate lyases are similar to the pectin lyases except that their substrate is pectic acid and their product, oligo galacturonides. This group of enzymes is commonly produced by pectolytic bacteria and the pectate lyases are the major cause of softening of potatoes attacked by soft rot bacteria (Mount *et al.* 1970; Hall and Wood 1973), although endo polygalacturonases may also be involved (Beraha *et al.* 1974).

The production of extracellular endo polygalacturonases and pectin lyases by fungi is generally induced by pectate or pectin in their environment. Cooper and Wood (1975) reported that these enzymes in the fungi *Verticillium* and *Fusarium* were induced by low concentrations of galacturonic acid rather than pectin or pectic acid and that the production of the enzymes was repressed when the inducer was in slight excess over that required for growth of the fungi. The extracellular pectate lyases produced by *Erwinia carotovora* are usually induced by pectin and, may or may not, be subject to catabolite repression, depending upon the individual isolate of the bacterium (Zucker *et al.* 1972).

Although microorganisms which cause postharvest disease must produce extracellular pectolytic enzymes, it does not follow that microorganisms that produce pectolytic enzymes can cause postharvest diseases. A number of nonpathogenic fungi and bacteria are capable of producing large amounts of enzymes which macerate slices of potatoes and other plant products (Ishii 1977). Both virulent and avirulent isolates of *Penicillium* spp. and *Erwinia carotovora* produce pectolytic enzymes in culture, indicating that the ability to attack native pectin in the host is not measured by pectolytic activity *in vitro*. However, avirulent strains of *Erwinia carotovora* showed lower activity of pectate lyase and endo polygalacturonase than did virulent strains (Beraha *et al.* 1974; Friedman 1962).

Host Interactions

Each type of fruit and vegetable may be attacked only by a relatively small and unique group of parasitic fungi, and possibly bacteria, which have nutritional requirements and enzymatic capabilities that permit them to develop extensively in the tissues of their host. *Penicillium digitatum* causes a postharvest disease of citrus fruit only, whereas *Penicillium expansum* is a serious pathogen of apples and pears, but not of citrus fruits. *Penicillium italicum* may attack a wide variety of fruit and vegetables. Many other examples could be cited to illustrate the specificity of parasitism possible in postharvest diseases. Cultivars of tomato (Bartz and Crill 1972), strawberries (Jennings and Carmichael 1975), and potatoes (Workman *et al.* 1976) vary in susceptibility to postharvest diseases. Indeed, even different tissues of the same fruit or vegetable may vary in susceptibility to the same isolate of a pathogen. In the carrot root, the pericyclic pyrenchyma beneath the surface periderm exhibits greater resistance to infection by *Centrospora acerina* than the underlying phloem pyrenchyma (Davies 1977). The outer leaves of cabbage are more resistant to attack by *Botrytis* than the inner leaves, and the adaxial surfaces of the leaves are more resistant than the abaxial surfaces (Yoder and Whalen 1975B).

The susceptibility of fruits and vegetables to postharvest decay is influenced dramatically by the maturity of the crop at the time of harvest and by subsequent physiological changes, collectively termed "ripening." The susceptibility of apples to blue mold *(Penicillium expansum)* increases with both maturity and ripeness, and appears to result from an increase in susceptibility of the flesh to bruise damage (Wright and Smith 1954). Likewise, Navel oranges become more susceptible to invasion by Penicillium molds with advance in maturity during the season. The resumption in activity of quiescent infections during ripening of fruits is a clear example of increasing disease susceptibility, rather than increased susceptibility of the host tissue to mechanical injury. Lenticel rotting of apples is virtually unknown at the time of harvest but develops in storage as the fruit ripens. Different pathogens appear at different stages of ripening of apples in storage. A striking increase in Alternaria stem end rot occurs when lemons pass a certain threshold of ripeness during storage. The lemon peel is highly resistant to invasion by *Alternaria citri* prior to this stage of development. Latent infections of *Colletotrichum* on banana, papaya and mango fruits seldom become a serious problem until the fruit approaches ripeness. A number of investigators have observed that potatoes become more susceptible to Fusarium dry rot during storage (Boyd 1972); the susceptibility of carrots to *Botrytis* and Centrospora at 5° C also increases with duration of the storage period (Goodliffe and Heale 1977; Davies 1977; Dennis 1977).

A few examples of disease susceptibility decreasing with maturity or storage of the crop are known. Several investigators have recorded that green tomatoes are more susceptible than red tomatoes to bacterial soft rot (Parsons and Spalding 1972). White potatoes become less susceptible to bacterial soft rot during the first few weeks of storage because the periderm layer matures during this period (Boyd 1972).

The intrinsic and variable resistance of fruits and vegetables to postharvest disease might be associated with one or a combination of several properties of the host: (1) pH, nutrient availability, or turgidity of the tissues; (2) inhibitors of microbial growth or of the action of pectolytic enzymes; (3) vulnerability of the cell wall to attack by macerating enzymes of the pathogen; and (4) ability of the host to form morphological or chemical barriers to the development of the pathogen.

pH, Nutrients and Water Status of the Host.—Acidity of the tissue of many fruits may be one of the most important reasons for their general resistance to bacteria that cause soft rot of many vegetable crops (Lund 1971). Tomatoes, peppers, cucumbers and pears are the few fruits known to be seriously affected by bacterial soft rot. Fruit tissues usually are below pH 5 which inhibits the growth of most bacteria capable of degrading plant tissues; vegetable tissues generally are less acid (Dennis 1977; Lund 1971). A decrease in acidity of the flesh of Bramley's Seedling apple during storage may be responsible for the decrease in resistance to *Nectria galligena* (Swinburne 1974). The increase in pH of the flesh associated with ripening in storage favors the dissociation of benzoic acid in the apple to the less toxic ionized form.

Clear-cut examples of unique nutrients or growth stimulants in susceptible varieties of fruits and vegetables are rare, but the possibility of their existence must nonetheless be considered. *Penicillium digitatum,* a unique pathogen on citrus fruits, requires a complex medium for rapid germination and vigorous hyphal growth. Citrus fruits contain both ascorbic acid and certan terpenes which are known to stimulate germination of *P. digitatum* spores (French *et al.* 1977; Pelser and Eckert 1977) as well as a complex of organic nutrients which sustain vigorous growth through the infection process.

Moisture Level.—Many fruits and vegetables are more susceptible to invasion by pathogens when their tissues are turgid. This is usually attributed to the presence of a water film associated with injuries which supports the growth of the pathogen during the infection process. Thus, slight desiccation reduces the susceptibility of citrus fruits to *Penicillium digitatum* (Eckert and Kolbezen 1963), potatoes to bacterial soft rot (Pérembelon and Lowe 1975) and carrots to *Centrospora acerina* (Davies 1977; Dennis 1977). In contrast, both *Botrytis* and *Rhizopus* preferentially infect wilted carrots (Dennis 1977; Goodliffe and Heale 1977; Thorne 1972). The susceptibility of carrots to infection by these fungi increases when the tissues exceed 8% water loss; more turgid roots are resistant. The increase in susceptibility of slightly flaccid carrots has been attributed to an increase in the intercellular spaces as the cells begin to separate at approximately 7% water loss (Thorne 1972). Cabbage also seems to be more resistant to *Botrytis* and other pathogens when stored at very high humidity which delays the onset of senescence of the outer leaves (Yoder and Whalen 1975A, B; Van den Burg and Lentz 1973).

Inhibitors of Microbial Growth and Pectolytic Enzymes.—Two types of microbial growth inhibitors have been recognized in plant tissues—preformed compounds and inhibitors which are synthesized by the host in response to attempted infection or other injuries. The latter are referred to as "phytoalexins." Examples of preformed inhibitors which are believed to offer resistance to infection are tannins and 3,4-dihydroxybenzaldehyde in bananas (Greene and Morales 1967; Mulvena *et al.* 1969), and benzylisothiocyanate in papaya (Patil *et al.* 1973). The concentration of these compounds decreases with ripening of the fruit and this is believed to play a role in the increase in susceptibility during this period.

The biosynthesis of antimicrobial compounds by plant tissue in response to injury is a rather common phenomenon (Kuć and Shain 1977). Several investigators have reported over the past few years that the production of 6-methoxymellein by the carrot is stimulated by the application of spores of *Rhizopus stolonifer, Botrytis cinerea,* and other storage pathogens to the injured surface of the roots (Menke *et al.* 1964; Heale and Sharman 1977; Heale *et al.* 1977; Coxon *et al.* 1973). This compound is believed to play a role in the resistance of freshly-harvested carrots to infection; loss of biosynthetic ability of the carrot to produce this compound during long-term cold storage is believed to be an important factor in the increase in susceptibility of carrot roots during this period (Goodliffe and Heale 1977; Heale *et al.* 1977). The development of resistance to *Rhizopus* by the injured surface of the carrot was shown to depend upon the metabolism of the cut surface of the carrot; surfaces which were not cleanly cut did not become resistant to infection after inoculation with spores of *Rhizopus stolonifer* (Menke *et al.* 1964). The antimicrobial terpenoid, rishitin, formed in potato tubers inoculated with *Erwinia carotovora* and stored in air, is thought to be a factor in restricting the expansion of lesions of bacterial soft rot, which developed under these conditions (Lyon *et al.* 1975). Benzoic acid accumulates in Bramley's Seedling apples in response to infection by *Nectria galligena* and other pathogens (Swinburne 1973). The ability of newly-harvested apples to produce this compound appears to be a significant factor in the resistance of fruit to decay during the initial period of storage.

The distribution of substances in plant tissues which may inhibit enzymes of the pathogen is probably more general than is usually appreciated. General enzyme inhibitors such as polyphenols are known to be present in the flesh of apples (Byrde 1963; Edney 1958); indeed, polyvinylpyrolidone must be added to the tissue to neutralize the polyphenols during homogenization prior to isolation of enzymes. Dicotyledenous plants have proteins associated with their cell walls which are capable of inhibiting the action of endo polygalacturonase secreted by plant pathogens. Albersheim and Anderson-Prouty (1975) suggest that a pathogen is incapable of attacking a plant unless the pathogen can secrete sufficient endo polygalacturonase to overcome the inhibitor present in the cell walls of the plant.

Increase in susceptibility of the host with ripeness.—Three factors may be involved in the observed increase in susceptibility of fruits and vegetables to disease during storage: (1) a decrease in ability of the host tissue to synthesize microbial inhibitors such as 6-methoxymellein and benzoic acid with age of the product in storage; (2) an increase in membrane permeability resulting in the release of nutrients and water into the intercellular spaces (Sacher 1973). [The apparent association of permeability increase and senescence has been critically evaluated by Simon (1977)]; and (3) increase in the susceptibility of the plant cell wall to attack by macerating enzymes of the pathogen.

The coherence of the flesh of fresh fruits and vegetables is dependent, to a large degree, upon the pectic substances in the middle lamella which functions as an adhesive, binding the cells together. The pectic materials are made up of linear chains of D-galacturonic acid, methylated to varying degrees (pectin), and with side chains of neutral sugars. As initially laid down by the cell, the pectic substances of the middle lamella are in a relatively insoluble form, protopectin. The insolubility of protopectin is a consequence of high molecular weight and an anchoring of the linear chains of pectin and pectic acid to the cellulose of the cell wall through bridges of neutral polysaccharides and perhaps proteins (Pilnik and Voragen 1970). As fruits and vegetables ripen, the bonds anchoring the pectic materials to the cell wall are broken, with the result that the pectic materials become more soluble and the tissue begins to soften. The increased solubility of the pectic substances and, perhaps, increased "openness" of the middle lamella makes this tissue more vulnerable to maceration by the pectolytic enzymes of the pathogen. The susceptibility of strawberries to attack by *Botrytis cinerea* is highly correlated with the soluble pectin content of the fruit (Hondelmann and Richter 1973).

Morphological barriers of the host to infection.—In response to injury, plant tissue may form protective barriers of tightly packed cells, provided that the tissue is still capable of cell division, or the cells surrounding the injury can deposit lignin and suberin in their walls to protect against the action of macerating enzymes of microorganisms. Lenticels of potatoes are normally blocked with a layer of suberized periderm which prevents the entry of soft rot bacteria into the cortex of the tuber (Fox *et al.* 1971; Adams 1975). Rupture of this periderm layer, either mechanically or by proliferation of underlying cells of the tuber in response to free water, makes the lenticels susceptible to invasion by soft rot bacteria (Pérembelon and Lowe 1975; Fox *et al.* 1971). A similar situation exists in the case of lenticels of immature apples (Edney 1958).

If potato tubers are superficially wounded and then placed at high humidity and moderate temperatures, suberin begins to form in the walls and intercellular spaces of the living cells surrounding the injury and, within several days, a suberized periderm forms beneath these suberized surface cells (Fox *et al.* 1971). These structures offer substantial resistance to the invasion of pathogens at wounded areas on the tuber. The surface cells of carrots also become suberized in response to wounding and this suberized surface layer is a barrier to infection by *Botrytis* (Goodliffe and Heale 1977; Heale and Sharman 1977; Heale *et al.* 1977).

Suberization of cells around injuries is probably the most important defense of the potato against infection by microorganisms, since this reaction takes place within 24 hr under optimal environmental conditions (Fox *et al.* 1971). The suberized periderm which follows in about four days is highly resistant to infection by microorganisms because it is composed of tightly packed cells, the walls of which are impregnated with suberin and have only a small amount of pectin (Fox *et al.* 1971). Periderm formation is most rapid at 21–27° C at high humidity, but storage temperatures of 16–21° C are usually recommended to reduce the rate of growth of *Erwinia carotovora* and *Fusarium* (Henricksen 1975). Smith and Smart (1955) found that the periderm formed on potato slices at 10° C in four days was adequate to protect against bacterial soft rot whereas slices held at 4.4° C formed only a slight amount of suberin, no periderm, and were highly susceptible to decay after transfer to 21° C. Low temperatures favor the development of gangrene *(Phoma)* on potatoes in storage because the pathogen continues to develop while the low temperature prevents the formation of barriers by the host (Boyd 1972). Carrots held at 22–26° C and high humidity for two days to permit wound healing before storage had less decay after long-term storage at 5° C than an uncured lot (Davies 1977). Curing of sweet potatoes and yams for 4–7 days at 26–32° C and 85–90% relative humidity allows for the tubers to form a suberized periderm to protect harvest injuries against invasion by *Rhizopus* and *Endoconidiophora* during subsequent storage of the crop (Steinbauer and Kushman 1971).

The ability of tissues of some plants to form a barrier of lignin around an injury is apparently an important mechanism of defense against invasion by a pathogen. Holding oranges at 30° C and 90–96% RH for several days after harvest reduces Penicillium decay because this environment is highly favorable to the synthesis of lignin, and the temperature is too high for the development of *Penicillium* (Table 7.1). The lignin barrier is not attacked by enzymes of the pathogen and probably prevents pectic enzymes of the pathogen from reaching the middle lamella of the host cells.

Environment

The best postharvest environment for maintenance of fresh fruits and vegetables is that which (1) maintains the product in the optimum condi-

TABLE 7.1
INFLUENCE OF TEMPERATURE AND RELATIVE HUMIDITY ON THE
DEVELOPMENT OF *PENICILLIUM DIGITATUM* IN ORANGES DURING
DEGREENING AND SUBSEQUENT STORAGE (BROWN 1973)

Degreening temp [1]	% of fruit infected with green mold after degreening at relative humidities of	
(° C)	90–96%	55–75%
27	89	71
30	63	93
33	21	97

[1] Fruit were degreened at each temperature at each range of relative humidity for 3 days and then stored at 25° C and 100% relative humidity for 4 days.

tion for consumption and (2) prevents invasion by microorganisms. Often the same environment satisfies both requirements; sometimes it does not, and some compromise must be made between these conflicting demands.

Temperature.—Exposure of fresh produce to moderately high temperatures after harvest, even briefly, can greatly increase their susceptibility to postharvest diseases even if physiological damage to the product is not evident. Exposure of harvested potatoes, yams and sweet potatoes to direct sunlight for more than a few hours is known to increase the severity of storage rots (Nielsen 1965; Nielsen and Johnson 1974). Soaking lemons in water at 48° C for 4 min, the recommended treatment for eradication of *Phytophthora* infections, predisposes some lots of lemons to excessive decay by *Penicillium* and other fungi during storage, without producing symptoms of heat damage to the fruit (Harding and Savage 1964). Hot water treatment has also predisposed tomatoes to *Alternaria* rot (Barkai-Golan 1973) and apples to Penicillium decay (Edney and Burchill 1967).

Low temperature storage is the most effective and practical means for delaying the development of decay in fruits and vegetables with deep-seated infections that cannot be eradicated by other postharvest treatments. The attack of the product by most microorganisms becomes very slow as the temperature is decreased below 5° C. The guiding principle of low temperature storage, therefore, should be to maintain the temperature as low as possible without injuring the commodity. With rare exceptions, low temperatures do not have a permanent effect upon the pathogen cells; their development is attenuated until late in the postharvest period when the temperature is once again suitable for their growth. Pathogens with relatively high temperature optima, such as *Geotrichum* and *Erwinia* may be inhibited almost indefinitely by storage near 0° C, but will rapidly produce a disease when the product is brought to a higher temperature for ripening for marketing. Other microorganisms such as *Botrytis, Penicillium, Sclerotinia, Centrospora* and *Cladosporium* grow slowly at near-freezing temperatures and limit the storage life of the crop if they are not controlled by other treatments (Eckert and Sommer 1967; Derbyshire and Crisp 1971). Low temperature delays the development of postharvest diseases by two mechanisms: (1) inhibition of ripening of the host, thereby prolonging the disease resistance associated with immaturity and (2) direct inhibition of the pathogen by a temperature unfavorable for growth. For each variety of fruit and vegetable there exists an optimum storage temperature which will maintain the desired quality of the product for the maximum period of time (Burton 1973). This optimum is near 0° C for many fruits and vegetables, *e.g.* apples, oranges, carrots and grapes. Storage at these low temperatures greatly retards the development of diseases as well as maintains the quality of the product. Other crops, typically of tropical origin, such as bananas, lemons, tomatoes, grapefruit and pineapples suffer chilling injury at storage temperatures of 10–15° C and become more susceptible to postharvest diseases if exposed to these temperatures for more than a few hours. Several investigators have observed that tomatoes and potatoes were more prone to attack by pathogens after exposure to low temperatures that did not produce visual damage (Bartz and Crill 1972; Daines 1970; McColloch and Worthington 1952; Workman *et al.* 1976). Exposure of sweet potatoes to 3–4° C for several hours before curing significantly increased Fusarium rot during storage due to an inhibition of wound healing by the low temperature treatment (Nielsen and Johnson 1974). Potatoes in England are normally stored at 10° C for 1 wk to allow wound healing before long term storage at a lower temperature. Gangrene rot *(Phoma)* can cause severe losses in potatoes held initially at 0–5° C, which inhibits effective wound healing, despite the fact that the optimum temperature for growth of *Phoma in vitro* is approximately 25° C (Burton 1973). Storage of freshly harvested potatoes at 4.4° C predisposes them to bacterial soft rot after storage because of inadequate wound healing (Smith and Smart 1955).

The optimum temperature for long term storage of grapefruit in Israel was reported as 12° C. Fruits stored at lower temperatures showed a higher decay level as a result of chilling injury which reduced the resistance of the fruit to Alternaria stem end rot (Schiffman-Nadel *et al.* 1971, 1975A).

Water and humidity.—Free water on the surface of fresh produce almost always increases decay by hydrating injuries, stomates, lenticels and by transporting pathogenic microorganisms into these and other avenues of infection. Free water on the surface of potato tubers may also cause "opening" of the lenticels and anaerobic conditions to develop in the tuber; both of these conditions greatly predispose potatoes to bacterial soft rot (Cromarty and Easton 1973; Fox *et al.* 1971; Lund 1971; Pérembelon and Lowe 1975). Lund and Kelman (1977) demonstrated that commercially-washed and dried potatoes had an inherently higher soft rot potential than unwashed potatoes. Water used to wash or cool fruits and vegetables should always be free of pathogenic microorganisms and preferably should contain an effective antimicrobial agent.

Washing fruits and vegetables has been found to reduce postharvest diseases in some rare instances. Washing seed potatoes is recognized to reduce Fusarium dry rot and gangrene *(Phoma)* during storage, presumably by removing pathogen propagules from the tuber surface (Boyd 1975). Washing citrus fruits before degreening with ethylene gas reduces the severity of anthracnose due to *Colletotrichum* by removing many of the appressoria of the pathogen from the surface of the fruit (Brown 1975; Smoot *et al.* 1971). Diplodia stem end rot is also reduced, presumably by washing propagules of the pathogen away from the stem end of the fruit (Smoot *et al.* 1971).

At temperatures above 5° C, a relative humidity in excess of 90% favors the development of postharvest diseases by permitting growth of microorganisms on the surface of fresh products and by maintaining wounds and other avenues of infection in a condition favorable for infection. Microorganisms generally do not grow on the surface of fruits and vegetables in equilibrium with an atmosphere of less than 90% relative humidity (Burton 1973). Superficial injuries on oranges, apples and bananas usually become resistant to infection if these fruits are held in an ambient environment for several days (Eckert and Kolbezen 1963; Greene and Goos 1963; Wright and Smith 1954). This may be attributed to mild desiccation of the injured tissue although wound defenses of the host might also be involved. Storage of commodities such as carrots and cabbage at 98–100% relative humidity and temperatures approaching 0° C provide excellent results, both in terms of product quality and low incidence of postharvest diseases (Van den Burg and Lentz 1973). Slight desiccation of cabbage during storage leads to senescence of outer leaves and increases their susceptibility to infection by *Botrytis* (Yoder and Whalen 1975A). Carrots also become more susceptible to both *Botrytis* and *Rhizopus* after a water loss of approximately 8% by weight and this phenomenon has been attributed to an increase in the intercellular spaces in the roots (Thorne 1972).

Moisture-impermeable plastic bags are an efficient means for maintaining a very high relative humidity around fresh produce stored near 0° C (Burton 1973). This practice should be approached with caution for fresh products stored at a higher temperature because of the growth of microorganisms on the surface of the product and the hazard of water condensation (Hardenburg 1971). A water film on the surface of potatoes and leafy vegetables greatly predisposes them to bacterial decay and may lead to fungal rots of other fruits and vegetables. Grierson (1969) examined the merits of film wraps for citrus fruits with regard to both weight loss and decay, concluding that ventilation-holes in the film are essential for humidity release to prevent excessive decay. An exception to these generalizations is the experimental practice in Australia of shipping green bananas in sealed plastic bags with an ethylene absorber at ambient temperatures. Ripening of the fruit during transit is prevented, but treatment of the fruit with a highly effective fungicide (e.g. thiabendazole) before bagging is absolutely essential (Scott et al. 1970).

In addition to carrots, cabbage and other vegetables, high humidity is essential to maintenance of resistance to disease in other fruits and vegetables. Citrus fruits produced in the humid subtropics are subject to a condition known as "stem end rind breakdown" if held at a relative humidity less than 90% for more than a few hours. The injured peel of fruit with this disorder is infected readily by Penicillium digitatum and P. italicum (McCornack 1973). High humidity (approximately 90%) is essential to the formation of suberin and periderm by sweet potatoes, white potatoes and other root tuber crops. Oranges held at 30° C and high humidity for several days develop a lignin barrier around superficial wounds in the peel which prevents later infection by Penicillium digitatum at these sites. Lignin formation is inhibited in oranges held under less humid conditions (Brown 1973).

The Atmosphere

Storage atmospheres, modified by altering the levels of O_2 and CO_2, can influence the development of postharvest disease either by direct inhibition of the pathogen or by altering the resistance of the host. Decay of strawberries and sweet cherries can be reduced significantly by raising the CO_2 level of the storage atmosphere to 20–30% (Smith 1963). These levels of carbon dioxide reduce decay by retarding the physiological deterioration of the berries (Sommer et al. 1973) and by directly inhibiting the growth of the pathogen (Wells and Uota 1970). High CO_2 storage for strawberries is most beneficial at temperatures above 5° C which do not completely inhibit the growth of Botrytis. The high CO_2 treatment has been utilized extensively in shipments of strawberries from California to distant markets. The CO_2 is generated by sublimation from solid carbon dioxide (dry ice) placed in the shipping container or in pallet stacks covered with polyethylene film (Harvey et al. 1971).

Modified atmospheres composed of O_2 and CO_2 levels less than 10% have been evaluated extensively as a means of extending the physiological life of fruits and vegetables in cold storage. In storages held near 0° C, it has been observed that the incidence of decay is somewhat less in modified atmospheres than in air (Edney 1964; Van den Berg and Lentz 1973). However, when these fruits and vegetables are transferred to air for ripening at ambient temperatures, decay develops to about the same extent regardless of the previous storage atmosphere (Smith and Anderson 1975). These results are explained by the effect of modified atmospheres on the resistance of the host to disease development. The usual modified atmosphere for fruit storage is unlikely to have a pronounced effect upon pathogens, since most fungi grow well in atmospheres containing more than 1% O_2 (Follstad 1966).

Florida avocados stored 45–60 days at 10° C in an atmosphere of 2% O_2 and 10% CO_2 showed a remarkable reduction in anthracnose compared to fruit stored in air at the same temperature. The 2% O_2–10% CO_2 atmosphere also prevented chilling injury, thereby permitting storage of the fruit at 7.2° C, a temperature which causes chilling injury to the fruit when stored in air (Spalding and Reeder 1975). Either 2% O_2 in nitrogen or 10% CO_2 in air was more effective than air alone, but the combined 2% O_2 plus 10% CO_2 was required for maximum decay control and increased cold tolerance. The 2% O_2–10% CO_2 treatment inhibited sporulation, but not growth, of *Colletotrichum,* indicating that control of decay on avocados by modified atmospheres was mediated through an increase in resistance of the fruit.

Decay of apples by *Nectria* was less in an atmosphere of 4% CO_2 than at higher concentrations of CO_2 or in air (Swinburne 1974). The accumulation of benzoic acid, a natural inhibitor of *Nectria* infections, was stimulated by concentrations of CO_2 up to 2.5% in the atmosphere.

Modified atmosphere storage does not always result in a reduction in postharvest diseases. Parsons and Spalding (1972) reported that mature green tomatoes stored in an atmosphere of 5% CO_2 and 3% O_2 had a higher incidence of bacterial soft rot than fruits stored in air. They attributed this to the fact that green tomatoes are uniquely more susceptible to this disease than are ripe fruit and the modified atmosphere delayed coloration of the fruit. Low O_2 (less than 5%) and high CO_2 (greater than 10%) greatly increases bacterial soft rot of potatoes, especially in the presence of free water, by inhibiting periderm formation over injured areas (Lipton 1967), and by increasing the susceptibility of the cortical tissues to *Erwinia carotovora* (Pérembelon and Lowe 1975; Cromarty and Easton 1973; Lund and Wyatt 1972). The severity of rot produced by *Geotrichum candidum* on tomatoes is increased by storage in low O_2 and high CO_2 atmospheres because the growth of the fungus is stimulated by this atmospheric environment (Wells and Spalding 1975).

Since susceptibility to disease increases with ripeness, it is not surprising that the application of ethylene gas to accelerate ripening has been observed to increase Alternaria rot on tomatoes (Segal *et al.* 1974), and stem end rot and anthracnose on citrus fruits (Brown and Barmore 1977; McCornack 1972; Smoot *et al.* 1971). McCornack (1972) demonstrated that ethylene degreening of oranges increased Diplodia stem end rot, but not Penicillium green mold, and that there was a direct relationship between the ethylene concentration (4–50 ppm) and stem end rot (Table 7.2). Five to ten parts per million ethylene is the minimum required for satisfactory coloring of the fruit. The ethylene treatment accelerates senescence and abscission of the fruit button, thereby stimulating the development of quiescent infections of *Diplodia* in this organ and providing an avenue for the fungus to enter the fruit (Brown and Wilson 1968).

Ethylene treatment of green tangerines stimulates appressoria of *Colletotrichum* present on the surface of the fruit to form infection hyphae which then penetrate the epidermal cells of the fruit (Brown 1975; Brown and Barmore 1977). Apparently the ethylene treatment simultaneously triggers an unknown resistance mechanism since the fruit become resistant to infection if they are treated with ethylene three days before inoculation with *Colletotrichum* spores.

The removal of low concentrations of ethylene, formed by ripening fruit or by fungi from the ambient atmosphere has reduced decay of lemons in modified atmosphere storage, presumably by maintaining the fruit in an immature, more resistant condition (Wild *et al.* 1976).

TABLE 7.2

EFFECT OF ETHYLENE ON ANTHRACNOSE, STEM END ROT, AND GREEN MOLD OF CITRUS FRUITS

Ethylene (μl-liter air)	Percent fruit decayed by		
	Anthracnose[1]	Stem end rot[2]	Green mold[2]
0	0	0	6
5	—	13	3
10	17	—	—
20	57	—	—
50	86	23	2

[1] Data of Brown and Barmore 1977. Anthracnose incited by *Colletotrichum gleosporioides*. Robinson tangerines exposed to ethylene for 2 days and then stored for 1 wk at 26° C and 100% relative humidity.
[2] Data of McCornack 1972. Stem end rot incited mostly by *Diplodia natalensis* and green mold by *Penicillium digitatum*. Valencia oranges exposed to ethylene for 2 days and then stored at 21° C for 4 wk.

CONTROL OF POSTHARVEST DISEASES

Several strategies for the control of postharvest diseases are feasible: (1) prevent infection of the product in the field before harvest; (2) reduce the level of pathogen inoculum in the postharvest environment to minimize the possibility of infection of injured fruit (sanitation); (3) maintain the resistance of the plant product to infection by low temperature storage and modified atmospheres; and (4) prevent, eradicate or attenuate the development of microorganisms in the host by postharvest treatment of the crop.

The principles underlying strategies 1–3 have been examined earlier in this paper and examples of practical applications and results have been summarized elsewhere (Burton 1973; Eckert 1975; Eckert and Sommer 1967). In the writer's opinion, most of the opportunities in these areas seem to have been fully exploited and, therefore, major advances in the control of postharvest diseases in the future should be realized in the application of chemical treatments to the crop to: (1) protect the surface of the product from subsequent infection; (2) prevent infection through harvest injuries; (3) eradicate or attenuate established infections or deep-seated inoculations in lenticels, stomates, and latent infections; and (4) retard senescence of the plant product, thereby retaining the resistance associated with immaturity.

Preharvest treatments

Several investigators have demonstrated that Rhizopus rot of peaches after harvest may be substantially reduced by spraying the fruit one week before harvest with dichloran (Luepschen et al. 1971; Ogawa et al. 1971). Similarly, orchard sprays of benomyl before harvest significantly reduced the incidence of brown rot which developed on peaches after harvest, but had little effect upon Rhizopus rot (Kable 1971; Ogawa et al. 1968). Oranges sprayed with 300–1000 ppm benomyl 30 days before harvest showed substantially less stem end rot and Penicillium mold 2 wk after harvest (Brown and Albrigo 1972). Orchard sprays of thiabendazole on pears produced a reduction in decay incited by *Penicillium expansum* and *Botrytis cinerea* (Ben-Arie and Guelfat-Reich 1973; Coyier 1970). As a means of preventing infection of harvest injuries, the application of fungicides in the field is a less desirable practice than treatment of the fruit after harvest because (1) only a fraction of the fungicide applied in the field is bound to the harvested product where it can protect subsequent injuries and (2) the deposit required for postharvest disease control may be removed by washing or waxing the fruit after harvest. Preharvest treatments appear justifiable in situations where substantial harvest injury is anticipated, but handling practices make postharvest treatment difficult to carry out. The application of preharvest sprays of fungicides to control Diplodia stem end rot of oranges during ethylene degreening and Rhizopus rot of peaches during controlled ripening or storage before processing, appear to be situations which justify the preharvest treatment (Ogawa *et al.* 1971). Mechanical harvesting invariably results in a substantial increase in decay arising in superficial injuries and preharvest treatments may be a desirable approach to this problem, especially if the crop cannot be treated soon after harvest. The fungicide chosen for preharvest application should be selected with care since there is a significant risk that residues from the preharvest treatment will permit the build-up of fungicide-resistant strains of the pathogen which would nullify all benefits of a postharvest treatment with the same fungicide. The wisdom of applying preharvest sprays to control postharvest diseases has been questioned by some investigators because of this potential problem. Postharvest applications of benomyl or another fungicide with similar properties by drenching of the fruit in field containers would provide the same measure of protection as the preharvest treatment without the disadvantages of the latter. Benomyl appears to penetrate to a limited extent, into the superficial tissues of the host and protects injuries made several weeks later against infection by *Penicillium digitatum* (Brown and Albrigo 1972; Gutter 1970; McCornack 1975).

Postharvest treatments

Prevention of infection through harvest injuries.—Injuries created during harvest and handling of the crop are the major sites of infection by "wound pathogens." Mechanical harvesting procedures are recognized to substantially increase such infections. Chemical treatments to control these infections should be applied as soon as possible after harvest, although in many cases a delay of several hours may significantly increase the effectiveness of the treatment (Fig. 7.2). The increase in effectiveness of chemical and physical treatments applied several hours after inoculation may be explained by the fact that germinating spores are more sensitive than dormant spores to these treatments (Eckert 1977). In practice, however, a few hours will necessarily elapse between harvest and treatment of the crop; therefore, the primary goal should be to apply the treatment as soon as possible before the pathogen penetrates too deeply into the host tissue. The maximum period of time which may elapse between harvest (inoculation) and successful treatment varies from about 10 hr at 25° C for a protective treatment applied to peaches inoculated with *Rhizopus* to about 2 wk for control of *Phoma* on potatoes by treatment with

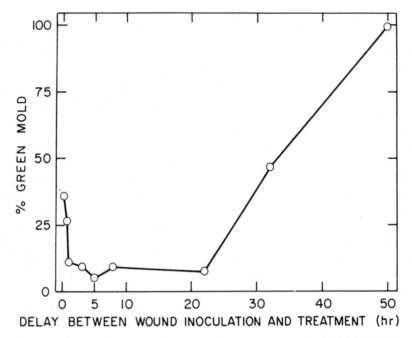

From Seberry (1969)

FIG. 7.2.   EFFECT OF TIME BETWEEN WOUND INOCULATION
AND TREATMENT UPON THE EFFECTIVENESS OF SODIUM
o-PHENYLPHENATE (SOPP) IN CONTROLLING GREEN MOLD
(*Penicillium digitatum*) ON ORANGES
Oranges were submerged in 2.0% SOPP solution pH 12 for 2 min and
then rinsed with water. Note that the treatment reaches maximum
effectiveness 1–2 hours after inoculation of fruit with spores of
Penicillium digitatum.

gaseous *sec*-butylamine (Graham *et al.* 1973). The principal factors influencing the allowable time between inoculation and treatment are:
temperature, growth rate of the pathogen, resistance of the host tissues,
and penetration of the chemical into the host tissue. In case of latent
infections, such as anthracnose of tropical fruits and lenticel rot of apples,
the pathogen is restricted to superficial tissues by resistance of the host so
that a systemic fungicide treatment may be very effective when applied
weeks or even months after inoculation.

Fungicides applied to prevent infection at harvest injuries may be classified into two categories according to their distribution on the treated
host: (1) nonionic water insoluble compounds (*e.g.*, thiabendazole) which
are applied in a manner which results in uniform coverage of the chemical

on the surface of the product; (2) water-soluble salts such as sodium *o*-phenylphenate and sodium carbonate which are applied in fairly concentrated solution (0.5–3%) in water (Eckert 1967). The injuries absorb the aqueous solution, whereas the intact cuticle is impermeable to the ionic substance. The product is then lightly rinsed with fresh water after treatment and most of the chemical is removed from the surface, except for a significant residue which remains in injury sites (Fig. 7.3). In the case of sodium *o*-phenylphenate, the water-insoluble *o*-phenylphenol is precipitated in acidic wound tissues which have a pH considerably below the pK a of *o*-phenylphenol (10.01). The advantage of uniform coverage of the fungicide is that the entire surface of the product is protected from injuries

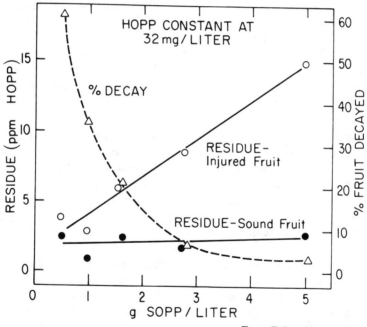

From Eckert et al. (1969)

FIG. 7.3. RELATIONSHIP BETWEEN RESIDUE OF *o*-PHENYLPHENOL (HOPP) ACCUMULATED IN SUPERFICIAL PEEL INJURIES AND THE CONTROL OF GREEN MOLD (*Penicillium digitatum*) ON VALENCIA ORANGES

Fruit were submerged for 4 min in solutions of sodium *o*-phenylphenate tetrahydrate at the indicated concentrations and rinsed with water. The concentration of HOPP in the solutions was held constant by slight adjustments in pH of the solutions.

incurred after the treatment. Compounds which may be phytotoxic to the product must be applied selectively to wounds and the remainder of the applied compound removed by rinsing. Rinsing greatly reduces the residue of the chemical on the treated product which is important in the case of sodium tetraborate which is used to treat oranges.

Solutions of sodium o-phenylphenate have been used extensively for control of postharvest diseases of citrus fruits, apples, pears, sweet potatoes and other perishable fruits and vegetables. Sodium o-phenylphenate is a broad-spectrum antimicrobial agent. Fortunately, the o-phenylphenate anion does not readily penetrate the cuticle of the host, since o-phenylphenate is toxic to living cells of the host as well as the pathogen. The benefits of treating fruits and vegetables with sodium o-phenylphenate are twofold: (1) fungus spores and bacteria on the surface of the fruit or introduced into the cleaning solution are killed, and (2) a residue of o-phenylphenol is deposited in harvest injuries which prevents infection at these sites during storage and marketing of the crop (Eckert *et al*. 1969). Krochta *et al*. (1977) demonstrated that a foam prepared from a solution of 0.5% sodium o-phenylphenate and 0.3% sodium laurylsulfate was highly effective as a cushion for mechanically harvested tomatoes dropped into a bin and also prevented decay of the fruit caused by *Geotrichum* and *Rhizopus*. Neither of these fungi are sensitive to the benzimidazole fungicides.

The fungicide *sec*-butylamine (Fig. 7.4, V) may be applied to the harvested crop as a fumigation treatment or as a solution of a salt of *sec*-butylamine. The fumigation treatment controls *Penicillium* on citrus fruits because (1) the injuries on the surface of the fruit become alkaline by absorption of the amine vapor and (2) fungistatic residues of neutral *sec*-butylamine salts persist in the injuries following the fumigation treatment (Eckert and Kolbezen 1970).

Seed potatoes in the UK may be fumigated with *sec*-butylamine to control several diseases in storage (Boyd 1975; Graham *et al*. 1973).

In California, oranges are drenched with a solution of 1% *sec*-butylamine (phosphate salt, pH ca.9) after harvest to protect the fruit injuries against infection by *Penicillium digitatum* and *P. italicum* during the subsequent degreening period or during a holding period of several days between harvesting and processing for shipment as fresh fruit (Eckert and Kolbezen 1964). The degreening operation is highly conducive to the development of Penicillium decay since it involves exposure of the fruit to a low concentration of ethylene gas for several days in a warm (23–25° C), humid environment. *sec*-Butylamine is added also to emulsion wax formulations applied to lemons before storage to control species of *Penicillium*, but the treatment has little effect upon diseases caused by *Alternaria*

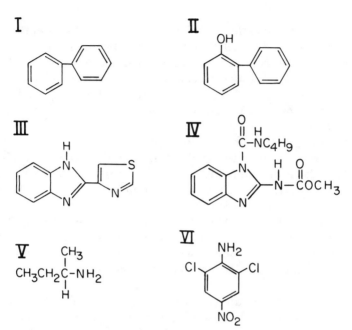

FIG. 7.4. CHEMICAL STRUCTURES OF FUNGICIDES
CURRENTLY IN COMMERCIAL USE TO CONTROL
POSTHARVEST DISEASES OF FRUITS AND
VEGETABLES
I—biphenyl; II—o-phenylphenol; III—thiabendazole;
IV—benomyl; V—sec-butylamine; VI—dicloran.

or *Geotrichum* which are resistant to *sec*-butylamine as well as to the
2-substituted benzimidazoles. The *sec*-butylamine treatment may be followed by a thiabendazole or benomyl treatment after storage since isolates
of *Penicillium* rarely exhibit resistance to both *sec*-butylamine and the
2-substituted benzimidazole fungicides.

The benzimidazole fungicides, thiabendazole and benomyl (Fig. 7.4, III
and IV), have been utilized extensively for control of postharvest diseases
since the late 1960's. They are used worldwide to control Penicillium decay
and stem-end rots of citrus fruits, crown rot and anthracnose on bananas,
Monilinia brown rot on stone fruits, Penicillium blue mold and lenticel
rots of apples, and anthracnose on several tropical fruits (Eckert 1977).

Benomyl is more active than thiabendazole and has provided outstanding activity against infections situated beneath the surface of the fruit,
such as anthracnose on bananas (Griffee and Burden 1973, 1974), papayas
(Bolkan *et al.* 1976), mangoes (Jacobs *et al.* 1973; Muirhead 1976), citrus

stem-end rots (Smoot and Melvin 1975; Wild *et al.* 1975) and lenticel rots of apples (Burchill and Edney 1972; Edney 1970). Benomyl is also superior to thiabendazole for control of Thielaviopsis stem-end rot of pineapples (Cho *et al.* 1977). The outstanding performance of benomyl in controlling these diseases appears to be due to the penetration of the compound into the living tissues of the host to encounter deep-seated infections (Ben-Arie 1975; Phillips 1975).

Both benomyl and thiabendazole penetrate the intact periderm of potato, making underlying tissue resistant to pathogenic fungi (Murdoch and Wood 1972; Tisdale and Lord 1973). Tests initiated in the late 1960's have shown that the benzimidazole fungicides can control dry rot *(Fusarium),* gangrene *(Phoma),* skin spot *(Oospora)* and silver scurf *(Helminthosporium)* of potatoes during storage (Boyd 1975; Meredith 1975). For reasons unknown, thiabendazole seems to be somewhat more effective than benomyl for control of dry rot and gangrene. Thiabendazole is currently applied to potatoes intended for seed, storage, and fresh pack for control of fungus diseases.

Thiabendazole is stable under most conceivable conditions of formulation and postharvest use. Even sublimation for application as a thermal dust has been recommended (Katchansky 1977). Thiabendazole is soluble in pure water to the extent of about 25 μg/ml, but owing to its amphoteric nature it is fairly soluble in both dilute acids and bases. In practice, however, thiabendazole is applied to fruits and vegetables only as a suspension in water because the acidic and basic conditions required for true solution could injure fresh produce.

Benomyl is highly insoluble in water and suspensions of 1000 mg/liter or greater are apparently stable for many hours in a neutral medium (Lowen 1977). Benomyl decomposes slowly, especially in acidic medium and at dilute concentrations, to methyl benzimidazolecarbamate (MBC) which is almost as fungitoxic as benomyl *in vitro.* In alkaline formulations or environments (pH greater than 8) the substituents attached to the benzimidazole nucleus of benomyl react to give 1,2,3,4-tetrahydro-3-butyl-2,4-dioxo-*s*-triazino[a]-benzimidazole (Lowen 1977; White *et al.* 1973), which has very weak if any antifungal properties (Fig. 7.5). For this reason, benomyl should never be incorporated into a fruit "wax" formulation, such as used to coat citrus and stone fruits, which has a pH greater than 8, or used in other alkaline formulations. This admonition may be applicable to the proposal to incorporate benomyl into a preharvest spray of lime sulfur intended to intensify the color of mandarins in Japan (Yamada *et al.* 1972). Benomyl is highly unstable in organic solvents, including those used in the preparation of fruit coatings, breaking down spontaneously to MBC (Chiba and Doornbos 1974).

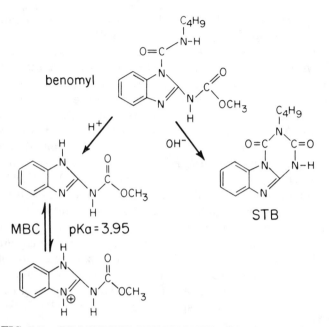

FIG. 7.5.   REACTIONS OF BENOMYL UNDER ACIDIC
AND BASIC CONDITIONS

The antifungal activity of MBC *in vitro* is almost equal to benomyl. MBC has been evaluated for control of postharvest diseases of bananas, citrus and stone fruits (Löcher and Hampel 1973). MBC is equal or slightly inferior to benomyl in preventing infection at harvest injuries (Koffman and Kable 1975; Rippon and Wild 1973), but is definitely inferior for applications which require deepest penetration into the host tissues (following section).

Thiophanate methyl, a progenitor of MBC (Vonk and Kaars-Sijpesteijn 1971) is about equal to MBC and benomyl for many applications, but inferior for others (Eckert 1977). Thiophanate methyl is most active fungicidally under alkaline conditions which increase its solubility and favor its conversion to MBC (Noguchi *et al.* 1972; Yamada *et al.* 1972).

The benzimidazole fungicides have two serious biological limitations: (1) all compounds in this group are inactive against several microorganisms responsible for important postharvest diseases: *Rhizopus* and other mucors, *Alternaria, Geotrichum, Phytophthora,* and all bacteria including *Erwinia* spp. which cause soft rot of potatoes and vegetable crops; (2) resistant individuals exist in populations of "benzimidazole sensitive fungi" which may proliferate when the benzimidazole fungicide is used continuously on a crop (Georgopoulos 1977; Wolfe 1975).

Relatively minor diseases incited by *Alternaria* and *Geotrichum* on citrus fruits have become the limiting factor in storage and handling of this crop after treatment with benzimidazole fungicides (Albrigo and Brown 1977). In fact, several investigators have observed that these problems have increased in severity in fruit lots treated with these fungicides (Kuramoto and Yamada 1976; McCornack 1975; Pelser 1974; Yamada *et al.* 1972). The selection of resistant strains present in a natural population of a "benzimidazole sensitive fungus" has become a serious problem in the control of Penicillium decay of citrus fruits (Gutter 1975; Harding 1972) and *Botrytis* on several fruits. The use of benomyl on strawberries in the presence of resistant strains of *Botrytis* has caused an increase in decay due to this fungus, apparently reflecting a selection of a benzimidazole-resistant strain with increased virulence or better ecological adaptation (Jordan and Richmond 1974, 1975). Benzimidazole-resistant strains of other pathogenic fungi have been isolated or induced by mutagenic agents, sounding the alarm to the potential hazards involved in the continuous use of benzimidazole fungicides on specific crops (Griffee 1973; Griffee and Burden 1974).

The limitations of the antimicrobial spectrum of the benzimidazole fungicides can be corrected by the addition of a second fungicide in some instances. Peaches, cherries and nectarines are seriously affected by both Monilinia brown rot and Rhizopus rot. The standard treatment today is a combination of benomyl and dicloran (Fig. 7.4 VI) applied in a wax formulation to the fruit (Koffman and Kable 1975; Wade and Gipps 1973; Wells and Bennett 1976). Thiabendazole or dicloran applied in a dust formulation has been recommended for the control of *Botrytis* on cabbage in storage. Alternation of the fungicide used in sequential crop years should minimize the hazard of the development of resistant strains of *Botrytis* (Brown *et al.* 1975). Captafol possesses considerable activity against *Alternaria* which is a serious problem in long term cold storage of many crops (Albrigo and Brown 1977). For several disease applications in the field, the EPA registration specifies that benomyl must be applied in combination with a broad-spectrum fungicide such as captan or maneb in order to discourage the development of resistant strains of the pathogen.

Imazalil (Fig. 7.6), a unique fungicide introduced in Europe in the early 1970's, is highly effective in preventing *Penicillium* decay at injury sites on citrus fruits, as well as the stem-end rots caused by *Diplodia* and *Phomopsis* (Harding 1976; Laville 1974; McCornack *et al.* 1977). Imazalil has only slight activity against sour rot caused by *Geotrichum*, but may be combined with 1% sodium *o*-phenylphenate to control most of the principal

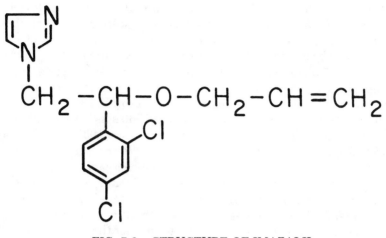

FIG. 7.6.   STRUCTURE OF IMAZALIL

postharvest diseases of citrus fruits (Harding 1976). The most significant property of imazalil is that it is fully active against strains of *Penicillium* which are resistant to both benzimidazole fungicides and *sec*-butylamine. Genetic studies of imazalil-resistant mutants of *Aspergillus nidulans,* revealed that resistance was based on a multigenic system and that all the single gene mutants showed a relatively low level of resistance (Van Tuyl 1977). These observations suggest that the development of pathogens with significant resistance to imazalil might be a rather rare event since it would require the combination of several resistant genes in a single strain of the fungus (Georgopoulos 1977). Trials conducted in many countries on the efficacy of imazalil on citrus fruits have been reviewed recently by Laville *et al.* (1977). Imazalil has not yet been approved by governmental agencies for treatment of citrus fruits or other crops on a commercial basis.

Several new fungicides recently have been reported to have applications in the postharvest field. Tests conducted with guazatine (1,17-diguani-dino-9-aza heptadecane acetate) in Japan (Kuramoto and Yamada 1976) showed that this compound provided good control of Geotrichum sour rot of satsuma mandarins in storage, as well as strains of *Penicillium* which were resistant to benzimidazole fungicides. Some activity against *Alternaria* was shown in *in vitro* tests, but guazatine did not control fruit decay initiated by this organism.

Several investigators have shown that postharvest treatment of stone fruits with triforine (N,N-[1,4-piperazinediylbis (2,2,2-trichloro-ethylidene)] bis formamide) controlled Monilinia brown rot at least as well as benomyl (Jones *et al.* 1973; Koffman and Kable 1975; Rohrbach 1973; Szkolnik 1975). Triforine is not active against *Rhizopus;* therefore, it should be combined with dicloran for broad-spectrum control of peach diseases (Rohrbach 1973). The experimental fungicide isopropyl carbamoyl-3-(3,5-dichlorophenyl)-hydantoin is highly effective against Rhizopus and Alternaria rots of stone fruits, although it is considerably less effective than benomyl against brown rot (Szkolnik 1975; Jones 1975).

Acetaldehyde was recognized in the early 1930's as a vapor phase inhibitor of postharvest diseases (Tomkins and Trout 1932), but has been evaluated intensively only in the past few years. *Botrytis* and *Rhizopus* on strawberries and raspberries and *Penicillium expansum* on apples have been controlled by exposure of the fruits to 0.25–1.0% (v/v) acetaldehyde for 1–2 hr (Prasad and Stadelbacher 1973, 1974; Stadelbacher and Prasad 1974). Interest in the development of acetaldehyde as a postharvest treatment may be attributed in large measure to the fact that this compound is a natural volatile constituent of fruits and vegetables and is permitted as a food additive in the United States.

All of the compounds discussed thus far in connection with the control of postharvest diseases are directly inhibitory to the growth of the pathogen *in vitro*. Other possible mechanisms for preventing infection of fruits and vegetables have not yet been fully explored. Three possibilities which may be suggested are: (1) agents which encourage the synthesis or retention of antimicrobial compounds or morphological barriers by the host, (2) agents which inhibit (repress) the induction of pectolytic enzymes by the pathogen, (3) agents which directly inhibit the action of pectolytic enzymes secreted by the pathogen in the host tissues.

The accumulation of benzoic acid, a natural antifungal compound in apples, can be stimulated by infusion of the apple flesh with a proteolytic enzyme extracted from the fungal pathogen *Nectria* or by increasing the CO_2 level in the atmosphere to 2.5% (Brown and Swinburne 1973; Swinburne 1973, 1974; Swinburne and Brown 1975). Many elicitors of antifungal compounds in plants are known (Kuć and Shain 1977). Cartwright *et al.* (1977) have described a derivative of cyclopropane dicarboxylic acid which may prevent the initiation of the rice blast disease in this manner. Physical barriers to infection consisting of suberin, lignin, and periderm are formed by a number of fruits and vegetables in response to injury; it may be possible to elicit these responses by other agents also.

The production of extracellular pectolytic enzymes by microorganisms responsible for postharvest disease are, for the most part, induced by low concentrations of their substrates (*e.g.* pectic substances); or more correctly, by low concentrations of the product of the enzyme reaction (*e.g.* galacturonic acid). Avirulent strains of pathogenic species produce little, if any, extracellular pectolytic enzymes (Beraha *et al.* 1974; Friedman 1962). Therefore, any agent which repressed enzyme synthesis should also interfere with the infection process. The synthesis of extracellular endo polygalacturonase and endo pectin lyase is controlled by the concentration of galacturonic acid in the environment (Cooper and Wood 1975). Low concentrations of galacturonic acid induce the synthesis of these pectic enzymes, whereas high concentrations of galacturonic acid repress their production. Thus, nonmetabolized structural analogs of galacturonic acid, if applied to the infection site at a sufficient concentration, could be expected to inhibit disease initiation by microorganisms which depended upon extracellular enzymes to supply an energy source for growth.

Although inhibitors of pectic enzymes are known to exist in plant tissues, and may be involved in disease resistance, little effort has been expended to exploit this phenomenon for disease control. Pathogens which depend upon very low concentrations of constitutive enzymes to generate the inducer (*e.g.*, galacturonic acid from pectic acid) for synthesis of extracellular enzymes should be vulnerable to inhibitors of the action of pectolytic enzymes. The effective application of an inhibitor of pectic enzyme action might be feasible at an early stage of infection when the concentration of the fungal enzyme in the host would be low. Attempting to inhibit a high concentration of induced enzyme would be futile. Finally, it might be possible to stimulate plant tissues to increase their production of naturally-occurring inhibitors of pectic enzymes (Albersheim and Anderson-Prouty 1975; Byrde 1963).

Treatments to Eradicate or Attenuate Established Infections

Latent infections can be eradicated in many instances by application of a "systemic fungicide" such as benomyl after harvest because the fungus pathogen has been restricted to the epidermal or lenticel region by the resistance of the host. Late season and postharvest infections, particularly those incited by wound-invading pathogens such as *Rhizopus*, *Geotrichum* and *Penicillium* are difficult, if not impossible, to eradicate by chemical treatment applied 48 hr after inoculation. However, past observations on the performance of both physical and chemical agents provide a basis for guarded optimism for new developments in this area.

Treatment of fruits and vegetables with heated water (*c.* 50° C) or gamma radiation has eradicated established infections which were uncontrollable by chemical agents. Dosages of both heat and radiation required for disease control are very close to the injury threshold for the host, resulting in delayed ripening, increased susceptibility to wound-invading pathogens and, in some cases, outright visible symptoms of injury (Dennison and Ahmed 1975; Maxie *et al.* 1971). Some interest has developed in recent years in treatments combining heat and fungicides. The temperatures employed are usually borderline for decay control, but well below the injury threshold for the host. The elevated temperature increases the effectiveness of a reduced dosage of the fungicide, resulting in acceptable disease control with lower chemical residues on the fruit and little risk of heat damage to the crop (Brodrick *et al.* 1972; Muirhead 1976; Smith and Anderson 1975).

Current investigations are focused on the development of fungicides which will penetrate several millimeters into the product with the expectation of: (1) eradicating lenticular, latent, and incipient infections; (2) reducing the expansion of decay lesions; (3) preventing surface growth and sporulation of the fungus pathogen; and (4) inducing resistance to future infections in superficial tissues of the host. Most fungicides that have been applied after harvest are very limited in their ability to penetrate host tissues and thus are ineffective against established infections. Nonetheless, the expansion of established lesions of *Rhizopus* have been retarded by treatment of peaches with dicloran several days after inoculation (Ogawa *et al.* 1963; Wells and Bennett 1976). Ravetto and Ogawa (1972) showed that dicloran penetrated 11 mm into the flesh of peaches, inhibiting the growth of *Rhizopus* at that depth. In similar fashion, benomyl inhibits the development of lesions of *Monilinia fructicola* on stone fruits. Phillips (1975) observed fungistatic residues of benomyl at a depth of 4 mm in the pericarp of treated peaches after 3 wk of storage. Several investigators have observed that benomyl prevented or at least delayed the development of latent infections of *Gloeosporium* in apples (Edney 1970), bananas (Griffee and Burden 1974) and mangoes and papayas (Bolkan *et al.* 1976; Muirhead 1976). Benomyl is able to penetrate the disc and calyx of the button of orange fruits, thereby halting the development of latent infections of *Diplodia* and *Phomopsis* (Brown and Albrigo 1972; McCornack 1975). Significant residues of benomyl have been found in the flavedo of treated oranges and may protect against subsequent infection by *Penicillium* for several weeks after treatment (Brown and Albrigo 1972; Gutter 1970; McCornack 1975). Fungistatic concentrations of thiabendazole have been observed in the flesh of treated pears (Ben-Arie 1975). Thiabendazole can penetrate at least 5 mm into the flesh of potato tubers to inhibit the development of 48 hr infections of *Fusarium* (Murdock and Wood 1972). Factors influencing the uptake of thiabendazole by potatoes have been analyzed in detail by Tisdale and Lord (1973).

For the past 40 yr, biphenyl has been added to the packaging material for citrus fruits in order to provide vapor-phase inhibition of sporulation of *Penicillium digitatum* and *Penicillium italicum* during shipment and marketing of the crop. Despite many years of successful use, biphenyl has three main problems which are well-recognized by the trade: (1) a transient chemical odor associated with the treated fruit; (2) the residue of biphenyl on treated fruit may exceed the tolerance of 70 ppm set by governmental agencies in Europe and Japan, especially if the packages are exposed to warm temperatures for more than a few days; and (3) the biphenyl treatment may be ineffective when used on fruit which were treated with o-phenylphenol and stored before packaging for shipment. Cross-resistance between these two fungicides might be anticipated in view of their structural similarity (Fig. 7.4) (Harding 1962). In recent years it has been observed that residues of 2–5 mg/kg of thiabendazole, benomyl, or imazalil can inhibit sporulation of *Penicillium* in the same manner as biphenyl (Eckert and Kolbezen 1977; Gutter *et al.* 1974; Harding 1976; McCornack *et al.* 1977). These treatments do not impart a chemical odor to the fruit and residues can be controlled to well within the legal tolerance for thiabendazole and benomyl. A tolerance has not yet been established for imazalil. Residues of 5–6 mg/kg fruit of either thiabendazole or MBC on the fruit are required for sporulation control equivalent to that provided by biphenyl, whereas 2 mg/kg benomyl provides equivalent sporulation control (Fig. 7.7). Thus, benomyl should be formulated in a manner which retards the breakdown to MBC in order to insure the highest effectiveness of the treatment. Imazalil is equivalent to benomyl in sporulation control.

Maintenance of host resistance.—In addition to lower temperatures and modified atmosphere storage, certain chemical growth regulators can retard ripening and senescence, significantly reducing postharvest disease. Most lemons produced in California are treated with the isopropyl ester of 2,4-D (2,4-dichlorophenoxyacetic acid) before storage for the purpose of delaying senescence of the button (calyx plus disc) which is the usual point of attack by the fungus *Alternaria citri* (Stewart *et al.* 1952). Oranges and grapefruit are treated similarly in Israel and South Africa to reduce *Diplodia* and *Alternaria* during storage and long distance shipment to northern Europe (Schiffmann-Nadel *et al.* 1972). Gibberellic acid may reduce Penicillium decay of navel oranges by delaying senescence of the peel, an event which renders the fruit more susceptible to injury and infection by wound-invading pathogens (Coggins and Lewis 1965).

Low temperatures inhibit physiological processes which lead to deterioration in quality as well as senescence of fresh fruits and vegetables. Therefore, there appears to be little justification for research on chemical agents to retard senescence of fruits and vegetables which can safely be stored at temperatures below 5° C. This conclusion might be tempered

FIG. 7.7. DISTRIBUTION OF SPORES OF *Penicillium italicum* THROUGHOUT A CARTON OF ORANGES (TOP)

Bottom: (C) Untreated oranges covered with dense sporulation of *P. digitatum*; (T) Orange with residue of 9.06 ppm thiabendazole; and (B) Orange with residue of 2.30 mg/kg benomyl. Orange on far left is not decayed.

somewhat for developing countries where adequate refrigeration may not be available. Substantial benefits should arise from the development of agents to delay the senescence of bananas, tomatoes, pineapples, lemons, mangoes, and other tropical fruits which must be stored at temperatures above 10° C in order to avoid chilling injury. In this connection, it is interesting that both thiabendazole and benomyl affect the physiology of grapefruit in an unknown manner which results in a reduction in chilling injury at temperatures which would normally cause significant low temperature damage to the fruit (Schiffmann-Nadel *et al.* 1975B; Wardowski et al. 1975).

SUCCESSES AND CHALLENGES OF POSTHARVEST DISEASE CONTROL

The technology is now available to satisfactorily control the principal postharvest diseases of the major fruit crops, provided that adequate refrigeration facilities are available and affordable. Dicloran, developed in the 1960's provided, for the first time, a highly effective treatment to control Rhizopus rot of peaches, nectarines, and sweet cherries and Botrytis rot of several vegetable crops during long term cold storage. *sec*-Butylamine, developed in the same period, afforded a means of treating citrus fruits immediately after harvest to prevent infection of injuries by *Penicillium* during degreening and storage. Thiabendazole and benomyl, extensively evaluated over the past decade, have solved the problem of crown rot of bananas which was created by the new practice of shipping bananas as hands rather than as stems. These compounds also provided a psychological breakthrough in postharvest pathology, in that they demonstrated for the first time that infections in the host tissues at harvest could be inactivated by chemical treatments applied to the surface of the product after harvest. Stem-end rots of citrus fruit, lenticel rots of apples, latent infections of *Colletotrichum* on bananas and other tropical fruit, and late season infection of *Monilinia* on peaches have all been controlled by benomyl to an extent thought impossible in earlier times. Surface deposits of thiabendazole and benomyl on citrus fruits, within legal residue tolerances, suppress the sporulation of *Penicillium* molds on the surface of the fruit more effectively than biphenyl, the standard of comparison for four decades. An active search for a substitute for biphenyl, recognized since its inception to impart an undesirable odor to treated fruit, has been underway for the past 20 yr. Thiabendazole has equalled or exceeded the organomercurial fungicides in controlling certain important diseases on stored seed potatoes, and this technology has now been extended to the reduction of postharvest losses due to fungal pathogens in warehouse and fresh-pack potatoes as well.

The euphoria associated with the numerous successful applications of the benzimidazole fungicides to problems of postharvest pathology was dampened with the realization by the mid-1970's that pathogenic fungi could develop resistance to this class of fungicides with comparative ease.

Genetic studies which followed (Van Tuyl 1977) demonstrated that single gene mutations could sufficiently alter the binding site for benzimidazole fungicides in the fungal cell so as to create new strains which retained their pathogenicity yet were practically resistant to the benzimidazole fungicides. This problem is a consequence of the single site of action of the benzimidazole fungicides, which provides a high degree of selective toxicity. Most fungicides developed in earlier times inactivated a wide variety of enzymes and cellular processes in target fungi. They did not injure the host because they were excluded by the impermeability of surface barriers of the host. Benomyl, thiabendazole and other "systemic fungicides" are valued for their ability to penetrate into the protoplasm of the plant cell in order to reach and inactivate deep-seated fungal infections. This compatibility with the protoplasm of the plant cell implies that the receptors which are responsible for the physiological action of these fungicides do not exist in the plant cell, or are protected by compartmentalization within the cell.

Although the development of systemic fungicides which would inactivate several processes in the fungal cell, without damaging the plant cell, is difficult to visualize at this moment, the problem of fungicide resistance can be approached by other means. The observation that experimentally produced mutants do not have a high level of resistance to imazalil and other compounds which inhibit the synthesis of ergosterol in fungi, suggests that such physiological functions are evolutionarily conservative, i.e., a single gene mutation cannot completely circumvent a block in the synthesis or function of cell membranes. Alternately, any mutation which alters this physiological function may also reduce the vigor or pathogenicity of the mutant strain. This contention has been borne out thus far with mutants which show partial resistance to triforine, imazalil and pimaricin, all of which interfere with the functions of the cell membrane (Georgopoulos 1977). Another contemplated strategy is the development of systemic fungicides which are inhibitors of respiration in fungi. Such compounds should be less subject to the development of resistance since they also inhibit the transcription and translation of the mutant code for fungicide resistance.

Until fungicides with greater immunity to the development of resistance are uncovered, the benzimidazole fungicides can continue to be utilized efficiently if the problem of resistance is recognized early and dealt with in a thoughtful manner by people in the commercial trade. Both biphenyl and sodium o-phenylphenate are currently utilized for control of decay of citrus fruits despite the recognized hazard of developing a Penicillium strain which has a high degree of resistance to both of these agents. The management of these problems in practice involves stringent sanitation in the postharvest environment, combined with the sequential application of fungicides with different mechanisms of action in multistage fruit handling operations (Dawson and Eckert 1977; Wolfe 1975).

Despite substantial progress over the past 15 yr in the development of highly effective fungicide treatments, there is as yet no chemical treatment which directly acts against several important postharvest diseases. Bacterial soft rot of vegetable crops, Geotrichum sour rot of citrus fruits and tomatoes, and Alternaria rot of several fruits and vegetables are all examples of diseases that cannot be controlled efficiently by treatment of the host with an antimicrobial agent. Control measures are limited to sanitation, refrigeration, and retarding the onset of host senescence. This situation appears to result from the insensitivity of the pathogens responsible for these diseases to existing biocides rather than a unique problem associated with the etiology of the disease. Thus, benomyl provides excellent control of Diplodia stem-end rot, but not Alternaria stem-end rot, of citrus. The sole reason appears to be the insensitivity of *Alternaria* to benzimidazole fungicides. Indications that bacterial soft rot could be controlled by a strongly antibacterial agent are apparent in earlier reports on the effectiveness of certain medicinal antibiotics against these diseases. If these assertions are correct, then we may anticipate the early development of effective chemical treatments for the control of these diseases.

The fresh fruit and vegetable industry is now in an era of dynamic innovation in the areas of mechanical harvesting, consumer packaging, and bulk handling and transportation. Most of the proposals that have been made in these areas will tend to intensify postharvest disease problems either by increasing the opportunity for wound infections or by creating an environment favorable for disease development. Mechanical harvesting in its current version invariably results in a high frequency of injury to fruit crops which must be processed, refrigerated, and treated with a fungicide within 24 hr after harvest to avoid serious decay losses. The ever-increasing interest in consumer packaging in moisture-impermeable plastic films can lead to serious problems in decay and superficial mold growth in the final stages of marketing. The limiting factor in all of these engineering and marketing innovations is the capability of controlling microbial deterioration. This will be the challenge to postharvest pathologists for the remainder of the century.

BIBLIOGRAPHY

ADAMS, M. J. 1975. Potato tuber lenticels: susceptibility to infection by *Erwinia carotovora* var.*atroseptica* and *Phytophthora infestans*. Ann. Appl. Biol. *79,* 275–282.

ALBERSHEIM, P. and ANDERSON-PROUTY, A. J. 1975. Carbohydrates, proteins, cell surfaces, and the biochemistry of pathogenesis. Ann. Rev. Plant Physiol. *26*, 31–52.

ALBRIGO, L. G. and BROWN, G. E. 1977. Storage studies with "Valencia" oranges. 1st World Congress of Citriculture *3*, 361–367.

BAKER, R. E. D. 1938. Studies in the pathogenicity of tropical fungi. 2. The occurrence of latent infections in developing fruits. Ann. Botany (N. S.) *2*, 919–931.

BARKAI-GOLAN, R. 1973. Postharvest heat treatment to control *Alternaria tenuis Auct.* rot in tomato. Phytopathol. Mediterr. *12*, 108–111.

BARTZ, J. A. and CRILL, J. P. 1972. Tolerance of fruit of different tomato cultivars to soft rot. Phytopathology *62*, 1085–1088.

BATEMAN, D. F. and MILLAR, R. L. 1966. Pectic enzymes in tissue degradation. Ann. Rev. Phytopath. *4*, 119–146.

BEN-ARIE, R. 1975. Benzimidazole penetration, distribution and persistence in postharvest treated pears. Phytopathology *65*, 1187–1189.

BEN-ARIE, R. and GUELFAT-REICH, S. 1973. Preharvest and postharvest applications of benzimidazoles for control of storage decay of pears. HortScience *8*, 181–183.

BERAHA, L., GARBER, E. D. and BILLETER, B. A. 1974. Enzyme profiles and virulence in mutants of *Erwinia carotovora.* Phytopath. Z. *81*, 15–22.

BOLKAN, H. A., CUPERTINO, F. P., DIANESE, J. C. and TAKATSU, A. 1976. Fungi associated with pre- and postharvest fruit rots of papaya and their control in central Brazil. Plant Disease Reptr. *60*, 605–609.

BOTHAST, R. J. 1978. Fungal deterioration and related phenomena in cereals, legumes and oilseeds. *In* Postharvest Biology and Biotechnology Symposium, H. O. Hultin (Editor). Food & Nutrition Press, Westport, CT.

BOYD, A. E. W. 1972. Potato storage diseases. Rev. Plant Pathol. *51*, 297–321.

BOYD, A. E. W. 1975. Fungicides for potato tubers. Proc. 8th British Insecticide Fungicide Conf. *3*, 1035–1044.

BOYD, M. R. and WILSON, B. J. 1972. Isolation and characterization of 4-ipomeanol, a lung-toxic furanoterpenoid produced by sweet potato (Ipomoea batatas). J. Agr. Food Chem. *20*, 428–430.

BRODRICK, H. T., JACOBS, C. J., SWARTZ, H. D. and MULDER, N. J. 1972. The control of storage diseases of papaws in South Africa. Citrus Grower and Subtropical Fruit J. *467*, 5, 7, 9, 21.

BROWN, G. E. 1973. Development of green mold in degreened oranges. Phytopathology *63*, 1104–1107.

BROWN, G. E. 1975. Factors affecting postharvest development of *Colletotrichum gloeosporioides* in citrus fruits. Phytopathology *65*, 404–409.

BROWN, G. E. and ALBRIGO, L. G. 1972. Grove application of benomyl and its persistence in orange fruit. Phytopathology *62*, 1434–1438.

BROWN, G. E. and BARMORE, C. R. 1977. The effect of ethylene on susceptibility of Robinson tangerines to anthracnose. Phytopathology *67*, 120–123.

BROWN, A. C., KEAR, R. W. and SYMONS, J. P. 1975. Fungicidal control of *Botrytis* on cold-stored white cabbage. Proc. 8th British Insecticide Fungicide Conf. *1*, 339–346.

BROWN, A. E. and SWINBURNE, T. R. 1973. Factors affecting the production of benzoic acid in Bramley's Seedling apples infected with *Nectria galligena.* Physiol. Plant Pathol. *3*, 91–99.

BROWN, G. E. and WILSON, W. C. 1968. Mode of entry of *Diplodia natalensis* and *Phomopsis citri* into Florida oranges. Phytopathology *58*, 736–739.

BUCHANAN, J. R., SOMMER, N. F., FORTLAGE, R. J., MAXIE, E. C., MITCH-ELL, F. G. and HSIEH, D. P. H. 1974. Patulin from *Penicillium expansum* in stone fruits and pears. J. Am. Soc. Hort. Sci. *99*, 262–265.

BURCHILL, R. T. and EDNEY, K. L. 1972. An assessment of some new treatments for the control of rotting of stored apples. Ann. Appl. Biol. *72*, 249–255.

BURTON, W. G. 1973. Environmental requirements in store as determined by potential deterioration. Proc. 7th British Insecticide Fungicide Conf. *3*, 1037–1055.

BUSH, D. A. and CODNER, R. C. 1970. Comparison of the properties of the pectin transeliminases of *Penicillium digitatum* and *Penicillium italicum*. Phytochemistry *9*, 87–97.

BYRDE, R. J. W. 1963. Natural inhibitors of fungal enzymes and toxins in disease resistance. *In* Perspectives of Biochemical Plant Pathology, S. Rich (Editor). Conn. Agr. Expt. Sta. Bull. *663*.

CARTWRIGHT, D., LANGCAKE, P., PRYCE, R. J. and LEWORTHY, D. P. 1977. Chemical activation of host defence mechanisms as a basis for crop protection. Nature *267*, 511–513.

CHIBA, M. and DOORNBOS, F. 1974. Instability of benomyl in various conditions. Bull. Environ. Contam. Toxicol. *11*, 273–274.

CHO, J. J., ROHRBACH, K. G. and APT, W. J. 1977. Induction and chemical control of rot caused by *Ceratocystis paradoxa* on pineapples. Phytopathology *67*, 700–703.

COGGINS, C. W. JR. and LEWIS, L. N. 1965. Some physical properties of the navel orange rind as related to ripening and to gibberellic acid treatments. Proc. Am. Soc. Hort. Sci. *86*, 272–279.

COOPER, R. M. and WOOD, R. K. S. 1975. Regulation of synthesis of cell wall degrading enzymes by *Verticillium albo-atrum* and *Fusarium oxysporum* f. sp. *lycopersici*. Physiol. Plant Pathol. *5*, 135–156.

COURSEY, D. G. and BOOTH, R. H. 1972. The postharvest phytopathology of perishable tropical produce. Rev. Plant Pathol. *51*, 751–765.

COXON, D. T., CURTIS, R. F., PRICE, K. R. and LEVETT, G. 1973. Abnormal metabolites produced by *Daucus carota* roots stored under conditions of stress. Phytochemistry *12*, 1881–1885.

COYIER, D. L. 1970. Control of storage decay in d'Anjou pear fruit by preharvest application of benomyl. Plant Disease Reptr. *54*, 647–650.

CROMARTY, R. W. and EASTON, G. D. 1973. The incidence of decay and factors affecting bacterial soft rot of potatoes. Am. Potato J. *50*, 398–407.

DAINES, R. H. 1970. Effects of temperature and a 2,6-dichoro-4-nitroaniline dip on keeping qualities of "Yellow Jersey" sweetpotatoes during the postharvest period. Plant Disease Reptr. *54*, 486–488.

DAVIES, W. P. 1977. Infection of carrot roots in cool storage by *Centrospora acerina*. Ann. Appl. Biol. *85*, 163–164.

DAWSON, A. J. and ECKERT, J. W. 1977. Problems of decay control in marketing citrus fruits: strategy and solutions, California and Arizona. Proc. Int. Soc. Citriculture *1*, 255–259.

DENNIS, C. 1977. Susceptibility of stored crops to microbial infection. Ann. Appl. Biol. *85*, 430–432.

DENNISON, R. A. and AHMED, E. M. 1975. Irradiation treatment of fruits and vegetables. *In* Postharvest Biology and Handling of Fruits and Vegetables, N. F. Haard and D. K. Salunkhe (Editors). Avi Publishing Co., Westport, CT.

DERBYSHIRE, D. M. and CRISP, A. F. 1971. Vegetable storage in East Anglia. Proc. 6th British Insecticide Fungicide Conf. *1*, 167–172.

ECKERT, J. W. 1967. Application and use of postharvest fungicides. *In* Fungicides. Vol. I, D. C. Torgeson (Editor). Academic Press, New York.

ECKERT, J. W. 1975. Postharvest diseases of fresh fruits and vegetables—etiology and control. *In* Postharvest Biology and Handling of Fruits and Vegetables, N. F. Haard and D. K. Salunkhe (Editors). Avi Publishing Co., Westport, CT.

ECKERT, J. W. 1977. Control of postharvest diseases. *In* Antifungal Compounds, M. R. Siegel and H. D. Sisler (Editors). Marcel Dekker, Inc., New York.

ECKERT, J. W. and KOLBEZEN, M. J. 1963. Control of Penicillium decay of oranges with certain volatile aliphatic amines. Phytopathology *53*, 1053–1059.

ECKERT, J. W. and KOLBEZEN, M. J. 1964. 2-Aminobutane salts for control of post-harvest decay of citrus, apple, pear, peach, and banana fruits. Phytopathology *54*, 978–986.

ECKERT, J. W. and KOLBEZEN, M. J. 1970. Fumigation of fruits with 2-aminobutane to control certain postharvest diseases. Phytopathology *60*, 545–550.

ECKERT, J. W. and KOLBEZEN, M. J. 1977. Influence of formulation and application method on the effectiveness of benzimidazole fungicides for controlling postharvest diseases of citrus fruits. Netherland J. Plant Pathol. *83* (Suppl. 1), 343–352.

ECKERT, J. W., KOLBEZEN, M. J. and KRAMER, B. A. 1969. Accumulation of *o*-phenylphenol by citrus fruits and pathogenic fungi in relation to decay control and residues. Proc. 1st Intern. Citrus Symp., Univ. Calif. (Riverside) *2*, 1097–1103.

ECKERT, J. W. and SOMMER, N. F. 1967. Control of diseases of fruits and vegetables by postharvest treatment. Ann. Rev. Phytopath. *5*, 391–432.

EDNEY, K. L. 1958. Observations on the infection of Cox's Orange Pippin apples by *Gloeosporium perennans* Zeller and Childs. Ann. Appl. Biol. *46*, 622–629.

EDNEY, K. L. 1964. The effect of the composition of the storage atmosphere on the development of rotting of Cox's Orange Pippin apples and the production of pectolytic enzymes by *Gloeosporium* spp. Ann. Appl. Biol. *54*, 327–334.

EDNEY, K. L. 1970. Some experiments with thiabendazole and benomyl as post-harvest treatments for the control of storage rots of apples. Plant Pathol. *19*, 189–193.

EDNEY, K. L. 1976. An investigation of persistent infection of stored apples by *Gloeosporium* spp. Ann. Appl. Biol. *82*, 355–360.

EDNEY, K. L. and BURCHILL, R. T. 1967. The use of heat to control the rotting of Cox's Orange Pippin apples by *Gloeosporium* spp. Ann. Appl. Biol. *59*, 389–400.

FOLLSTAD, M. N. 1966. Mycelial growth rate and sporulation of *Alternaria tenuis, Botrytis cinerea, Cladosporium herbarum,* and *Rhizopus stolonifer* in low oxygen atmospheres. Phytopathology *56*, 1098–1099.

FOX, R. T. V., MANNERS, J. G. and MYERS, A. 1971. Ultrastructure of entry and spread of *Erwinia carotovora* var. *atroseptica* into potato tubers. Potato Res. *14*, 61–73.

FRENCH, R. C., LONG, R. K., LATTERELL, F. M., GRAHAM, C. L., SMODT, J. J. and SHAW, P. E. 1978. Effect of n-nonanal, citral, and citrus oils on germination of conidia of *Penicillium digitatum* and *P. italicum*. Phytopathology (In Press).

FRIEDMAN, B. A. 1962. Physiological differences between a virulent and a weakly virulent radiation-induced strain of *Erwinia carotovora*. Phytopathology *52*, 328–332.

GEORGOPOULOS, S. G. 1977. Development of fungal resistance to fungicides. *In* Antifungal Compounds. Vol 2, M. R. Siegel and H. D. Sisler (Editors). Marcel Dekker, Inc., New York.

GOODLIFFE, J. P. and HEALE, J. B. 1977. Factors affecting the resistance of cold-stored carrots to *Botrytis cinerea*. Ann. Appl. Biol. *87*, 17–28.

GRAHAM, D., HAMILTON, G. A., QUINN, C. E. and RUTHVEN, A. D. 1973. Use of 2-aminobutane as a fumigant for control of gangrene, skin spot and silver scurf diseases of potato tubers. Potato Res. *16*, 109–125.

GREENE, G. L. and GOOS, R. D. 1963. Fungi associated with crown rot of boxed bananas. Phytopathology *53*, 271–275.

GREENE, G. L. and MORALES, C. 1967. Tannins as the cause of latency in anthracnose infections of tropical fruits. (*Gloeosporium musarum* in bananas). Turrialba *17*, 447–449.

GRIERSON, W. 1969. Consumer packaging of citrus fruits. Proc. 1st Intern. Citrus Symp., Univ. Calif. (Riverside). *3*, 1389–1401.

GRIFFEE, P. J. 1973. Resistance to benomyl and related fungicides in *Colletotrichum musae*. Trans. British Mycol. Soc. *60*, 433–439.

GRIFFEE, P. J. and BURDEN, O. J. 1973. Banana diseases in the Windward Islands. Proc. 7th British Insecticide Fungicide Conf. *3*, 887–897.

GRIFFEE, P. J. and BURDEN, O. J. 1974. Incidence and control of *Colletotrichum musae* on bananas in the Windward Islands. Ann. Appl. Biol. *77*, 11–16.

GUTTER, Y. 1970. Influence of application time on effectiveness of fungicides for green mold control in artificially inoculated oranges. Plant Disease Reptr. *54*, 325–327.

GUTTER, Y. 1975. Interrelationship of *Penicillium digitatum* and *P. italicum* in thiabendazole-treated oranges. Phytopathology *65*, 498–499.

GUTTER, Y., YANKO, U., DAVIDSON, M. and RAHAT, M. 1974. Relationship between mode of application of thiabendazole and its effectiveness for control of green mold and inhibiting fungus sporulation on oranges. Phytopathology *64*, 1477–1478.

HALL, J. H. and WOOD, R. K. S. 1973. The killing of plant cells by pectolytic enzymes. *In* Fungal Pathogenicity and the Plant's Response, R. J. W. Byrde and C. V. Cutting (Editors). Academic Press, New York.

HARDENBURG, R. E. 1971. Effect of in-package environment on keeping quality of fruits and vegetables. HortScience *6*, 198–201.

HARDING, P. R. JR. 1962. Differential sensitivity to sodium *o*-phenylphenate by biphenyl-sensitive and biphenyl-resistant strains of *Penicillium digitatum*. Plant Disease Reptr. *46*, 100–104.

HARDING, P. R. JR. 1972. Differential sensitivity to thiabendazole by strains of *Penicillium italicum* and *P. digitatum*. Plant Disease Reptr. *56*, 256–260.

HARDING, P. R. JR. 1976. R23979, a new imidazole derivative effective against postharvest decay of citrus by molds resistant to thiabendazole, benomyl and 2-aminobutane. Plant Disease Reptr. *60*, 643–646.

HARDING, P. R. JR. and SAVAGE, D. C. 1964. Investigation of possible correlation of hot-water washing with excessive storage decay in coastal California packing houses. Plant Disease Reptr. *48*, 808–810.

HARPER, K. A., BEATTIE, B. B., PITT, J. I. and BEST, D. J. 1972. Texture changes in canned apricots following infection of the fresh fruit with *Rhizopus stolonifer*. J. Sci. Food Agr. *23*, 311–320.

HARVEY, J. M., HARRIS, C. M. and PORTER, F. M. 1971. Air transport of California strawberries: pallet covers to maintain modified atmospheres and reduce market losses. U.S. Dept. of Agr., Marketing Res. Rept. *920*.

HEALE, J. B., HARDING, V., DODD, K. and GAHAN, P. B. 1977. Botrytis infection of carrot in relation to the length of the cold storage period. Ann. Appl. Biol. *85*, 453–457.

HEALE, J. B. and SHARMAN, S. 1977. Induced resistance to *Botrytis cinerea* in root slices and tissue cultures of carrots (*Daucus carota* L.). Physiol. Plant Pathol. *10*, 51–61.

HENRIKSEN, J. B. 1975. Prevention of gangrene and Fusarium dry rot by physical means and with thiabendazole. Proc. 8th British Insecticide Fungicide Conf. *2*, 603–608.

HONDELMANN, W. R. E. and RICHTER, E. 1973. On the susceptibility of strawberry clones to *Botrytis cinerea* Pers. in relation to pectin quantity and quality in the fruit. Gartenbauwissenschaft *38*, 311–314. (German).

ISHII, S. 1977. Purification and characterization of a factor that stimulates tissue maceration by pectolytic enzymes. Phytopathology *67*, 994–1000.

JACOBS, C. J., BRODRICK, H. T., SWARTS, H. D. and MULDER, N. J. 1973. Control of postharvest decay of mango fruit in South Africa. Plant Disease Reptr. *57*, 173–176.

JARVIS, W. R. 1962. The infection of strawberry and raspberry fruits by *Botrytis cinerea* Fr. Ann. Appl. Biol. *50*, 569–575.

JENNINGS, D. L. and CARMICHAEL, E. 1975. Resistance to grey mold (*Botrytis cinerea* Fr.) in red raspberry fruits. Hort. Res. *14*, 109–115.

JONES, A. L. 1975. Control of brown rot of cherry with a new hydantoin fungicide and with selected fungicide mixtures. Plant Disease Reptr. *59*, 127–129.

JONES, A. L., BURTON, C. L. and TENNES, B. R. 1973. Postharvest fungicide and heat treatments for brown rot control on stone fruits. Michigan State Univ. Agr. Expt. Sta. Res. Rept. *209*.

JORDAN, V. W. L. and RICHMOND, D. V. 1974. The effects of benomyl on sensitive and tolerant isolates of *Botrytis cinerea* infecting strawberries. Plant Pathol. *23*, 81–83.

JORDAN, V. W. L. and RICHMOND, D. V. 1975. Perennation and control of benomyl-insensitive *Botrytis* affecting strawberries. Proc. 8th British Insecticide Fungicide Conf. *1*, 5–13.

KABLE, P. F. 1971. Significance of short term latent infections in control of brown rot in peach fruits. Phytopathol. Z. *70*, 173–176.

KATCHANSKY, M. 1977. Fumigation of produce storage rooms with thiabendazole thermal dusting tablets for control of pathogenic fungi. Proc. 1st World Congress Citriculture *3*, 343–346.

KOFFMAN, W. and KABLE, P. F. 1975. Improved control of brown rot in harvested sweet cherries by triforine dip treatments. Plant Disease Reptr. *59*, 586–590.

KROCHTA, J. M., CARLSON, R. A., OGAWA, J. M. and MANJI, B. T. 1977. Harvesting into foam reduces tomato losses. Food Technol. *313*, 42–46.

KUC, J. and SHAIN, L. 1977. Antifungal compounds associated with disease resistance in plants. *In* Antifungal Compounds, M. R. Siegel and H. D. Sisler (Editors). Marcel Dekker, Inc. New York.

KURAMOTO, T. and YAMADA, S. 1976. DF-125, a new experimental fungicide for control of satsuma mandarin postharvest decays. Plant Disease Reptr. *60*, 809–812.

LAVILLE, E. 1974. Action of R23979 (imazalil) on Penicillium decays of oranges. Meded Fac. Landbouwwet., Rijksuniv. Gent *39*, 1121–1126.

LAVILLE, E., HARDING, P. R., DAGAN, Y., KRAGHT, A. J. and RIPPON, L. E. 1977. Studies on imazalil as potential treatment for control of citrus fruit decay. Proc. Int. Soc. Citriculture *1*, 259–263.

LIPTON, W. J. 1967. Some effects of low-oxygen atmospheres on potato tubers. Am. Potato J. *44*, 292–299.

LÖCHER, F. and HAMPEL, M. 1973. Control of postharvest diseases of bananas, pineapples and citrus with carbendazim. Proc. 7th British Insecticide Fungicide Conf. 1, 301–308.
LOWEN, W. K. 1977. Terminal residues of Benlate benomyl fungicides. Proc. 1st World Congress Citriculture 3, 583–593.
LUEPSCHEN, N. S., ROHRBACH, K. G., JONES, A. C. and PETERS, C. L. 1971. Methods of controlling Rhizopus decay and maintaining Colorado peach quality. Colorado State Univ. Expt. Sta. Bull. 547S.
LUND, B. M. 1971. Bacterial spoilage of vegetables and certain fruits. J. Appl. Bacter. 34, 9–20.
LUND, B. M. and KELMAN, A. 1977. Determination of the potential for development of bacterial soft rot of potatoes. Am. Potato J. 54, 211–225.
LUND, B. M. and WYATT, G. M. 1972. The effect of oxygen and carbon dioxide concentrations on bacterial soft rot of potatoes. 1. King Edward potatoes inoculated with Erwinia carotovora var. atroseptica. Potato Res. 15, 174–179.
LYON, G. D., LUND, B. M., BAYLISS, C. E. and WYATT, G. M. 1975. Resistance of potato tubers to Erwinia carotovora and formation of rishitin and phytuberin in infected tissue. Physiol. Plant Pathol. 6, 43–50.
MAXIE, E. C., SOMMER, N. F. and MITCHELL, F. G. 1971. Infeasibility of irradiating fresh fruits and vegetables. HortScience 6, 202–204.
McCOLLOCH, L. P. and WORTHINGTON, J. T. 1952. Low temperature as a factor in the susceptibility of mature green tomatoes to Alternaria rot. Phytopathology 42, 425–427.
McCORNACK, A. A. 1972. Effect of ethylene degreening on decay of Florida citrus fruit. Proc. Florida State Hort. Soc. 84, 270–272.
McCORNACK, A. A. 1973. Factors affecting decay and peel injury in Temples. Proc. Florida State Hort. Soc. 85, 232–235.
McCORNACK, A. A. 1975. Control of citrus fruit decay with postharvest application of Benlate. Proc. Florida State Hort. Soc. 87, 230–233.
McCORNACK, A. A., BROWN, G. E. and SMOOT, J. J. 1977. R23979, an experimental postharvest citrus fungicide with activity against benzimidazole-resistant Penicilliums. Plant Disease Reptr. 61, 788–791.
MENKE, G. H., PATEL, P. N. and WALKER, J. C. 1964. Physiology of Rhizopus stolonifer infection on carrot. Z. Pflanzenkrank. 71, 128–140.
MEREDITH, D. S. 1975. Control of fungal diseases of seed potatoes with thiabendazole. Proc. 8th British Insecticide Fungicide Conf. 2, 581–587.
MOUNT, M. S., BATEMAN, D. F. and BASHAM, H. G. 1970. Induction of electrolyte loss, tissue maceration and cellular death of potato tissue by an endo polygalacturonate trans-eliminase. Phytopathology 60, 924–931.
MUIRHEAD, I. F. 1976. Postharvest control of mango anthracnose with benomyl and hot water. Austral. J. Exptl. Agr. Animal Husbandry 16, 600–603.
MULVENA, D., WEBB, E. C. and ZERNER, B. 1969. 3,4-Dihydroxy-benazaldehyde, a fungistatic substance from green Cavendish bananas. Phytochemistry 8, 393–395.
MURDOCH, A. W. and WOOD, R. K. S. 1972. Control of Fusarium solani rot of potato tubers with fungicides. Ann. Appl. Biol. 72, 53–62.
NIELSEN, L. W. 1965. Harvest practices that increase sweet potato surface rot in storage. Phytopathology 55, 640–644.
NIELSEN, L. W. and JOHNSON, J. T. 1974. Postharvest temperature effects on wound healing and surface rot in sweet potato. Phytopathology 64, 967–970.
NOGUCHI, T., OHKUMA, K. and KOSAKA, S. 1972. Relation of structure and

activity in thiophanates. *In* Pesticide Chemistry. Vol. V, A. S. Tahori (Editor). Gordon and Breach, New York.

OGAWA, J. M., LEONARD, S., MANJI, B. T., BOSE, E. and MOORE, C. J. 1971. Monilinia and Rhizopus decay control during controlled ripening of freestone peaches for canning. J. Food Sci. *36*, 331–334.

OGAWA, J. M., MANJI, B. T. and BOSE, E. 1968. Efficacy of fungicide 1991 in reducing fruit rot of stone fruits. Plant Disease Reptr. *52*, 722–726.

OGAWA, J. M., MATHRE, J. H., WEBER, D. J. and LYDA, S. D. 1963. Effects of 2,6-dichloro-4-nitroaniline on Rhizopus species and comparison with other fungicides on control of Rhizopus rot of peaches. Phytopathology *53*, 950–955.

PARSONS, C. S. and SPALDING, D. H. 1972. Influence of a controlled atmosphere, temperature and ripeness on bacterial soft rot of tomatoes. J. Am. Soc. Hort. Sci. *97*, 297–299.

PATIL, S. S., TANG, C. S. and HUNTER, J. E. 1973. Effect of benzyl isothiocyanate treatment on the development of postharvest rots in papayas. Plant Disease Reptr. *57*, 86–89.

PEACOCK, B. C. 1973. Effect of *Colletotrichum musae* infection on the preclimacteric life of bananas. Queensland J. Agr. Animal Sci. *30*, 239–249.

PELSER, P. du T. 1974. Influence of thiabendazole and benomyl on Alternaria rot in stored grapefruit. Citrus Grower Sub-Trop. Fruit J. *486*, 15–17.

PELSER, P. du T. and ECKERT, J. W. 1977. Constituents of orange juice that stimulate the germination of conidia of *Penicillium digitatum*. Phytopathology *67*, 747–754.

PEROMBELON, M. C. M. and LOWE, R. 1975. Studies on the initiation of bacterial soft rot in potato tubers. Potato Res. *18*, 64–82.

PHILLIPS, D. J. 1975. Detection and translocation of benomyl in postharvest-treated peaches and nectarines. Phytopathology *65*, 255–258.

PILNIK, W. and VORAGEN, A. G. J. 1970. Pectic substances and other uronides. *In* The Biochemistry of Fruits and their Products. Vol. 1, A. C. Hulme (Editor). Academic Press, New York.

PRASAD, K. and STADELBACHER, G. J. 1973. Control of postharvest decay of fresh raspberries by acetaldehyde vapor. Plant Disease Reptr. *57*, 795–797.

PRASAD, K. and STADELBACHER, G. J. 1974. Effect of acetaldehyde vapor on postharvest decay and market quality of fresh strawberries. Phytopathology *64*, 948–951.

RAVETTO, D. J. and OGAWA, J. M. 1972. Penetration of peach fruit by benomyl and 2,6-dichloro-4-nitroaniline fungicides. Phytopathology *62*, 784.

RIPPON, L. E. and WILD, B. L. 1973. Comparison of four systemic fungicides for post-harvest treatment of citrus. Australian J. Exptl. Agr. Animal Husbandry *13*, 724–726.

ROHRBACH, K. U. 1973. Postharvest control of brown rot on stone fruits by triforine. Proc. 7th British Insecticide Fungicide Conf. *2*, 631–636.

ROMBOUTS, F. M. and PILNIK, W. 1972. Research on pectin depolymerases in the sixties—a literature review. Critical Rev. Food Technol. *3*, 1–26.

SACHER, J. A. 1973. Senescence and postharvest physiology. Ann. Rev. Plant Physiol. *24*, 197–224.

SCHIFFMANN-NADEL, M., CHALUTZ, E., WAKS, J. and DAGAN, M. 1975B. Reduction of chilling injury in grapefruit by thiabendazole and benomyl during long-term storage. J. Am. Soc. Hort. Sci. *100*, 270–272.

SCHIFFMANN-NADEL, M., LATTAR, F. S. and WAKS, J. 1971. The response of grapefruit to different storage temperatures. J. Am. Soc. Hort. Sci. *96*, 87–90.

SCHIFFMANN-NADEL, M., LATTAR, F. S. and WAKS, J. 1972. The effect of

2,4-D applied in waxes on the preservation of "Marsh Seedless" grapefruit and "Valencia" orange during prolonged storage. HortScience 7, 120–121.

SCHIFFMANN-NADEL, M., WAKS, J. and CHALUTZ, E. 1975A. Frost injury predisposes grapefruit to storage rots. Phytopathology 65, 630.

SCOTT, K. J., McGLASSON, W. B. and ROBERTS, G. A. 1970. Potassium permanganate as an ethylene absorbent in polyethylene bags to delay ripening of bananas during storage. Australian J. Exptl. Agr. Animal Husbandry 10, 237–240.

SEBERRY, J. A. 1969. Comparison of various fungicides for control of postharvest rots of Australian citrus fruits. Proc. 1st Intern. Citrus Symp., Univ. Calif. (Riverside) 3, 1309–1315.

SEGALL, R. H. 1967. Bacterial soft rot, bacterial necrosis, and Alternaria rot of tomatoes as influenced by field washing and postharvest chilling. Plant Disease Reptr. 51, 151–152.

SEGALL, R. H., GERALDSON, C. M. and EVERETT, P. H. 1974. The effects of cultural and postharvest practices on postharvest decay and ripening of two tomato cultivars. Proc. Florida State Hort. Soc. 86, 246–249.

SIMON, E. W. 1977. Membranes in ripening and senescence. Ann. Appl. Biol. 85, 417–421.

SMITH, H. W. 1963. The use of carbon dioxide in the transport and storage of fruits and vegetables. In Advances in Food Research. Vol 12, C. O. Chichester, E. M. Mrak and G. F. Stewart (Editors). Academic Press, New York.

SMITH, W. L. JR. and ANDERSON, R. E. 1975. Decay control of peaches and nectarines during and after controlled atmosphere and air storage. J. Am. Soc. Hort. Sci. 100, 84–86.

SMITH, W. L. JR. and SMART, H. F. 1955. Relation of soft rot development to protective barriers in Irish potato slices. Phytopathology 45, 649–654.

SMOOT, J. J. and MELVIN, C. F. 1975. Decay control of oranges with benomyl by three methods of postharvest application. Proc. Florida State Hort. Soc. 87, 234–236.

SMOOT, J. J., MELVIN, C. F. and JAHN, O. L. 1971. Decay of degreened oranges and tangerines as affected by time of washing and fungicide application. Plant Disease Reptr. 55, 149–152.

SOMMER, N. F., BUCHANAN, J. R. and FORTLAGE, R. J. 1974. Production of patulin by Penicillium expansum. Appl. Microbiol. 28, 589–593.

SOMMER, N. F., FORTLAGE, R. J., MITCHELL, F. G. and MAXIE, E. C. 1973. Reduction of postharvest losses of strawberry fruits from gray mold. J. Am. Soc. Hort. Sci. 98, 285–288.

SPALDING, D. H. 1969. Toxic effect of macerating action of extracts of sweetpotatoes rotted by Rhizopus stolonifer and its inhibition by ions. Phytopathology 59, 685–692.

SPALDING, D. H. and REEDER, W. F. 1975. Low-oxygen high-carbon dioxide controlled atmosphere storage for control of anthracnose and chilling injury of avocados. Phytopathology 65, 458–460.

STADELBACHER, G. J. and PRASAD, K. 1974. Postharvest decay control of apple by acetaldehyde vapor. J. Am. Soc. Hort. Sci. 99, 364–368.

STANGHELLINI, M. E. and ARAGAKI, M. 1966. Relation of periderm formation and callous deposition to anthracnose resistance in papaya fruit. Phytopathology 56, 444–449.

STEINBAUER, C. E. and KUSHMAN, L. J. 1971. Sweetpotato culture and disease. U.S. Dept. Agr. Handbook 388.

STEWART, W. S., PALMER, J. E. and HIELD, H. Z. 1952. Packing-house experiments on the use of 2,4-dichlorophenoxyacetic acid and 2,4,5-trichlorophenoxyacetic acid to increase storage life of lemons. Proc. Am. Soc. Hort. Sci. *59*, 327–334.

STOVER, R. H. 1972. Banana, Plantain and Abaca Diseases. Commonwealth Mycological Institute, Kew, Surrey, England.

SWINBURNE, T. R. 1973. The resistance of immature Bramley's Seedling apples to rotting by *Nectria galligena* Bres. *In* Fungal Pathogenicity and the Plant's Response, R. J. W. Byrde and C. V. Cutting (Editors). Academic Press, New York.

SWINBURNE, T. R. 1974. The effect of storage conditions of the rotting of apples, cv. Bramley's Seedling, by *Nectria galligena*. Ann. Appl. Biol. *78*, 39–84.

SWINBURNE, T. R. and BROWN, A. E. 1975. The biosynthesis of benzoic acid in Bramley's Seedling apples infected by *Nectria galligena* Bres. Physiol. Plant Pathol. *6*, 259–264.

SZKOLNIK, M. 1975. Control of Rhizopus rot on peach with a new hydantoin fungicide. Proc. Am. Phytopath. Soc. *2*, 108.

TANI, T. 1967. The relation of soft rot caused by pathogenic fungi to pectic enzyme production by the host. *In* The Dynamic Role of Molecular Constituents in the Plant-Parasite Interaction, C. J. Mirocha and I. Uritani (Editors). Am. Phytopath. Soc.

THORNE, S. M. 1972. Studies on the behavior of stored carrots with respect to their invasion by *Rhizopus stolonifer* Lind. J. Food Technol. *7*, 139–151.

TISDALE, M. J. and LORD, K. A. 1973. Uptake and distribution of thiabendazole by seed potatoes. Pesticide Sci. *4*, 121–130.

TÓIBÍN, M. 1974. Pectic enzymes in sour rot of orange caused by *Geotrichum candidum* var. *citri-aurantii* (Ferr.). M.Sc. Thesis, National University of Ireland, Dublin.

TOMKINS, R. G. 1939. Treated wraps for the prevention of fungal rotting. Gr. Brit., Dept. Sci. Ind. Res., Food Invest. Board Rept., 1938.

TOMKINS, R. G. and TROUT, S. A. 1932. The prevention of decay in stored fruit by the use of volatile compounds. Gr. Brit., Dept. Sci. Ind. Res., Food Invest. Board Rept., 1931.

U.S. DEPT. AGR. 1965. Losses in Agriculture. U.S. Dept. Agr., Agr. Handbook *291*.

VAN DEN BERG, L. and LENTZ, C. P. 1973. High humidity storage of carrots, parsnips, rutabagas and cabbage. J. Am. Soc. Hort. Sci. *98*, 129–132.

VAN TUYL, J. M. 1975. Genetic aspects of acquired resistance to benomyl and thiabendazole in a number of fungi. Meded. Fac. Landbouwwet. Rijksuniv., Gent. *40*, 691–697.

VAN TUYL, J. M. 1977. Genetic aspects of resistance to imazalil in *Aspergillus nidulans*. Netherlands J. Plant Pathol. *83* (Suppl. 1). (In Press)

VERHOEFF, K. 1974. Latent infections by fungi. Ann. Rev. Phytopath. *12*, 99–110.

VONK, J. W. and KAARS-SIJPESTEIJN, A. 1971. Methyl benzimidazol-2-yl-carbamate, the fungitoxic principle of thiophanate methyl. Pesticide Sci. *2*, 160–164.

WADE, N. L. and GIPPS, P. G. 1973. Postharvest control of brown rot and Rhizopus rot in peaches with benomyl and dicloran. Australian J. Exptl. Agr. Animal Husbandry *13*, 600–603.

WARDOWSKI, W. F., ALBRIGO, L. G., GRIERSON, W., BARMORE, C. R. and WHEATON, T. A. 1975. Chilling injury and decay of grapefruit as affected by thiabendazole, benomyl and CO_2. HortScience *10*, 381–383.

WELLS, J. M. and BENNET, A. H. 1976. Hydrocooling and hydaircooling with fungicides for reduction of postharvest decay of peaches. Phytopathology *66*, 801–805.

WELLS, J. M. and SPALDING, D. H. 1975. Stimulation of *Geotrichum candidum* by low oxygen and high carbon dioxide atmosphere. Phytopathology *65*, 1299–1302.

WELLS, J. M. and UOTA, M. 1970. Germination and growth of five fungi in low-oxygen and high-carbon dioxide atmospheres. Phytopathology *60*, 50–53.

WHITE, E. R., BOSE, E. A., OGAWA, J. M., MANJI, B. T. and KILGORE, W. W. 1973. Thermal and base-catalyzed hydrolysis products of the systemic fungicide, benomyl. J. Agr. Food Chem. *21*, 616–618.

WILD, B. L., McGLASSON, W. B. and LEE, T. H. 1976. Effect of reduced ethylene levels in storage atmospheres on lemon keeping quality . HortScience *11*, 114–115.

WILD, B. L., RIPPON, L. E. and SEBERRY, J. A. 1975. Comparison of thiabendazole and benomyl as post-harvest fungicides for wastage control in long term lemon storage. Australian J. Exptl. Agr. Animal Husbandry *15*, 108–111.

WOLFE, M. S. 1975. Pathogen response to fungicide use. Proc. 8th British Insecticide Fungicide Conf. *3*, 813–822.

WORKMAN, M., KERSCHNER, R. and HARRISON, M. 1976. The effect of storage factors on membrane permeability and sugar content of potatoes and decay by *Erwinia carotovora* var. *atroseptica* and *Fusarium roseum* var. *sambucinum*. Am. Potato J. *53*, 191–204.

WRIGHT, T. R. and SMITH, E. 1954. Relation of bruising and other factors to blue mold decay of Delicious apples. U.S. Dept. Agr., Circular *935*.

YAMADA, S., KURAMOTO, T. and TANAKA, H. 1972. Studies on the control of citrus fruit decay in storage. 7. Effect of the fungicide grove spray to satsuma mandarin (*Citrus unshiu* Marcovitch) fruit. Bull. Hort. Res. Station Ser. B (Okitsu) *12*, 207–228.

YODER, O. C. and WHALEN, M. L. 1975A. Factors affecting postharvest infection of stored cabbage tissues by *Botrytis cinerea*. Can. J. Botany *53*, 691–699.

YODER, O. C. and WHALEN, M. L. 1975B. Variations in susceptibility of stored cabbage tissue to infection by *Botrytis cinerea*. Can. J. Botany *53*, 1972–1977.

ZUCKER, M., HANKIN, L. and SANDS, D. 1972. Factors governing pectate lyase synthesis in soft rot and non-soft rot bacteria. Physiol. Plant Pathol. *2*, 59–67.

ACKNOWLEDGEMENT

The dedicated assistance of Mrs. Kathleen Eckard and Ms. Sandhya Patel in compiling the references and preparing the manuscript is gratefully acknowledged.

CHAPTER 8

FUNGAL DETERIORATION AND RELATED PHENOMENA IN CEREALS, LEGUMES AND OILSEEDS

R. J. BOTHAST

Northern Regional Research Center
Agricultural Research Service
U.S. Department of Agriculture
Peoria, IL 61604

ABSTRACT

Fungi are a major cause of postharvest deterioration of cereals, legumes and oilseeds. Concern over fungal invasion of these grains has intensified over the last 17 yr because mycotoxins (toxic metabolites produced by filamentous fungi) are being detected in some commodities. Apparently, the harvesting of more and more grain at moisture levels too high for safe storage has contributed to the mycotoxin problem. The fungi that invade grains are generally grouped into two categories: (1) field fungi which include species of Alternaria, Cladosporium, Fusarium, *and* Helminthosporium; *and (2) storage fungi which are predominantly species of* Aspergillus *and* Penicillium. *This division is based primarily upon moisture requirements. In addition to moisture, other factors such as temperature, O_2–CO_2 atmosphere, aeration, pH, and grain condition interact to affect fungal growth. Fungal deterioration of grain is a dynamic ecological process which often involves a succession of microorganisms, the breakdown of organic matter to yield CO_2 and H_2O, and the generation of heat. Nutrients are lost because of changes in carbohydrates, protein, lipids and vitamins. Germinability is lost and aesthetic changes occur which include discoloration, caking and abnormal odors. Also, mycotoxins such as aflatoxin, ochratoxin, citrinin, zearalenone, T-2 and vomitoxin may be produced and are capable of eliciting a toxic response when ingested. Usually these deteriorative changes in grains are prevented by reducing the moisture content to a level too low for fungi to grow. High-temperature drying has been feasible in the past, but as fuel becomes more expensive and less available, alternate methods to control fungal deterioration will be required. Low-temperature drying, solar drying, chilling, controlled atmosphere storage, chemical preservatives and various combinations of these control methods have application.*

INTRODUCTION

A major problem of agricultural production is loss of grain during and after harvest. Microorganisms, insects, and rodents contribute greatly to these postharvest losses. A commonly quoted estimate provided by the FAO for worldwide losses for all cereals, leguminous seeds, and oilseeds is 10% (Janicki and Green 1976). Christensen and Kaufmann (1974) state that fungi are the major cause of spoilage in stored grains and seeds in the technologically advanced countries, because insects and rodents are effectively controlled. Over the past 3 yr, the Agricultural Stabilization and Conservation Service (1976) reported that mold deterioration of the containers occurred in 25% of the shipments of Corn Soya Milk (CSM) under the Food for Peace Program. This spoilage resulted in losses estimated at 3% of the total amount of CSM shipped.

Consequently, it is well established that fungi destroy food and feed. However, the basic problem remains of implementing effective measures to reduce fungal losses.

This review presents a general discussion of the kinds of molds, factors, and processes involved in fungal deterioration of plant material. The effects of fungal invasion of grains on nutritional quality, aesthetic value, and mycotoxins are discussed. Finally, attention is focused on control methods to reduce losses and conserve resources.

FUNGI AND GRAIN

The microflora of cereals (including oilseeds) is made up of a wide variety of fungi, bacteria and actinomycetes. These microorganisms are the same as those found in soil, air, and on or in living or dead plants and animals (Semeniuk 1954). Christensen and Kaufmann (1969) reported that more than 150 species of fungi have been isolated from within seeds or kernels. Fungi are generally ubiquitous and omnivorous (Wallace 1973). However, the exact mechanism and time of fungal invasion of seeds is unclear. Christensen (1957) grouped fungi that invade cereals into two categories: (1) field fungi and (2) storage fungi. This division is not taxonomically valid but is based primarily upon moisture requirements. Field fungi attack developing and mature seeds which contain at least 20% moisture or are at an equilibrium relative humidity (RH) of 90–100%; storage fungi are usually encountered when grain is stored after harvest at moisture levels of 13–20% or in equilibrium with RH of 70–90%. The major field fungi are species of *Alternaria, Cladosporium, Fusarium,* and *Helminthosporium*, although species of *Curvularia, Stemphylium, Epicoc-*

cum, and *Nigrospora* infect seed at or near harvest and are included in this group. These fungi may discolor the grain, weaken or kill the embryo, or cause seedling blight, scab, or other disease. A few species of *Fusarium* and *Alternaria* may produce toxins in the invaded grain (Christensen *et al.* 1968). Nevertheless, species of *Alternaria, Helminthosporium,* and *Cladosporium* commonly occur in freshly harvested seeds.

Storage fungi are predominantly species of *Aspergillus* and *Penicillium.* "Species" of *Aspergillus* are not always well defined and are sometimes referred to as "groups." The major storage fungi consist of five or six groups of *Aspergillus*, plus several species of *Penicillium* which are common until deterioration is well advanced (Christensen and Kaufmann 1974). Certain other species of *Penicillium* are considered field fungi (Mislivec and Tuite 1970). Wallace (1973) lists 26 species of *Aspergillus* and 66 species of *Penicillium* which have been isolated from stored grain and grain products.

Over the past 20 yr, Christensen and Kaufmann (1974) have tested thousands of samples of cereal grains from commercial bins in the United States, Mexico, South America, and several European countries. Two groups consistently associated with beginning or incipient deterioration have been *A. restrictus* and *A. glaucus*. In grain where the equilibrium RH is less than 78–80%, these are the only species that can grow. However, in grain with an equilibrium RH above 80%, an ecological succession usually occurs; *A. restrictus* and *A. glaucus* appear first and may be followed by *A. candidus, A. flavus, A. ochracus, A. versicolor,* and *Penicillium* spp. Each species has a rather sharp lower limit of moisture equilibrium, below which it cannot grow. Table 8.1 reflects the minimum equilibrium RH at which different common storage fungi can grow.

In contrast to the Aspergilli, the Penicillia require more moisture and vary little in color, ranging through shades of blue to green and grey. The Penicillia are frequently referred to as "blue" or "green" molds or *Penicillium* spp. because they are difficult to identify (Wallace 1973).

Other molds included in the storage fungi category are *Absidia, Mucor, Rhizopus, Chaetomium, Scopulariopsis, Paecilomyces,* and *Neurospora.* *Absidia, Mucor,* and *Rhizopus* are generally associated with spoilage under moist conditions because they require a minimum RH of 88% for growth; consequently, they are not usually initiators of grain deterioration in storage (Wallace 1973).

For the most part, the above categories are accurate; however, exceptions exist. *A. flavus* can invade in the field and *Fusarium* can continue to decay grain in storage if the moisture is high enough (Lillehoj *et al.* 1975A, 1976; Caldwell and Tuite 1974).

TABLE 8.1
APPROXIMATE MINIMUM EQUILIBRIUM RELATIVE HUMIDITY FOR
GROWTH OF COMMON STORAGE FUNGI

Mold	RH Limit	Reference
Aspergillus halophilicus	68	Christensen and Kaufmann (1974)
A. restrictus group	70	Christensen and Kaufmann (1974)
A. glaucus group	73	Christensen and Kaufmann (1974)
A. chevalieri	71	Ayerst (1969)
A. repens	71	Ayerst (1969)
A. candidus group	80	Christensen and Kaufmann (1974)
A. candidus	75	Ayerst (1969)
A. ochraceus group	80	Christensen and Kaufmann (1974)
A. flavus group	85	Christensen and Kaufmann (1974)
A. flavus	78	Ayerst (1969)
A. nidulans	78	Ayerst (1969)
A. fumigatus	82	Ayerst (1969)
Penicillium, depending on species	80–90	Christensen and Kaufmann (1974)
P. cyclopium	82	Ayerst (1969)
P. martensii	79	Ayerst (1969)
P. islandicum	83	Ayerst (1969)

FACTORS INFLUENCING FUNGAL GROWTH AND DETERIORATION OF GRAIN

The principal factors which control fungal growth and deterioration are moisture, temperature, atmosphere, aeration, pH and condition of the grain. All of these factors interact as deterioration progresses, but moisture and temperature are probably most important.

Moisture

If the moisture content is maintained at a sufficiently low level, grain can be stored for many years with little deterioration even under otherwise unfavorable storage conditions (Pomeranz 1974). However, modern farming practices and harvesting techniques yield grain at moisture levels too high for safe storage. This grain must be dried, protected by airtight storage, or preserved if fungal growth is to be prevented.

The most useful measure of the availability of water to fungi is the ratio of the vapor pressure of the water in the substrate to that of pure water at the same temperature and pressure. This figure is referred to as water activity (a w) or when it is expressed as a percentage, as the equilibrium RH. Pomeranz (1974) reports that an RH of 75% is about minimum for the germination of most mold spores at ordinary temperatures and that the equilibrium moisture content of different grains (Table 8.2) at 75% RH may vary markedly because of differences in composition. Consequently, the critical moisture level for a particular grain is the percentage moisture when the seed is in equilibrium with an atmospheric RH of approximately 75%. The equilibrium RH of grain is more important than the moisture content for controlling fungal deterioration.

Despite the fact that grain may be uniform and within what is normally considered a safe moisture limit at the outset of storage, fungal deterioration may still result because of excessive moisture. Differences in temperature between different portions of a grain bulk can result in a rapid transfer of moisture from warmer to cooler regions. RH of interstitial air in stored grain tends to remain in equilibrium with moisture in the grain. At any RH, the actual amount of water vapor per cubic foot of air increases with rising temperature. Thus, when warm air reaches a cool region, it gives up moisture to the grain to maintain equilibrium. This moisture interchange usually takes place in the vapor phase; but with extreme temperature differences, the warm air may be cooled below the dew point and water will condense on the surface of the grain or bin. Christensen and Drescher (1954) reported a range in moisture contents from a low of 10% to a high of 18% in a bin of wheat in which the moisture content of a representative sample was 13.2%. Ramstad and Geddes (1942) found soybeans with a moisture content of 28% in a bin where the average moisture content of the beans was 15%. Numerous other examples of moisture transfer are in the literature (Johnson 1957; Holman 1950; Christensen

TABLE 8.2

EQUILIBRIUM MOISTURE CONTENT OF DIFFERENT GRAINS AT 75% RH
AND 25–28° C

Grain	Moisture Content[1]
Sorghum	15.3
Rye	14.9
Wheat	14.6
Barley	14.4
Corn	14.4
Rice	14.4
Oats	13.9
Soybeans	13.5[2]
Cottonseed	11.4
Sunflower seed	10.5[2]
Peanuts	10.5
Flaxseed	10.0

[1]Adapted from data of Milner and Geddes (1954).
[2]Adapted from Christensen (1972).

1970). Consequently, the actual moisture content of the grain in localized areas of the bin is important rather than the moisture content of the grain when loaded into the bin.

Temperature

Growth of fungi is accelerated by an increase in temperature until such factors as thermal inactivation of enzymes, exhaustion of substrate, oxygen or moisture depletion, or accumulation of carbon dioxide become limiting. The interrelationships of these factors are so complex that determination of minimum, optimum, and maximum temperatures for growth of fungi is approximate. Nevertheless, Christensen and Kaufmann (1974) have summarized the temperature limits for growth of common storage fungi (Table 8.3).

There is a balance between safe moisture content and safe temperature in the grain bulk. Within limits, low temperature can be substituted for low moisture in prolonging the storage of grain. Burrell (1974B) reported that *A. glaucus*, several species of *Penicillium, Cladosporium, Fusarium*, and *Mucor*, and some yeasts grow at $-5°$ C to $-8°$ C, and in some instances sporulate at temperatures below freezing. Also, fungal toxins can be produced on moldy grain at low temperature (Joffe 1965; Kurtzman and Ciegler 1970). Nevertheless, at a given moisture content, grain can be stored longer at lower temperatures. Fig. 8.1 from the work of Steele *et al.* (1969) gives an example of the length of time that corn can be stored under various conditions of temperature and moisture before fungi destroy 0.5% dry matter. However, extrapolation of the data of Steele *et al.* (1969) to field recommendations is risky because of the specific conditions (aeration, RH, and mechanical damage) under which their experiments were conducted.

TABLE 8.3

APPROXIMATE MINIMUM, OPTIMUM, AND MAXIMUM TEMPERATURES FOR GROWTH OF COMMON STORAGE FUNGI ON GRAINS[1]

Fungus	Temperature for Growth, ° C		
	Minimum	Optimum	Maximum
Aspergillus restrictus	5–10	30–35	40–45
A. glaucus	0–5	30–35	40–45
A. candidus	10–15	45–50	50–55
A. flavus	10–15	40–45	45–50
Penicillium	−5–0	20–25	35–40

[1]From Christensen and Kaufmann (1974).

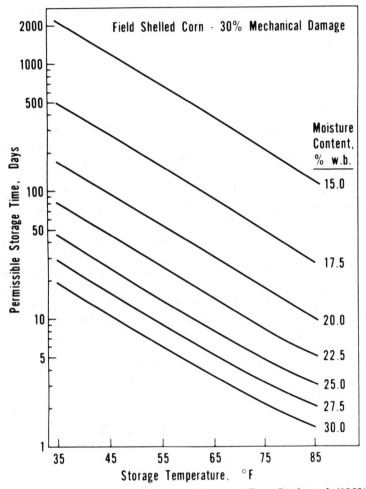

FIG. 8.1. PERMISSIBLE STORAGE TIME UNDER VARIOUS
CONDITIONS OF TEMPERATURE AND MOISTURE BEFORE 0.5%
DRY MATTER IS DESTROYED

Atmosphere

Generally, it is accepted that fungi are aerobic microorganisms and that
oxygen depletion or carbon dioxide build-up limits their activity. Detroy *et
al.* (1971) state that reducing oxygen from 5 to 1% dramatically inhibits
growth of *A. flavus* and aflatoxin formation, while Christensen and Kaufmann (1974) reflect that a carbon dioxide concentration above 14% is

detrimental to mold growth. Nevertheless, some fungi can grow at low oxygen concentrations. Tabak and Cooke (1968) tested a number of fungi, including *Fusarium*, under conditions "as anaerobic as could be maintained in the laboratory" and all fungi grew to some extent. Peterson *et al.* (1956) reported that growth of storage fungi was reduced with decreasing oxygen concentration, but some growth occurred down to about 0.2% oxygen. Below a minimum oxygen level, fungal growth will cease, but this does not mean that the fungus is killed.

Aeration

Aeration of grain consists of blowing or drawing ambient air at a low rate through the grain mass. This practice almost seems self-defeating because a favorable O_2–CO_2 environment is ensured for fungal growth. However, the main function of aeration is to establish and maintain a moderately low and uniform temperature throughout the bulk. Burrell (1974A) reported that fungal growth rarely occurs during storage periods up to 8 months in bulks of cooled grain at 5°–10° C if the moisture content is below about 18%. Also, Burrell (1974A, B) states that no change in fungal counts occurred during aerated storage of wheat and barley up to 18% moisture. However, it is clear that aeration as a means for slowing fungal attack is only effective in cool climates where the grain is not over 18% moisture content. In warmer climates, fairly rapid drying to 14% moisture or less is necessary to prevent fungal deterioration.

The Grain and Its Condition

Substrate is a determining factor in fungal deterioration. Most grains, leguminous seeds, and oilseeds support mold growth. Nutritional investigations have shown that simple carbon and nitrogen sources, plus selected mineral salts, provide the necessary nutrients for growth of most fungi (Detroy *et al.* 1971). Most molds can grow over a wide pH range (2–8.5) but an acid pH usually favors growth (Frazier 1967). However, species of *Scopulariopsis* grow best under alkaline conditions (Bothast *et al.* 1975B).

Condition or soundness of grain can be critical in establishing fungal invasion. Condition of grain is influenced by the environment during growth and maturation, by the degree of fungal invasion, by the maturity at harvest, by methods of harvesting, and by handling prior to and after storage, e.g., during drying and during international shipment, stress cracks and mechanical damage or breakage predispose grain to fungal invasion (Thompson and Foster 1963; Steele *et al.* 1969). Also, varietal differences may influence fungal attack. Softer types of wheat respire more rapidly than harder types at similar moisture levels and temperatures (Pomeranz 1974). With so many interacting factors affecting fungal growth, it is clear why it is difficult to establish an absolute maximum moisture limit for the safe storage of grain.

RESPIRATION AND HEATING

When fungal deterioration of grains occurs, energy is required. This energy is normally produced in the presence of oxygen by the respiratory process. Respiration involves the breakdown of organic matter (carbohydrate-hexose) to yield carbon dioxide, water, and the generation of heat. However, in the absence of oxygen, the process is less complete (less CO_2, H_2O, and heat) and other compounds such as acetic acid or ethyl alcohol are produced. The direct effects of respiration on grain are loss in dry matter, gain in moisture content, increase in CO_2 in the intergranular air, and a rise in temperature of the grain (Milner and Geddes 1954).

Early research on respiration of moist grain did not distinguish between respiration of the grain itself and respiration of the microflora on and within the grain (Bailey and Gurjar 1918). However, later work by Ramstad and Geddes (1942) and Milner and Geddes (1945) showed that fungi are primarily responsible for respiration and heating of moist soybeans. Hummel *et al.* (1954) found in a study with mold-free wheat that respiration at moisture contents of 15–31% and a temperature of 35° C was low and constant. Consequently, the evidence shows that the respiratory activity of seed or grain itself usually plays only a minor role, whereas fungi are often solely responsible for the initial heating of moist grains and moist plant material.

The succession of microorganisms during the heating of moist (30% moisture) agricultural materials has been studied extensively (Carlyle and Norman 1941; Norman *et al.* 1941; Wedberg and Rettger 1941; Gregory *et al.* 1963; Gray *et al.* 1971). At the start of the process, the indigenous mesophilic flora (molds, yeasts, bacteria-growing best between 20–40° C) multiply and the temperature rises. At approximately 40° C, the activity of the mesophiles diminishes and degradation continues by thermophiles. At 60° C, the thermophilic fungi die and the process is kept going by certain spore-forming bacteria and thermophilic actinomycetes to maximum of 70° C. Thereafter, heating cannot be attributed to microorganisms but to chemical processes (Milner and Geddes 1946; Milner *et al.* 1947; Currie and Festenstein 1971).

These investigations are not directly applicable to stored grain because respiration and heating of grain begin at lower moisture levels (*i.e.*, a moisture content in equilibrium with a relative humidity of 70–75%)

where only certain molds can grow. Christensen and Kaufmann (1974) effectively described the succession of fungi during the heating of grain: "Either some of the grain is moist enough when stored, or through moisture transfer later acquires a high enough moisture content, so that *Aspergillus restrictus* and *A. glaucus* can grow. *A. restrictus* grows so slowly that it probably does not increase either the temperature or moisture content of the grain appreciably. *A. glaucus*, however, if growing rapidly, can increase the temperature of the grain at least to 30–40° C. This results in some increase in moisture content in the grain where the fungus is growing, and a greater increase in the grain just above. Once the moisture content of the grain exceeds 15.0–15.5%, *A. candidus* can grow; and given optimal conditions, it can increase the moisture content and temperature of the grain rapidly. Once the moisture content of the grain reaches that in equilibrium with a relative humidity of 85% (18.5% moisture in the cereal seeds), *A. flavus* can grow. *A. candidus* and *A. flavus* together can increase the temperature of the grain to 55° C and hold it there for weeks. Depending on whether the metabolic and distillation water from the activities of these fungi is carried off or whether it accumulates in the grain, the heating may gradually subside, or may pass into the next stage in which thermophilic bacteria plus perhaps a variety of thermophilic fungi may be involved."

Milner and Geddes (1946) made simultaneous measurements of seed viability, mold infection, chemical changes, respiratory exchange, and temperature increase in aerated soybeans containing 22.8% moisture under near adiabatic conditions. Fig. 8.2 shows the course of temperature increase, CO_2 evolution, and fungal infection. The first heating stage is a result of fungal metabolism and ends when the thermal death range (50–55° C) of certain molds (predominantly *A. glaucus* and *A. flavus*) is reached. The second heating stage may initially involve thermotolerant fungi (*A. fumigatus*, *Mucor pusillus*, etc.), but as the temperature increases the molds and seeds are killed. The rapid heating and CO_2 production after 14 days is due to chemical oxidation which follows death of microorganisms.

As more high-moisture grain is harvested and alternatives to costly high-temperature drying are adapted, the heating process in grain may be modified. For example, in our studies on preservation of high-moisture corn (Bothast *et al.* 1975A), the temperature of 1,500 bu (38 metric tons) of ammonia-treated corn (27% moisture content) increased from 25–60° C (Fig. 8.3). Chemical heating was probably responsible for the initial rise in temperature, but subsequent heating was by bacterial respiration. Apparently, fungi contributed little to the heating of this grain.

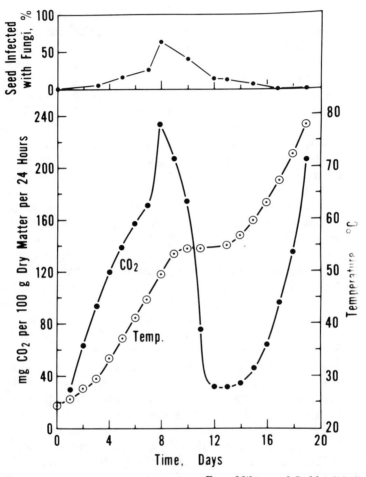

From Milner and Geddes (1946)

FIG. 8.2.   THE COURSE OF SPONTANEOUS ADIABATIC HEAT-
ING, RESPIRATION, AND FUNGAL INFECTION EXHIBITED BY
ILLINI SOYBEANS, UNDER ADIABATIC CONDITIONS, AT 22.8%
MOISTURE. THE AERATION WAS DOUBLED ON THE SEVENTH
DAY

NUTRITIVE CHANGES

Fungal activity can cause changes during storage of grain and grain
products that are detrimental to nutritive value (Zeleny 1954). Specifically, nutrients are lost because of changes in carbohydrates, protein, lipids,
and vitamins (Semeniuk 1954). Pomeranz (1974) has reviewed some of
these changes.

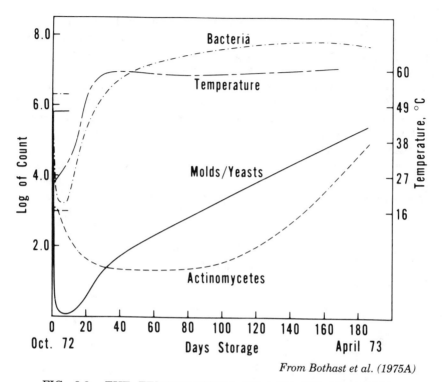

From Bothast et al. (1975A)

FIG. 8.3. THE TEMPERATURE OF 1,500 BU OF AMMONIA TREATED CORN (27% MOISTURE CONTENT) INCREASED FROM 25–60° C AND APPARENTLY FUNGI CONTRIBUTED LITTLE TO THE HEATING

Carbohydrates

Conditions that favor fungal activity lead to carbohydrate decomposition. Sugars are consumed and converted into CO_2 and H_2O. At moisture levels of approximately 15%, grain loses both starch and sugar and the dry weight decreases.

Ramstad and Geddes (1942) found a marked increase in reducing sugars in soybeans stored at more than 15% moisture. The increase was followed by an equally significant decrease in nonreducing sugars. Milner and Geddes (1946) demonstrated that sugars in stored soybeans disappear during the biological phase of heating, but that reducing substances increase when the heating has advanced to chemical oxidation. Bottomley *et al*. (1950, 1952) reported a marked disappearance of nonreducing sugars in corn stored at high moisture levels. The nonreducing sugars are converted into reducing sugars by the action of invertase or similar enzymes, which are produced by molds and by amylase which is normally present in corn.

Fig. 8.4 shows the relationship between the number of viable mold spores in the corn and the content of nonreducing sugars.

Glass *et al.* (1959) studied aerobic and anaerobic storage of wheat in the laboratory. When damp wheat was stored in air, extensive mold growth occurred and the increase in reducing sugars was only about one-fourth as great as the decrease in nonreducing sugars. The difference in sugars was attributed to metabolism by molds.

In an atmosphere of nitrogen, where mold growth was prevented, marked changes still occurred in the sugars. The decrease in nonreducing sugars was almost exactly compensated for by the increase in reducing

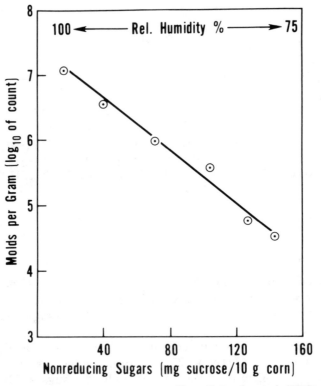

From Bottomley et al. (1950)

FIG. 8.4. RELATIONSHIP BETWEEN MOLD COUNT AND NONREDUCING SUGAR CONTENT OF CORN STORED UNDER RELATIVE HUMIDITIES OF 75–100%

sugars. Lynch *et al.* (1962) stored 20% moisture wheat under various atmospheres for 8 wk at 30° C. In the sample stored in air, reducing sugars remained unchanged while sucrose decreased (Table 8.4). In wheat stored under nitrogen or CO_2, maltose remained unchanged, whereas fructose and glucose each increased threefold and galactose four to fivefold. The increase in galactose reveals that sucrose is not the only nonreducing sugar hydrolyzed during storage. It appears that the reducing sugars produced from the breakdown of nonreducing sugars in air are used by fungi.

In our studies (Lancaster and Bothast 1976) on preservation of high-moisture corn, *Scopulariopsis brevicaulis* invaded ammonia-treated corn and altered the sugar composition markedly. Sucrose decreased from 2.4% to 0.6% while reducing sugars increased from 0.21% to 0.77%.

Protein

The total protein content of grain as calculated from its nitrogen content is generally assumed to be constant during storage. However, as fungal deterioration advances and carbohydrate is used in the respiratory processes, protein increases mathematically. Daftary *et al.* (1970) demonstrated this by finding that the protein content (determined by the Kjeldahl method) was slightly, but consistently, higher in flours from mold-damaged samples than in corresponding flours from sound wheat.

Proteolytic enzymes produced by fungi can modify the proteins in grains by hydrolyzing them into polypeptides and amino acids. Subsequently fungi can convert these materials into fungal protein which can be nutri-

TABLE 8.4

CHANGES IN THE MONO- AND DISACCHARIDES OF WHEAT STORED UNDER VARIOUS ATMOSPHERES FOR 8 WK AT 30° C AND 20% MOISTURE[1]

Sugar	Concentration of Sugars (mg/10g dry matter)			
	Control	Stored in Air	Stored in Nitrogen	Stored in Carbon Dioxide
Fructose	6	5	18	16
Glucose	8	7	24	23
Galactose	2	3	9	9
Sucrose	54	21	39	36
Maltose	5	1	4	3
Total reducing sugars, as maltose	41	41	117	117
Total nonreducing sugars, as sucrose	190	43	100	115

[1]Data of Lynch *et al.* (1962).

tionally beneficial to animals. Although these effects are only significant at advanced stages of deterioration, several investigators have reported qualitative transformations of free amino acids. DeVay (1952) observed changes in concentration of gamma-aminobutyric acid in hard red spring wheat stored at 19.5% moisture. Linko and Milner (1959) showed a considerable change in the composition of free amino acids of wheat which had been wetted.

Bothast et al. (1975B) reported that lysine, methionine, and proline increased in proportion to total amino acids while glutamic acid, leucine, alanine, and phenylalanine decreased when S. brevicaulis was grown on ammonia-treated corn. In contrast, Trolle and Pedersen (1971) reported that the lysine content of barley decreased during storage damage. Kao and Robinson (1972) investigated the changes in amino acid content of molded wheat. Arginine, cysteine, lysine, and histidine decreased while methionine increased.

Lipids

Because most molds have a high lipolytic activity, fats and oils in grain are readily broken down into free fatty acids and partial glycerides during the fungal deterioration of grains. These changes are greatly accelerated when moisture and temperature are favorable for fungal growth (Goodman and Christensen 1952; Loeb and Mayne 1952). Nagel and Semeniuk (1947) grew pure cultures of nine fungi on steam-sterilized corn containing 32% moisture and found that all fungi increased the free fatty acid content of the corn. Christensen and Dorworth (1966) reported that invasion of soybeans by storage fungi was accompanied by increases in free fatty acids. Consequently, the free fatty acid content of grain has been used as an index for estimating grain deterioration (Zeleny 1954).

Pomeranz et al. (1956) reported a 20% reduction in free lipids of wheat when the mold count increased a thousandfold. Daftary and Pomeranz (1965) studied changes in lipids of soft and hard wheat stored at moistures and temperatures conducive to mold growth. A decrease of 40% in total lipid content was accompanied by an increase in mold count from 1,000 to about 2,000,000 per gram. Nonpolar lipids decreased about 25% and damaged wheat contained only one-third as much polar lipids as did sound wheat. Also, grain deterioration was accompanied by rapid disappearance of glycolipids and phospholipids. The breakdown of polar lipids was more rapid and extensive than formation of free fatty acids or disappearance of triglycerides. Subsequently, Daftary et al. (1970) investigated the effects of temperature (23, 30 and 37° C) on the composition of wheat flours stored for 16 weeks at 18% moisture. Mold counts increased up to 10,000-fold.

Aspergillus niger, A. candidus, and *A. vesicolor* were predominant. Free lipids and the polar components of bound lipids decreased during storage.

Ramstad and Geddes (1942) stored soybeans in excess of 15% moisture in glass jars for 1 yr. All samples were moldy and showed decreased iodine numbers. Zeleny (1954) concluded that the oil from damaged soybeans is likely to be of inferior quality. Krober and Collins (1948) reported that the free fatty acid increase in damaged soybeans gives rise to high refinery losses. In response to this problem, List *et al.* (1977) developed methods to improve the quality of oil from damaged beans by increasing the deodorization temperature and subsequently removing high molecular weight flavor bearing compounds.

Recently we (Bothast *et al.*, unpublished data) studied the effect of specific fungi on flavorful carbonyl compounds in soybeans. Species of *Aspergillus, Penicillium, Mucor, Rhizopus,* and *Candida* increased the total carbonyl and the monocarbonyl content of crude oil during controlled fermentations (Table 8.5).

Vitamins

Cereal grains and their products are important sources of vitamins in food and feed (Zeleny 1954). Generally cereal grains are good sources of thiamine, niacin, pyridoxine, inositol, biotin, pantothenic acid and vitamin E. Vitamin A activity of yellow corn, although low and unstable, is

TABLE 8.5

TOTAL CARBONYL AND TOTAL MONOCARBONYL CONTENT OF CRUDE SOYBEAN OIL AFTER FERMENTATION[1]

Culture[2]	μMoles per 1.0g of Extract[3]	
	Total Carbonyls	Total Monocarbonyls
Aspergillus chevalieri	8.89	1.85
A. amstelodami	11.12	1.79
A. niger	2.70	1.57
A. flavus	5.65	3.62
A. candidus	5.51	4.90
Penicillium meleagrinum	5.57	4.25
Rhizopus oligosporous	7.18	3.87
Mucor pusillus	8.17	1.87
Candida lipolytica	3.85	1.53
Phycomyces blakesleeanus	4.48	2.02
Autoclaved control	2.71	0.90
Nonautoclaved control	3.00	1.31

[1]Unpublished data of Bothast *et al.*
[2]Fermentation was conducted at 28° C for 12 days.
[3]Mean values of two trials.

important in animal feeds and may also be of significance in human nutrition. No other grain has appreciable vitamin A activity (Pomeranz 1974). Consequently, losses in vitamin content that occur during storage are of considerable practical importance.

Bayfield and O'Donnell (1945) showed that wheat lost approximately 30% of its thiamine in a 5-month storage period at 17% moisture. Kao and Robinson (1972) reported near 50% losses in thiamine in molded wheat. In contrast, this same wheat increased in riboflavin, vitamin B6, vitamin B12, and pantothenic acid. For the most part, it is believed that the B vitamins, with the possible exception of pantothenic acid, are rather stable and are not readily destroyed under normal storage conditions.

Trolle and Pedersen (1971) demonstrated that the tocopherol content of barley decreased after storage damage, e.g., 30.5 γ/g in mold damaged barley compared to 54.5–81.0 γ/g in sound barley.

Minerals

Except under unusual conditions, the mineral content of grain or its products does not change during storage (Zeleny 1954). However, it is possible for the percentage of total mineral matter in grain, as determined by ash content, to increase as a result of the loss of other constituents such as carbohydrate. This type of change may be measured in grain that has undergone extensive fungal deterioration. An example of this was cited by Zeleny (1954):

"A sample of barley was taken from an excavation in Asia Minor and claimed to be from 3,000 to 5,000 years old. Its ash content on a moisture-free basis was 17.2% as compared to 3% for normal barley ash. The barley was black in color, very light in weight, and had obviously lost much of its organic substance."

GERMINATION AND AESTHETIC CHANGES

Numerous reports (Dorworth and Christensen 1968; Qasem and Christensen 1960; Christensen 1955; Armolik et al. 1956; Papavizas and Christensen 1957; Christensen 1962; Fields and King 1962; and Lopez and Christensen 1967) confirm that invasion of seeds by storage fungi can result in loss of germinability.

Briefly, samples of soybeans, wheat, barley, corn, sorghum and peas stored at moisture contents and temperatures favorable for the growth of storage fungi, but kept free of fungi, retained a germinability of 95–100% for a few months. On the other hand, similar samples invaded with storage fungi were reduced to zero or near zero germinability. Most of the storage fungi invade the embryo of the seed preferentially as shown in Fig. 8.5. Thus, it is not surprising that germinability is reduced. An example of the effect of storage fungi on germination of wheat is illustrated in Fig. 8.6.

Aesthetic changes that occur when grain deteriorates in storage include discoloration, caking, and abnormal odors. With increasing fungal invasion, grain loses its natural luster and becomes rather dull and lifeless in appearance. General appearance alone is considered a quality factor in the routine inspection and grading of barley, oats, grain sorghums, and soybeans. According to Christensen (1955), Papavizas and Christensen (1957), and Schroeder and Sorenson (1961), it is highly probable that

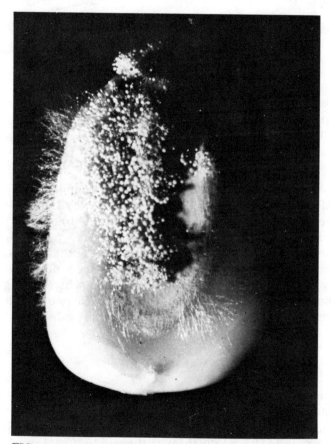

FIG. 8.5. *Aspergillus flavus* GROWING FROM THE EMBRYO OF A SURFACE-STERILIZED CORN KERNEL ON MALT AGAR

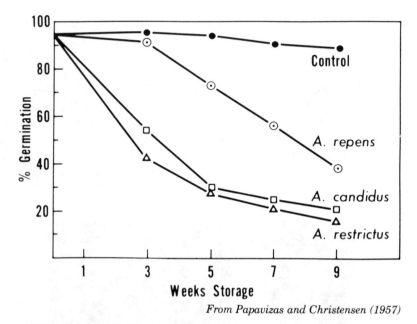

From Papavizas and Christensen (1957)

FIG. 8.6. LOSS OF GERMINABILITY OF WILLET WHEAT IN-
OCULATED WITH VARIOUS SPECIES OF *Aspergillus* AND
STORED AT 85% RELATIVE HUMIDITY AND 25° C

invasion of the germ or embryo of seeds by storage fungi is a major cause of
discoloration. Damaged kernels can be identified by the brown to black
color of the germ. Wheat damaged in this manner is frequently referred to
as "sick wheat" (Pomeranz 1974).

Abnormal odors, such as musty or sour odors, and caking are associated
with grain that has reached a fairly advanced stage of deterioration. Sour
odors may be produced by bacterial fermentation, whereas fungi are usually responsible for the musty odors, which may carry over into products
made from deteriorated grain.

MYCOTOXINS

Toxic metabolites (mycotoxins) produced by filamentous fungi in grains
are capable of producing a toxic response (mycotoxicosis) when ingested.
Despite the fact that numerous poisonings from the consumption of moldy
grains were reported from 1826 through the mid 1900's (Mayer 1953; Bilay
1960; Dounin 1926; Joffe 1971; Forgacs 1972), mycotoxin problems were

not fully recognized until it was discovered that aflatoxins were responsible for the deaths of a large number of turkey poults in England in 1960 and that the toxins were potent carcinogens in laboratory animals (Lancaster *et al.* 1961). Since 1960, hundreds of scientific reports and reviews have been published on mycotoxins (Goldblatt 1969; Lillehoj *et al.* 1970; Ciegler *et al.* 1971; Kadis *et al.* 1971, 1972; Purchase 1971; Rodricks 1976).

Cereal grains, peanuts and cottonseed may be contaminated with mycotoxins (Detroy *et al.* 1971). Mycotoxin production can occur in the field or during harvest, processing, storage and shipment. However, only a few mycotoxins have definitely been implicated in mycotoxicoses (Ciegler 1975). Although the diagnosed cases are usually more acute and dramatic, the most important mycotoxicoses probably involve subacute doses; e.g., livestock exhibit poor weight gains and lowered feed efficiency; humans may contract hepatomas and suffer from degeneration of the hematopoietic system. I will briefly summarize the major mycotoxins implicated in natural outbreaks of mycotoxicoses from grains (Table 8.6).

Aflatoxin

The aflatoxins (B_1, B_2, M_1, M_2, G_1 and G_2) are the most studied mycotoxins. They appear to constitute a contamination problem primarily in peanuts and peanut products, cottonseed meal, and corn; however, many other grains have also been reported to be contaminated (Ciegler 1975). The occurrence of aflatoxins is usually associated with poor storage conditions, although more and more evidence indicates that these compounds are also produced in the field.

Lillehoj *et al.* (1975A) determined that 2.5% of 5,000 test ears of field corn from southeast Missouri and east Central Illinois contained aflatoxin B_1 at levels exceeding 20 μg/kg. In a subsequent study of field corn in northeastern South Carolina, Lillehoj *et al.* (1975B) demonstrated a 49% incidence of aflatoxin in samples collected from 184 fields. Aflatoxin B_1 levels were less than 80 μg/kg in 80% of the samples containing detectable aflatoxin. Lillehoj *et al.* (1977) detected extensive *A. flavus* infection at harvest in 214 Iowa corn samples but only four contained 20 μg/kg or more aflatoxin B_1. According to comprehensive studies by Shotwell *et al.* (1969A, B) on samples mainly from the corn belt, only nine out of 1,368 samples of wheat, grain sorghum and oats, and 35 out of 1,311 corn samples contained small amounts of aflatoxins (up to 19 μg/kg). A second survey of U.S. corn for aflatoxin showed a similarly low incidence of aflatoxins at levels up to 25 μg/kg (Shotwell *et al.* 1970). Most of the contamination was in the lower grades.

TABLE 8.6
MAJOR MYCOTOXINS FOUND IN NATURAL OUTBREAKS OF MYCOTOXICOSES FROM GRAINS

Mycotoxin	Producing Fungi	Animals Affected	Biological Effects	Reference
Aflatoxin	*Aspergillus flavus*, *A. parasiticus*	Mammals, fish, birds	Hepatotoxin, cancer	Detroy *et al.* (1971)
Ochratoxin A	*Penicillium viridicatum*, *A. ochraceous*	Swine, man?	Nephrotoxin	Ciegler *et al.* (1971)
Citrinin	*P. viridicatum* *P. citrinum*	Swine	Nephrotoxin	Ciegler *et al.* (1971)
Zearalenone or F-2	*Fusarium graminearum* (*Gibberella zeae*)	Swine	Volvovaginitis, abortion, enlarged mammary glands	Christensen *et al.* (1965)
Trichothecenes T-2	*F. tricinctum*	Cattle	Dermal necrosis, hemorrhage	Hsu *et al.* (1972)
Vomitoxin	*F. graminearum*	Swine, man	Refusal vomiting	Vesonder *et al.* (1973, 1976)

In 1969 and 1970, Shotwell *et al.* (1973) expanded their survey to southern corn of all grades except U.S. No. 1. Incidence of contamination was near 30% and ranged from 5 μg/kg to 308 μg/kg. Only 16% of the positive samples contained more than 100 μg/kg.

In general, incidence and levels of aflatoxins in grains and grain products appear to be low (Scott 1973).

In addition to being a potent toxin, aflatoxin B_1 is the most carcinogenic mold metabolite known. Liver damage is the usual symptom. Wogan (1968) reported that a dietary level of 15 ppb aflatoxin B_1 was sufficient to cause 100% incidence of liver tumors in male rats after 68 wk. Susceptibility of domestic animals to aflatoxin varies widely (Keyl and Booth 1971). Pigs can tolerate up to 233 ppb. No toxic effects were observed at levels of 300 ppb or lower in beef steers fed for 4.5 months. In dairy cows, weekly intakes of 67–200 mg of aflatoxin B_1 produced 70–154 ppb aflatoxin M_1 in lyophilized milk. No adverse effects were discernible in broilers fed from 1 day to 8 wk of age a ration containing 400 ppb aflatoxin. Sheep were resistant for 3 yr to 17,500 ppb aflatoxin B_1 in the feed, while liver tumors were found in ducklings fed a diet containing 30 ppb aflatoxin (Scott 1973).

There is no established toxic dose for humans; but circumstantial evidence from Southeast Asia, India, and Africa, plus a suspect case in Germany, indicates that aflatoxins have been involved in human deaths, particularly among children (Ciegler 1975).

Ochratoxin A

Merwe *et al.* (1965) first isolated ochratoxin A in South Africa from maize on which *Aspergillus ochraceus* had grown. Since then the toxin has been experimentally produced by this fungus and *Penicillium viridicatum* on a number of substrates (Scott 1973). Shotwell *et al.* (1969C) detected the first natural occurrence of ochratoxin A in corn, but since then it has been found in wheat, oats, barley, peanuts, white beans, and mixed feed grain from Canada (Scott *et al.* 1972) and in coffee (Levi *et al.* 1974). However, the incidence of ochratoxin A in U.S. corn was low (Shotwell *et al.* 1970).

Ochratoxin A is a highly toxic metabolite that produces both liver and kidney damage. The oral LD_{50} in male rats is 22 mg/kg (Purchase and Theron 1968). Chronic toxicity studies carried out with laying hens from 14 through 52 wk of age showed that up to 4 ppm ochratoxin A in the diet caused high mortality, depressed growth, delayed sexual maturity, low egg production, and poor egg shell quality (Choudhury *et al.* 1970). Krogh *et al.* (1974) studied the effect of ochratoxin in swine at levels of 200, 1,000 and 4,000 μg/kg of feed for about 4 months. These levels correspond to levels found commonly in animal feeds in Denmark. Changes in kidney structure was a characteristic symptom, and ochratoxin A was detected in the kidney, liver, adipose, and muscular tissue of the experimental animals.

Citrinin

Citrinin was originally isolated from *Penicillium citrinum* by Hetherington and Raistrick (1931). Citrinin was considered responsible for the yellow color of rice imported into Japan from Thailand around 1951 (Scott 1973). However, the occurrence of citrinin as a contaminant of feedstuffs has been associated with *P. viridicatum* and always as a co-contaminant with ochratoxin. Scott *et al.* (1972) found citrinin at 0.07–80 mg/kg levels in 13 of 18 samples which also contained ochratoxin. Citrinin caused kidney damage in experimental animals (Krogh *et al.* 1970), comparable to that caused by feeding barley contaminated with *P. viridicatum*. Thus, it is possible that both ochratoxin A and citrinin are involved in the mycotoxicosis which has affected up to 7% of the pigs in Denmark (Krogh 1969). Hesseltine (1976) reported that citrinin occurs naturally in wheat, rye, barley, and oats in both Canada and Denmark.

Zearalenone or F-2

Various reports since 1928 indicate that the estrogenic syndrome in swine is associated with consumption of moldy corn (Ciegler 1975). Subsequently, Stob *et al.* (1962) isolated an anabolic uterotropic compound from corn infected with *Gibberella zeae* and partially characterized the toxin. Later, Christensen *et al.* (1965) and Mirocha *et al.* (1967) isolated the estrogenic substance and labeled it F-2. Urry *et al.* (1966) determined the structure of this compound and gave it the name, zearalenone.

The estrogenic syndrome is one of the best understood mycotoxicosis because zearalenone or F-2 has been found naturally in feeds in sufficient amounts to cause the disease. In swine, the estrogenic syndrome involves development of a swollen vulvae in the females, shrunken testes in young males, enlarged mammary glands in the young of both sexes, and breeding and abortions in females (Mirocha *et al.* 1968). The authors demonstrated experimentally that 1 mg of F-2 fed daily to gilts weighting 27 kg produced swollen vulvae within 5 days and in some cases atrophy of the ovaries after 8 days. Usually, levels of 1–5 ppm zearalenone in the feed are enough to induce these physiological responses in swine. At lower levels, zearalenone and various derivatives actually stimulate growth in farm animals (Hesseltine 1976).

The occurrence of zearalenone is related to low temperatures and the invasion of grain by various species of *Fusarium*, particularly *F. graminearum* or *F. roseum* (perfect stage, *G. zeae*). These organisms invade developing corn at silking stage in periods of heavy rainfall and proliferate on mature grains that have not dried because of wet weather at harvest or on grains that are stored wet (Tuite *et al.* 1974; Caldwell and Tuite 1970, 1974).

In two general surveys of U.S. corn for zearalenone, Shotwell *et al.* (1970, 1971) found the mycotoxin in about 1% of the samples examined at an average level of 625 µg/kg (range 450–800 µg/kg). Subsequently, the FDA (Eppley *et al.* 1974) conducted a concentrated survey in an area where there was evidence of *F. roseum* damage. Zearalenone was found in 17% of 223 samples assayed, at an average level of 0.9 mg/kg (range 0.1–5.0 mg/kg). Zearalenone has also been found in feed grains from Finland, Denmark, France, England, Mexico, and Yugoslavia (Stoloff 1976; Hesseltine 1976).

Trichothecenes

The 12,13-epoxy-Δ^9-trichothecenes have been implicated in a variety of mycotoxicoses involving both humans and animals on a large scale; diseases include alimentary toxic aleukia, stachybotryotoxicosis, moldy corn toxicosis, and the refusal-vomition phenomenon (Ciegler 1975).

In the only natural isolation reported, Hsu *et al.* (1972) identified T-2 toxin as the cause of a lethal toxicosis in Wisconsin dairy cattle. These cattle had consumed feed containing 60% corn molded with *F. tricinctum.* The cows had extensive hemorrhaging on the serosal surface of all internal viscera, typical of previously reported cases of moldy corn poisoning (Smalley 1973).

Reports of vomiting in animals and humans caused by the consumption of moldy wheat, barley, and flour go back beyond the early 1900's (Ciegler 1975; Hesseltine 1976). Curtin and Tuite (1966) demonstrated that extracts of corn naturally infected with *G. zeae* caused emesis in pigs and also possibly caused refusal. Both effects have been observed naturally in the midwestern U.S. However, the causative agent eluded detection until 1973 when Vesonder *et al.* (1973) isolated a new trichothecene, vomitoxin, from corn infected in the field with *F. graminearum*. From barley invaded with Fusaria, Yoshizawa and Morooka (1973) isolated deoxynivalenol which is structurally the same as vomitoxin. Subsequently, Vesonder *et al.* (1976) demonstrated that vomitoxin was responsible for refusal. Vomitoxin does not appear to cause hemorrhaging and is less potent in causing dermal necrosis than is T-2 toxin.

CONTROL OF FUNGAL DETERIORATION IN GRAIN

During storage of grain, the primary aim is to prevent deterioration in quality. Generally, this is done by reducing the moisture content (drying) to a level too low for fungi to grow. However, reduced temperatures in combination with low moisture are even more effective for preventing fungal deterioration. A low, uniform moisture content coupled with a low, uniform temperature (aeration) reduces the possibility of moisture transfer within the bulk and adds to the storage life of grain (Christensen and Kaufmann 1974). Other control methods include limiting the O_2 content or increasing the CO_2 content of the atmosphere, chilling, treating with chemicals, and combinations of these methods. In addition, time is an important consideration because fungal deterioration of stored grain is a dynamic process and control becomes more complicated as length of storage increases.

High-temperature drying to prevent deterioration has closely paralleled the growth of mechanical harvesting (combining) of high-moisture grain (Foster 1973). However, as fuel and electrical power for operating driers become less available and more expensive, alternate processes will be required. Fortunately, there appear to be some alternatives. Shove (1973) demonstrated that a 7° C temperature rise is sufficient to dry shelled corn to 13–15% moisture in Illinois from late October through the middle of December. The data indicate that the energy requirement for low-temperature drying was considerably less than for normal drying procedures. Nevertheless, there are risks involved in this type of drying. When unheated air is blown upward through the grain from a perforated floor, the drying front moves upward more slowly than with heated air. If the drying front reaches the top and all grain is dried before any spoils, the process is a success. A bad effect of drying grain with unheated air is that fungal deterioration may occur because of failure to quickly reach a low enough moisture level.

Application of solar energy to low-temperature grain drying is currently receiving much attention and appears to be feasible (Foster and Peart 1976). However, its availability depends on weather. Several alternate systems, differing mostly in the design of collectors, are available for drying farm products with solar energy.

Refrigerated or chilled storage of grain is another alternative, but this requires considerable energy, and the risks are appreciable (Burrell 1974B). Ensilation or storage of high-moisture grain in air-tight systems involves organic acid-producing fermentations and is exemplified by the Harvestore®[1]. This system is successfully being used for animal feeds (Hyde 1974).

Still another alternative for minimizing fungal losses in storage and

---

[1]The mention of firm names or trade products does not imply that they are endorsed or recommended by the U.S. Department of Agriculture over other firms or similar products not mentioned.

energy use in drying is the preservation of high-moisture grain with chemicals (Sauer and Burroughs 1974). Propionic acid is widely used in Europe to prevent mold and bacterial activity on damp grain in open storage. Feeding trials with cattle, pigs, sheep, and poultry were satisfactory (Hutson 1968). In the U.S., this procedure is also being used and several companies are marketing propionic acid and combinations of acetic, propionic, and formaldehyde for treating high-moisture grain to be fed on-farm. The feedlot performance of beef and dairy cattle consuming acid-treated grain has been excellent (Lane 1972; Perry 1972; Wilson 1972).

A novel approach for the storage of high-moisture grain, which I feel has considerable potential for reducing storage losses at a low cost, involves the introduction of a small amount of gaseous or liquid preservative into grain to control microbial growth during ambient air drying. In our initial test (Nofsinger et al. 1977) intermittent application of gaseous ammonia to 560 bu of high-moisture corn permitted low flow (1.8 m³/min/1,000 kg) ambient drying (Fig. 8.7). Moisture content was reduced from 23.3% to

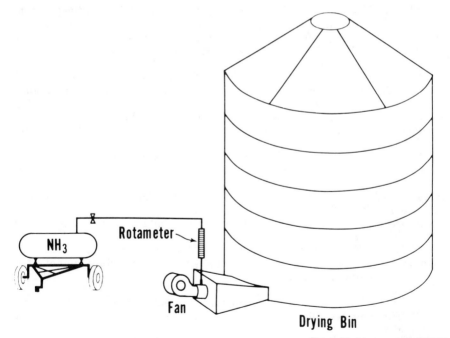

From Nofsinger et al. (1977)

FIG. 8.7. DIAGRAM OF GASEOUS AMMONIA EMPLOYED IN AMMONIA-SUPPLEMENTED AMBIENT TEMPERATURE DRYING OF HIGH-MOISTURE CORN

17.7% in 56 days and mold growth was effectively controlled throughout 6 months' storage. Feedlot steers gained fast (2.86 lb/day) and efficiently (6.57 lb feed/lb gain) on a ration formulated with this corn. Currently, larger scale tests using ammonia, formaldehyde, and methylene-*bis*-propionate to supplement ambient air drying of high-moisture corn are encouraging from both a cost and an energy conservation viewpoint.

BIBLIOGRAPHY

AGRICULTURAL STABILIZATION AND CONSERVATION SERVICE. 1976. Report "Plan of operation for test shipment of CSM to India, MV Jalamani" prepared by Charles M. McGuire, Deputy Director, Programs, U.S. Department of Agriculture, ASCS, Commodity Office, Shawnee Mission, KS.

ARMOLIK, N., DICKSON, J. G. and DICKSON, A. D. 1956. Deterioration of barley in storage by microorganisms. Phytopathology *46*, 457–461.

AYERST, G. 1969. The effects of moisture and temperature on growth and spore germination in some fungi. J. Stored Prod. Res. *5*, 127–141.

BAILEY, C. H. and GURJAR, A. M. 1918. Respiration of stored wheat. J. Agric. Res. *12*, 685–713.

BAYFIELD, E. G. and O'DONNELL, W. W. 1945. Observations on the thiamine content of stored wheat. Food Res. *10*, 485–488.

BILAY, V. I. 1960. Mycotoxicoses of man and agricultural animals. Kiev, Translation distributed by Office of Technical Services, U.S. Department of Commerce, Washington, D.C.

BOTHAST, R. J., ADAMS, G. H., HATFIELD, E. E. and LANCASTER, E. B. 1975A. Preservation of high-moisture corn: A microbiological evaluation. J. Dairy Sci. *58*, 386–391.

BOTHAST, R. J., GREGG, G. and LIST, G. R. Unpublished data. The effect of fungal fermentation on the carbonyl compounds of soybean oil. To be submitted to J. Food Sci.

BOTHAST, R. J., LANCASTER, E. B. and HESSELTINE, C. W. 1975B. *Scopulariopsis brevicaulis*: Effect of pH and substrate on growth. Eur. J. Appl. Microbiol. *1*, 55–66.

BOTTOMLEY, R. A., CHRISTENSEN, C. M. and GEDDES, W. F. 1950. Grain storage studies. 9. The influence of various temperatures, humidities, and oxygen concentrations on mold growth and biochemical changes in stored yellow corn. Cereal Chem. *27*, 271–296.

BOTTOMLEY, R. A., CHRISTENSEN, C. M. and GEDDES, W. F. 1952. Grain storage studies. 10. The influence of aeration, time, and moisture content on fat acidity, nonreducing sugars, and mold flora of stored yellow corn. Cereal Chem. *29*, 53–64.

BURRELL, N. J. 1974A. Aeration. *In* Storage of Cereal Grain and Their Products. C. M. Christensen (Editor). American Association of Cereal Chemists, Inc., St. Paul, MN.

BURRELL, N. J. 1974B. Chilling. *In* Storage of Cereal Grain and Their Products. C. M. Christensen (Editor). American Association of Cereal Chemists, Inc., St. Paul, MN.

CALDWELL, R. W. and TUITE, J. 1970. Zearalenone production in field corn in Indiana. Phytopathology *60*, 1696–1697.

CALDWELL, R. W. and TUITE, J. 1974. Zearalenone in freshly harvested corn. Phytopathology *64*, 752–753.

CARLYLE, R. E. and NORMAN, A. G. 1941. Microbial thermogenesis in the decomposition of plant materials. Part 2. Factors involved. J. Bacteriol. *41*, 699–724.

CHOUDHURY, H., CARLSON, C. W. and SEMENIUK, G. 1970. Toxic effects of ochratoxin on laying hens. Poult. Sci. *49*, 1374–1375 (Abstr).

CHRISTENSEN, C. M. 1955. Grain storage studies. 21. Viability and moldiness of commercial wheat in relation to the incidence of germ damage. Cereal Chem. *32*, 507–518.

CHRISTENSEN, C. M. 1957. Deterioration of stored grain by fungi. Bot. Rev. *23*, 108–134.

CHRISTENSEN, C. M. 1962. Invasion of stored wheat by *Aspergillus ochraceus*. Cereal Chem. *39*, 100–106.

CHRISTENSEN, C. M. 1970. Moisture content, moisture transfer, and invasion of stored sorghum seeds by fungi. Phytopathology *60*, 280–283.

CHRISTENSEN, C. M. 1972. Microflora and seed deterioration. *In* Viability of Seeds. E. H. Roberts (Editor). Chapman and Hall Ltd., London.

CHRISTENSEN, C. M. and DORWORTH, C. E. 1966. Influence of moisture content, temperature, and time on invasion of soybeans by storage fungi. Phytopathology *56*, 412–418.

CHRISTENSEN, C. M. and DRESCHER, R. F. 1954. Grain storage studies. 14. Changes in moisture content, germination percentage, and moldiness of wheat samples stored in different portions of bulk wheat in commercial bins. Cereal Chem. *31*, 206–216.

CHRISTENSEN, C. M. and KAUFMANN, H. H. 1969. Grain storage: The role of fungi in quality loss. University of Minnesota Press, Minneapolis, MN.

CHRISTENSEN, C. M. and KAUFMANN, H. H. 1974. Microflora. *In* Storage of Cereal Grain and Their Products. C. M. Christensen (Editor). American Association of Cereal Chemists, Inc., St. Paul, MN.

CHRISTENSEN, C. M., NELSON, G. H. and MIROCHA, C. J. 1965. Effect on the white rat uterus of a toxic substance isolated from *Fusarium*. Appl. Microbiol. *13*, 653–659.

CHRISTENSEN, C. M., NELSON, G. H., MIROCHA, C. J. and BATES, F. 1968. Toxicity to experimental animals of 943 isolates of fungi. Cancer Res. *28*, 2293–2295.

CIEGLER, A. 1975. Mycotoxins: Occurrence, chemistry, biological activity. Lloydia *38*, 21–35.

CIEGLER, A., KADIS, S. and AJL, S. J. 1971. Microbial Toxins. 6. Fungal Toxins. Academic Press, New York.

CURRIE, J. A. and FESTENSTEIN, G. N. 1971. Factors defining spontaneous heating and ignition of hay. J. Sci. Food Agric. *22*, 223–230.

CURTIN, T. M. and TUITE, J. 1966. Emesis and refusal of feed in swine associated with *Gibberella zeae*-infected corn. Life Sci. *5*, 1937–1944.

DAFTARY, R. D. and POMERANZ, Y. 1965. Changes in lipid composition in wheat during storage deterioration. J. Agric. Food Chem. *13*, 442–446.

DAFTARY, R. D., POMERANZ, Y. and SAUER, D. B. 1970. Changes in wheat flour damaged by mold during storage. Effects on lipid, lipoprotein, and protein. J. Agric. Food Chem. *18*, 613–616.

DETROY, R. W., LILLEHOJ, E. B. and CIEGLER, A. 1971. Aflatoxin and related compounds. *In* Microbial Toxins. A. Ciegler, S. Kadis and S. J. Ajl (Editors), Vol.

6. Academic Press, New York.

DeVAY, J. E. 1952. A note on the effect of mold growth and increased moisture content on the free amino acids in hard red spring wheat. Cereal Chem. *29*, 309–311.

DORWORTH, C. E. and CHRISTENSEN, C. M. 1968. Influence of moisture content, temperature, and storage time upon changes in fungus flora, germinability, and fat acidity values of soybeans. Phytopathology *58*, 1457–1459.

DOUNIN, M. 1926. The Fusariosis of cereal crops in Europian Russia in 1923. Phytopathology *16*, 305–308.

EPPLEY, R. M., STOLOFF, L., TRUCKSESS, M. W. and CHUNG, C. W. 1974. Survey of corn for *Fusarium* toxins. J. Assoc. Off. Anal. Chem. *57*, 632–635.

FIELDS, R. W. and KING, T. H. 1962. Influence of storage fungi on deterioration of stored pea seed. Phytopathology *52*, 336–339.

FORGACS, J. 1972. Stachybotryotoxicosis. *In* Microbial Toxins. Vol. 8. Fungal Toxins. S. Kadis, A. Ciegler, and S. J. Ajl (Editors). Academic Press, New York.

FOSTER, G. H. 1973. Heated-air grain drying. *In* Grain Storage: Part of a System. R. N. Sinha and W. E. Muir (Editors). Avi Publishing Co., Westport, CT.

FOSTER, G. H. and PEART, R. M. 1976. Solar grain drying. USDA Agriculture Information Bulletin, No. 401, p. 14.

FRAZIER, W. C. 1967. Food Microbiology, 2nd ed. McGraw-Hill Book Company, New York.

GLASS, R. L., PONTE, J. G., JR., CHRISTENSEN, C. M. and GEDDES, W. F. 1959. Grain storage studies. 28. The influence of temperature and moisture level on the behavior of wheat stored in air or nitrogen. Cereal Chem. *36*, 341–356.

GOLDBLATT, L. A. 1969. Aflatoxin: Scientific Background, Control and Implications. Academic Press, New York.

GOODMAN, J. J. and CHRISTENSEN, C. M. 1952. Grain storage studies. 11. Lipolytic activity of fungi isolated from stored corn. Cereal Chem. *29*, 299–308.

GRAY, K. R., SHERMAN, K. and BIDDLESTONE, A. J. 1971. A review of composting. Part 1. Process Biochem. *6*(6): 32–36.

GREGORY, P. H., LACEY, M. E., FESTENSTEIN, G. N. and SKINNER, F. A. 1963. Microbial and biochemical changes during the moulding of hay. J. Gen. Microbiol. *33*, 147–174.

HESSELTINE, C. W. 1976. Mycotoxins other than aflatoxins. *In* Proc. 3rd Int. Biodegradation Symp. J. M. Sharpley and A. M. Kaplan (Editors). Applied Science Publishers Ltd., London.

HETHERINGTON, A. C. and RAISTRICK, H. 1931. Studies on biochemistry of microorganisms. 11. On the production and chemical constitution of a new yellow coloring matter, citrinin, produced from glucose by *Penicillium citrinum* Thom. Phil. Trans. Roy. Soc. Ser. B. *220*, 269–297.

HOLMAN, L. H. 1950. Handling and storage of soybeans. *In* Soybeans and Soybean Products. K. S. Markley (Editor). Vol. 1. Interscience Publishers, Inc., New York, NY.

HSU, I. C., SMALLEY, E. B., STRONG, F. M. and RIBELIN, W. B. 1972. Identification of T-2 toxin in moldy corn associated with a lethal toxicosis in dairy cattle. Appl. Microbiol. *24*, 684–690.

HUMMEL, B. C. W., CUENDET, L. S., CHRISTENSEN, C. M. and GEDDES, W. F. 1954. Grain storage studies. 12. Comparative changes in respiration, viability, and chemical composition of mold-free and mold-contaminated wheat upon storage. Cereal Chem. *31*, 143–150.

HUTSON, J. J. 1968. Cereal preservation with propionic acid. Process Biochem. *3*(11), 31–32.

HYDE, M. B. 1974. Airtight storage. *In* Storage of Cereal Grain and Their Products. C. M. Christensen (Editor). American Association of Cereal Chemists, Inc., St. Paul, MN.

JANICKI, L. J. and GREEN, V. E. JR. 1976. Rice losses during harvest drying and storage. Il. Riso *25*, 333–338.

JOFFE, A. Z. 1965. Toxin production by cereal fungi causing toxic alimentary aleukia in man. *In* Mycotoxins in Foodstuffs. G. N. Wogan (Editor). Massachusetts Institute of Technology Press, Cambridge, MA.

JOFFE, A. Z. 1971. Alimentary toxic aleukia. *In* Microbial Toxins. 7. Algal and Fungal Toxins. S. Kadis, A. Ciegler and S. J. Ajl (Editors). Academic Press, New York.

JOHNSON, H. E. 1957. Cooling stored grain by aeration. Agric. Eng. *38*, 597–601.

KADIS, S., CIEGLER, A. and AJL, S. J. 1971. Microbial Toxins. 7. Algal and Fungal Toxins. Academic Press, New York.

KADIS, S., CIEGLER, A. and AJL, S. J. 1972. Microbial Toxins. 8. Fungal Toxins. Academic Press, New York.

KAO, C. and ROBINSON, R. J. 1972. *Aspergillus flavus* deterioration of grain: Its effect on amino acids and vitamins in whole wheat. J. Food Sci. *37*, 261–263.

KEYL, A. C. and BOOTH, A. N. 1971. Aflatoxin effects in livestock. J. Am. Oil Chem. Soc. *48*, 599–604.

KROBER, O. A. and COLLINS, F. I. 1948. Effect of weather damage on the chemical composition of soybeans. J. Am. Oil Chem. Soc. *25*, 296–298.

KROGH, P. 1969. The pathology of mycotoxicoses. J. Stored Prod. Res. *5*, 259–264.

KROGH, P., AXELSEN, N. H., ELLING, F., GYRD-HANSEN, N., HALD, B., HYLDGAARD-JENSEN, J., LARSEN, A. F., MADSEN, A., MORTENSEN, H. P., MOLLER, T., PETERSON, O. K., RAVNSKOV, U., ROSTGAARD, M. and AALAND, O. 1974. Experimental porcine nephropathy. Acta Pathol. Microbiol. Scand. Sect. A Suppl. *246*, 1–21.

KROGH, P., HASSELAGER, E. and FRIIS, P. 1970. Studies on fungal nephrotoxicity. 2. Isolation of two nephrotoxic compounds from *Penicillium viridicatum* Westling: Citrinin and oxalic acid. Acta Pathol. Microbiol. Scand. B. *78*, 401–413.

KURTZMAN, C. P. and CIEGLER, A. 1970. Mycotoxin from a blue-eye mold of corn. Appl. Microbiol. *20*, 204–207.

LANCASTER, E. B. and BOTHAST, R. J. 1976. Treating maize with ammonia—a controlled storage experiment. J. Stored Prod. Res. *12*, 171–175.

LANCASTER, M. C., JENKINS, F. P. and McL. PHELP, J. 1961. Toxicity associated with certain samples of groundnuts. Nature *192*, 1095–1096.

LANE, G. T. 1972. Preventing mold growth in high moisture grain. *In* Master Manual on Molds and Mycotoxins. G. L. Berg (Editor). Farm Technol. Agri-Fieldman *28*(5), 34a–41a.

LEVI, P., TRENK, H. L. and MOHR, H. K. 1974. Study of the occurrence of ochratoxin A in green coffee beans. J. Assoc. Off. Anal. Chem. *57*, 866–870.

LILLEHOJ, E. B., CIEGLER, A. and DETROY, R. W. 1970. Fungal toxins. *In* Essays in Toxicology, Vol. 2, F. R. Blood (Editor). Academic Press, New York.

LILLEHOJ, E. B., KWOLEK, W. F., FENNELL, D. I. and MILBURN, M. S. 1975A. Aflatoxin incidence and association with bright greenish-yellow fluorescence and insect damage in a limited survey of freshly harvested high-moisture corn. Cereal Chem. *52*, 403–412.

LILLEHOJ, E. B., KWOLEK, W. F., SHANNON, G. M., SHOTWELL, O. L. and HESSELTINE, C. W. 1975B. Aflatoxin occurrence in 1973 field corn. 1. A limited survey in the southeastern U.S. Cereal Chem. *52*, 603–611.

LILLEHOJ, E. B., FENNELL, D. I. and KWOLEK, W. F. 1977. Aflatoxin and *Aspergillus flavus* occurrence in 1975 corn at harvest from a limited region of Iowa. Cereal Chem. *54*, 366–372.

LINKO, P. and MILNER, M. 1959. Free amino and keto acids of wheat grains and embryos in relation to water content and germination. Cereal Chem. *36*, 280–294.

LIST, G. R., EVANS, C. D., WARNER, K., BEAL, R. E., KWOLEK, W. F., BLACK, L. T. and MOULTON, K. J. 1977. Quality of oil from damaged soybeans. J. Am. Oil Chem. Soc. *54*, 8–14.

LOEB, J. R. and MAYNE, R. Y. 1952. Effect of moisture on the microflora and formation of free fatty acids in rice bran. Cereal Chem. *29*, 163–175.

LOPEZ, L. C. and CHRISTENSEN, C. M. 1967. Effect of moisture content and temperature on invasion of stored corn by *Aspergillus flavus*. Phytopathology *57*, 588–590.

LYNCH, B. T., GLASS, R. L. and GEDDES, W. F. 1962. Grain storage studies. 32. Quantitative changes occurring in the sugars of wheat deteriorating in the presence and absence of molds. Cereal Chem. *39*, 256–262.

MAYER, C. F. 1953. Endemic Panmyelotoxicosis in the Russian Grain Belt. Mil. Surg. *113*, 295–315.

MERWE, K. J., van der, STEYN, P. S. and FOURIE, L. 1965. Mycotoxins. 2. The constitution of ochratoxins A, B, and C. metabolites of *Aspergillus ochraceus* Wilh. J. Chem. Soc. 7083–7088.

MILNER, M., CHRISTENSEN, C. M. and GEDDES, W. F. 1947. Grain storage studies. 6. Wheat respiration in relation to moisture content, mold growth, chemical deterioration, and heating. Cereal Chem. *24*, 182–199.

MILNER, M. and GEDDES, W. F. 1945. Grain storage studies. 2. The effect of aeration, temperature, and time on the respiration of soybeans containing excessive moisture. Cereal Chem. *22*, 484–501.

MILNER, M. and GEDDES, W. F. 1946. Grain storage studies. 4. Biological and chemical factors involved in the spontaneous heating of soybeans. Cereal Chem. *23*, 449–470.

MILNER, M. and GEDDES, W. F. 1954. Respiration and heating. *In* Storage of Cereal Grains and Their Products. J. A. Anderson and A. W. Alcock (Editors). American Association of Cereal Chemists, St. Paul, MN.

MIROCHA, C. J., CHRISTENSEN, C. M. and NELSON, G. H. 1967. Estrogenic metabolite produced by *Fusarium graminearum* in stored corn. Appl. Microbiol. *15*, 497–503.

MIROCHA, C. J., CHRISTENSEN, C. M. and NELSON, G. H. 1968. Physiologic activity of some fungal estrogens produced by *Fusarium*. Cancer Res. *28*, 2319–2322.

MISLIVEC, P. B. and TUITE, J. 1970. Species of *Penicillium* occurring in freshly harvested and in stored dent corn kernels. Mycologia *62*, 67–74.

NAGEL, C. M. and SEMENIUK, G. 1947. Some mold-induced changes in shelled corn. Plant Physiol. *22*, 20–33.

NOFSINGER, G. W., BOTHAST, R. J., LANCASTER, E. B. and BAGLEY, E. B. 1977. Ammonia-supplemented ambient temperature drying of high-moisture corn. Trans. ASAE. *20*, 1151–1154, 1159.

NORMAN, A. G., RICHARDS, L. A. and CARLYLE, R. E. 1941. Microbial thermogenesis in the decomposition of plant materials. 1. An adiabatic fermentation apparatus. J. Bacteriol. *41*, 689–697.

PAPAVIZAS, G. C. and CHRISTENSEN, C. M. 1957. Grain storage studies. 25. Effect of invasion by storage fungi upon germination of wheat seed and upon development of sick wheat. Cereal Chem. *34*, 350–359.

PERRY, T. W. 1972. Improving feed efficiency with organic acids. *In* Master Manual on Molds and Mycotoxins. G. L. Berg (Editor). Farm Technol. Agri-Fieldman *28*(5), 42a–45a.

PETERSON, A., SCHLEGEL, V., HUMMEL, B., CUENDET, L. S., GEDDES, W. F. and CHRISTENSEN, C. M. 1956. Grain storage studies. 22. Influence of oxygen and carbon dioxide concentrations on mold growth and grain deterioration. Cereal Chem. *33*, 53–66.

POMERANZ, Y. 1974. *In* Storage of Cereal Grain and Their Products. C. M. Christensen (Editor). American Association of Cereal Chemists, Inc., St. Paul, MN.

POMERANZ, Y., HALTON, P. and PEERS, F. G. 1956. The effects on flour dough and bread quality of molds grown in wheat and those added to flour in the form of specific cultures. Cereal Chem. *33*, 157–169.

PURCHASE, I. H. F. 1971. Symposium on Mycotoxins in Human Health. Macmillan Press Ltd., London.

PURCHASE, I. F. H. and THERON, J. J. 1968. The acute toxicity of ochratoxin A to rats. Food Cosmet. Toxicol. *6*, 479–483.

QASEM, S. A. and CHRISTENSEN, C. M. 1960. Influence of various factors on the deterioration of stored corn by fungi. Phytopathology *50*, 703–709.

RAMSTAD, P. E. and GEDDES, W. F. 1942. The respiration and storage behavior of soybeans. Minn. Agric. Exp. Sta. Tech. Bull. 156.

RODRICKS, J. V. 1976. Mycotoxins and Other Fungal Related Food Problems. Advances in Chemistry Series 149. American Chemical Society, Washington, D.C.

SAUER, D. B. and BURROUGHS, R. 1974. Efficacy of various chemicals as grain mold inhibitors. Trans. ASAE *17*, 557–559.

SCHROEDER, H. W. and SORENSON, J. W. JR. 1961. Mold development of rough rice as affected by aeration during storage. Rice J. *64*, 8–10, 12, 21–23.

SCOTT, P. M. 1973. Mycotoxins in stored grain, feeds, and other cereal products. *In* Grain Storage: Part of a System. R. N. Sinha and W. E. Muir (Editors). Avi Publishing Co., Westport, Conn.

SCOTT, P. M., van WALBEEK, W., KENNEDY, B. and ANYETI, D. 1972. Mycotoxins (ochratoxin A, citrinin, and sterigmatocystin) and toxigenic fungi in grains and other agricultural products. J. Agric. Food Chem. *20*, 1103–1109.

SEMENIUK, G. 1954. Microflora. *In* Storage of Cereal Grains and Their Products. J. A. Anderson and A. W. Alcock (Editors). American Association of Cereal Chemists, St. Paul, MN.

SHOTWELL, O. L., HESSELTINE, C. W., BURMEISTER, H. R., KWOLEK, W. F., SHANNON, G. M. and HALL, H. H. 1969A. Survey of cereal grains and soybeans for the presence of aflatoxin. 1. Wheat, grain sorghum, and oats. Cereal Chem. *46*, 446–454.

SHOTWELL, O. L., HESSELTINE, C. W., BURMEISTER, H. R., KWOLEK, W. F., SHANNON, G. M. and HALL, H. H. 1969B. Survey of cereal grains and soybeans for the presence of aflatoxin. 2. Corn and soybeans. Cereal Chem. *46*, 454–463.

SHOTWELL, O. L., HESSELTINE, C. W. and GOULDEN, M. L. 1969C. Note on the natural occurrence of ochratoxin A. J. Assoc. Off. Anal. Chem. *52*, 81–83.

SHOTWELL, O. L., HESSELTINE, C. W. and GOULDEN, M. L. 1973. Incidence of aflatoxin in southern corn, 1969–1970. Cereal Sci. Today *18*, 192–195.

SHOTWELL, O. L., HESSELTINE, C. W., GOULDEN, M. L. and VANDEGRAFT, E. E. 1970. Survey of corn for aflatoxin, zearalenone, and ochratoxin. Cereal Chem. *47*, 700–707.

SHOTWELL, O. L., HESSELTINE, C. W., VANDEGRAFT, E. E. and GOULDEN, M. L. 1971. Survey of corn from different regions for aflatoxin, ochratoxin, and zearalenone. Cereal Sci. Today *16*, 266–273.

SHOVE, G. C. 1973. New techniques in grain conditioning. *In* Grain Storage: Part of a System. R. N. Sinha and W. E. Muir (Editors). Avi Publishing Co., Westport, Conn.

SMALLEY, E. B. 1973. T-2 toxin. J. Am. Vet. Med. Assoc. *163*, 1278–1281.

STEELE, J. L., SAUL, R. A. and HUKILL, W. V. 1969. Deterioration of shelled corn as measured by carbon dioxide production. Trans. ASAE *12*, 685–689.

STOB, M., BALDWIN, R. S., TUITE, J., ANDREWS, F. W. and GILLETTE, K. B. 1962. Isolation of an anabolic, uterotrophic compound from corn infected with *Gibberella zeae*. Nature *196*, 1318.

STOLOFF, L. 1976. Occurrence of mycotoxins in foods and feeds. *In* Mycotoxins and Other Fungal Related Food Problems. J. V. Rodricks (Editor). Advances in Chemistry Series 149. American Chemical Society, Washington, D.C.

TABAK, H. A. and COOKE, W. B. 1968. Growth and metabolism of fungi in an atmosphere of nitrogen. Mycologia *60*, 115–140.

THOMPSON, R. A. and FOSTER, G. H. 1963. Stress cracks in artificially dried corn. U.S. Dep. Agric. Mark. Res. Rep. *631*, 1–24.

TROLLE, B. and PEDERSEN, H. 1971. Grain quality research committee under the Danish Academy of Technical Sciences. Summary of a report on the activities of the committee. J. Inst. Brew. (London) *77*, 338–348.

TUITE, J., SHANER, G., RAMBO, G., FOSTER, J. and CALDWELL, R. W. 1974. The *Gibberella* ear rot epidemics of corn in Indiana in 1965 and 1972. Cereal Sci. Today *19*, 238–241.

URRY, W. H., WEHRMEISTER, H. L., HODGE, E. B. and HIDY, P. H. 1966. The structure of zearalenone. Tetrahedron Lett. *1966*, 3109–3114.

VESONDER, R. F., CIEGLER, A. and JENSEN, A. H. 1973. Isolation of the emetic principle from *Fusarium*-infected corn. Appl. Microbiol. *26*, 1008–1010.

VESONDER, R. F., CIEGLER, A., JENSEN, A. H., ROHWEDDER, W. K. and WEISLEDER, D. 1976. Co-identity of the refusal and emetic principle from *Fusarium*-infected corn. Appl. Environ. Microbiol. *31*, 280–285.

WALLACE, H. A. H. 1973. Fungi associated with stored grain. *In* Grain Storage: Part of a System. R. N. Sinha and W. E. Muir (Editors). Avi Publishing Co., Westport, Conn.

WEDBERG, S. E. and RETTGER, L. F. 1941. Factors influencing microbial thermogenesis. J. Bacteriol. *41*, 725–743.

WILSON, L. L. 1972. What is the nutritional value of organic acids? *In* Master Manual on Molds and Mycotoxins. G. L. Berg (Editor). Farm Technol. Agri-Fieldman *28*(5), 50a–53a.

WOGAN, G. N. 1968. Aflatoxin risks and control measures. Fed. Proc. *27*, 932–938.

YOSHIZAWA, T. and MOROOKA, N. 1973. Deoxynivalenol and its monoacetate:

New mycotoxins from *Fusarium roseum* and moldy barley. Agric. Biol. Chem. *37*, 2933–2934.

ZELENY, L. 1954. Chemical, physical, and nutritive changes during storage. *In* Storage of Cereal Grains and Their Products. J. A. Anderson and A. W. Alcock (Editors). American Association of Cereal Chemists, St. Paul, MN.

Native Structure of Edible Plant Tissues and Effects of Disruption

CHAPTER 9

STRUCTURE OF CEREAL GRAINS AS RELATED TO END-USE PROPERTIES

Y. POMERANZ

U.S. Grain Marketing Research Center
Agricultural Research Service

and

D. B. BECHTEL

U.S. Department of Agriculture
Manhattan, KS 66502

ABSTRACT

Kernel structure is of great significance in producing, storing, marketing and processing of cereal grains. A general outline of the kernel structure, of the hull and bran layers, the germ (embryonic axis and scutellum), and of cereal starches is given. The implications of kernel structure in minimizing damage during grain harvest, drying, storage, and handling in marketing channels; in optimizing milling of wheat and rice, in germination and in malting, and in enhancing the nutritional value and end-use properties of cereal grains are reviewed.

For optimum utilization of cereal grains knowledge of their structures and compositions is required. The practical implications of kernel structure are numerous (Pomeranz 1976A). They relate to the various stages of grain production, harvest, storage, marketing, and use. Some of the implications are listed in Table 9.1.

KERNEL STRUCTURE—GENERAL

The cereal grain is a one-seeded fruit called a caryopsis, in which the fruit coat is adherent to the seed. As the fruit ripens, the pericarp (fruit wall) becomes firmly attached to the wall of the seed proper. The pericarp, seed coats, nucellus, and aleurone cells form the bran. The embryo occupies only a small part of the seed. The bulk of the seed is taken up by the endosperm, which constitutes a food reservoir (Pomeranz and MacMasters 1968, 1970).

The floral envelopes (modified leaves known as lemma and palea), or chaffy parts, within which the caryopsis develops, persist to maturity in

244

TABLE 9.1
SOME IMPLICATIONS OF KERNEL STRUCTURE

Significance in	Parameter	Effect	Commodity
Threshing	Germ damage or skinning	Reduced germin-ability, impaired storability	All cereal grains
Drying	Cracks, fissures, and breakage; hardening	Reduced commercial value; lowered grade, impaired storability, dust formation reduced starch yield	Mainly corn and rice
	Discoloration	Reduced commercial value, lowered grade	Mainly rice
Marketing	Breakage	Reduced commercial value in food processing	Mainly corn and rice
General use	High husk: caryopsis ratio or high pericarp: endosperm ratio	Reduced nutritional value—as food or feed	All cereal grains
General use	Kernel shape and dimensions propor-tions of tissues in the kernel, distri-bution of nutrients in the tissues	Yield of food prod-ucts; nutritional value of cereal (or cereal prod-ucts) as food or feed	All cereal grains
Malting	Germ damage, skinning, or inadequate husk adherence	Reduced germinabili-ty, uneven malting	Mainly barley
Milling	Uneven surface, deep crease or uneven aleurone	Reduced milling yield	Mainly wheat and rice
Milling	Steely texture	Increased power re-quirements, starch damage, high water absorption, difficulty in air-classification	Wheat and malt milling
Germination-Malting	Starch granule size	Uneven degradation	All cereal grains
Consumption-Nutrition	Distribution and composition of proteins	Change in nutritional value	All cereal grains

the grass family. If the chaffy structures envelope the caryopsis so closely that they remain attached to it when the grain is threshed (as in rice and most varieties of oats and barley), the grain is considered to be covered. However, if the caryopsis readily separates from the floral envelopes on threshing, as with common wheats, rye, hull-less barleys, and the common varieties of corn, the grain is considered to be naked.

The structure of the wheat kernel is shown in Fig. 9.1. The dorsal side of

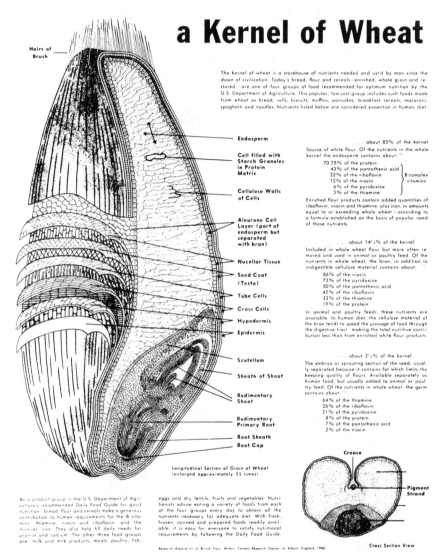

a Kernel of Wheat

Hairs of Brush

The kernel of wheat is a storehouse of nutrients needed and used by man since the dawn of civilization. Today's bread, flour and cereals - enriched, whole grain and restored - are one of four groups of food recommended for optimum nutrition by the U.S. Department of Agriculture. This popular, low-cost group includes such foods made from wheat as bread, rolls, biscuits, muffins, pancakes, breakfast cereals, macaroni, spaghetti and noodles. Nutrients listed below are considered essential in human diet.

Endosperm

Cell filled with Starch Granules in Protein Matrix

Cellulose Walls of Cells

Aleurone Cell Layer (part of endosperm but separated with bran)

Nucellar Tissue

Seed Coat (Testa)

Tube Cells

Cross Cells

Hypodermis

Epidermis

Scutellum

Sheath of Shoot

Rudimentary Shoot

Rudimentary Primary Root

Root Sheath

Root Cap

. . . about 83% of the kernel
Source of white flour. Of the nutrients in the whole kernel the endosperm contains about:

    70-75% of the protein
    43% of the pantothenic acid
    32% of the riboflavin   } B-complex
    12% of the niacin   } vitamins
    6% of the pyridoxine
    3% of the thiamine

Enriched flour products contain added quantities of riboflavin, niacin and thiamine, plus iron, in amounts equal to or exceeding whole wheat — according to a formula established on the basis of popular need of those nutrients.

. . . about 14½% of the kernel
Included in whole wheat flour but more often removed and used in animal or poultry feed. Of the nutrients in whole wheat, the bran, in addition to indigestible cellulose material contains about:

    86% of the niacin
    73% of the pyridoxine
    50% of the pantothenic acid
    42% of the riboflavin
    33% of the thiamine
    19% of the protein

In animal and poultry feeds, these nutrients are available. In human diet, the cellulose material of the bran tends to speed the passage of food through the digestive tract - making the total nutritive contribution less than from enriched white flour products

. . . about 2½% of the kernel
The embryo or sprouting section of the seed, usually separated because it contains fat which limits the keeping quality of flours. Available separately as human food, but usually added to animal or poultry feed. Of the nutrients in whole wheat, the germ contains about:

    64% of the thiamine
    26% of the riboflavin
    21% of the pyridoxine
    8% of the protein
    7% of the pantothenic acid
    2% of the niacin

Longitudinal Section of Grain of Wheat (enlarged approximately 35 times)

Crease

Pigment Strand

Cross Section View

As a product group in the U.S. Department of Agriculture's recommended Daily Food Guide for good nutrition, bread, flour and cereals make a generous contribution to human requirements for the B vitamins, thiamine, niacin and riboflavin and the mineral, iron. They also help fill daily needs for protein and calcium. The other three food groups are milk and milk products, meats, poultry, fish, eggs and dry lentils; fruits and vegetables. Nutritionists advise eating a variety of foods from each of the four groups every day to obtain all the nutrients necessary for adequate diet. With fresh, frozen, canned and prepared foods readily available, it is easy for everyone to satisfy nutritional requirements by following the Daily Food Guide.

Research Association of British Flour Millers Cereals Research Station, St. Albans, England, 1960

From Wheat Flour Institute, Chicago, IL

FIG. 9.1. LONGITUDINAL AND CROSS SECTION OF A WHEAT KERNEL
Enlarged approximately 35 ×

the wheat grain is rounded, while the ventral side has a deep groove or crease along the entire longitudinal axis. At the apex or small end (stigmatic end) of the grain is a cluster of short, fine hairs known as brush hairs. The pericarp, or dry fruit coat, consists of four layers: the epidermis, hypodermis, cross cells, and tube cells. The remaining tissues of the grain are the inner bran (seed coat and nucellar tissue), endosperm, and embryo (germ). The aleurone layer consists of large rectangular, heavy-walled, supposedly starch-free cells. Botanically, the aleurone is the outer layer of the endosperm, but as it tends to remain attached to the outer coats during wheat milling, it is shown in the diagram as the innermost bran layer.

The embryonic axis consists of the plumule and radicle, which are connected by the mesocotyl. The scutellum, serves as an organ for food storage. The outer layer of the scutellum, the epithelium, may function as either a secretory or an absorption organ. In a well-filled wheat kernel, the germ comprises ~ 2–3% of the kernel, the bran 13–17%, and the endosperm the remainder. The inner bran layers (the aleurone) are high in protein, whereas, the outer bran (pericarp, seed coats and nucellus) is high in cellulose, hemicelluloses, and minerals; biologically, the outer bran functions as a protective coating and remains practically intact when the seed germinates. The germ is high in proteins, lipids, sugars (chiefly sucrose), and minerals; the endosperm consists largely of starch grains surrounded by protein.

Grains of other cereals are similar in structure to wheat. The corn grain is the largest of all cereals (Fig. 9.2). The kernel is flattened, wedge-shaped, and broader at the apex than at its attachment to the cob. The aleurone cells contain much protein and oil and also contain the pigments that make certain varieties appear blue, black, or purple. Two types of starchy endosperms—horny and floury—are found beneath the aleurone layer. The horny endosperm is harder and contains a higher level of protein. In dent corn varieties, the horny endosperm is found on the sides and back of the kernel and bulges in toward the center at the sides. The floury endosperm fills the crown (upper part) of the kernel, extends downward to surround the germ, and shrinks as dent corn matures; the shrinking causes an indentation at the top of the kernel. In a typical dent corn, the pericarp comprises ~ 6%, the germ 11%, and the endosperm 83% of the kernel. Flint corn varieties contain more horny than floury endosperm. A small amount of floury endosperm extends around the top of the kernel, with the bulk of the floury portion around the embryo.

The common varieties of oats have the fruit (caryopsis) enveloped by a hull composed of certain floral envelopes. Naked or hull-less oat varieties are not extensively grown. In light thin oats, hulls may comprise as much as 45% of the grain; in very heavy or plump oats, they may represent only 20%. The hull normally makes up ~ 30% of the grain. Oat kernels with the hulls removed are called groats.

Rice is a covered cereal; in the threshed grain (or rough rice), the kernel is enclosed in a tough siliceous hull, which renders it unsuitable for human consumption. When this hull is removed during milling, the kernel (or

a Kernel of Corn

The kernel of corn is a fruit enclosing a single seed—a type known botanically as a caryopsis. The corn plant is from the grass family Gramineae, genus Zea. Corn kernels vary considerably in size and shape among the different types of the corn plant, and even from the same cob. Their color, too, ranges from white to orange, cherry-red, red, dark red, brown or variegated.

In milling corn, the kernels are separated into three main parts: hull, endosperm and germ—by a process similar to that used for milling wheat into flour. The germ, containing most of the oil, is difficult to separate. Wet milling (for industrial purposes) separates about 70 percent of the oil. Dry milling (for food purposes) removes only about 28 percent.

Hull

- Epidermis
- Mesocarp
- Cross cells
- Tube cells
- Seed Coat (Testa)
- Aleurone Layer (part of endosperm but separated with bran)
- Horny Endosperm
- Floury Endosperm
- Cells filled with Starch Granules in Protein Matrix
- Walls of Cells
- Scutellum
- Plumule or Rudimentary Shoot and Leaves
- Radicle or Primary Root

Longitudinal Section of a Grain of Corn (enlarged approximately 30 times)

Tip Cap

PERICARP

These outer layers of the kernel constitute the fruit coat, which protects the seed. The pericarp constitutes about six percent of the whole kernel, and consists largely (73 percent) of insoluble non-starch carbohydrates, with 16 percent fiber, 7 percent protein and 2 percent oil.

ENDOSPERM

The corn miller seeks the separation and grinding of this portion to obtain corn meal. The endosperm comprises about 80 to 84 percent of the kernel, and contains 85 percent starch, 12 percent protein. Kernels of corn have both hard (horny), outer endosperm, as well as soft, inner endosperm. Classification of corn into types is based on characteristics of the endosperm, proportions of soft and hard, as well as kind of carbohydrate contained.

GERM

The germ is found in the lower portion of the endosperm, and comprises about 10 to 14 percent of the kernel. Most of the oil in the corn kernel (81-86 percent) is in the germ, although this portion also has some protein and carbohydrates.

Corn—a Versatile Grain

Corn is an increasingly important grain in the world, because of its high yields per acre, its versatility as food for man and animals, and its importance as a source of starch, oil and sweeteners. Corn has proved itself highly susceptible to genetic manipulation and control by agronomists—affecting the nutrients it contains, its yields, its growth periods, and the climatic conditions in which it will flourish.

The U.S. leads the world in corn production, with a harvest of over 99 million tons (over 4 billion bushels) in 1965, or almost half of the total world production.

Milled for human food, the coarsest grind is popularly known as "grits"—also as corn grits and hominy grits—consisting of large fragments of endosperm relatively free of germ and hulls.

Corn meal, more finely ground, has a wide variation in granulation, or size of endosperm chunks, depending on regional preferences. Corn flour is a fine grind, which can be produced instead of, or together with grits or meal. In processing corn, there is almost no waste—there are uses and demands for all parts of the kernel. Corn can be called the most efficient and economical of the cereal grains.

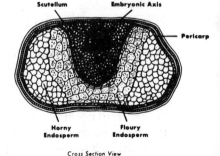

Scutellum | Embryonic Axis | Pericarp | Horny Endosperm | Floury Endosperm

Cross Section View

FIG. 9.2.   LONGITUDINAL AND CROSS SECTION OF A CORN KERNEL
Enlarged approximately 30 ×

caryopsis), comprised of the pericarp (outer bran) and the seed proper (inner bran, endosperm, and germ), is known as brown rice or sometimes as unpolished rice. Brown rice is in little demand as a food. Unless stored under very favorable conditions, it tends to become rancid and is more subject to insect infestation than the various forms of milled white rice. When brown rice is subjected to further milling processes, the bran and germ are removed and the purified endosperms are marketed as white rice or polished rice. Milled rice is classified as head rice (whole endosperm), and various classes of broken rice known as second head, screenings, and brewers' rice, in order of decreasing size.

HULL AND BRAN LAYERS

The structure and adherence of the hull are important in protecting the germinating grain and in the malting process. One reason that barley is uniquely suited for malting is that a cementing layer is present between the hull and the caryopsis. The hull restricts excessive seedling growth without adversely affecting the desirable enzymic degradation of insoluble high molecular weight materials. The adhering hull also protects the seedling from mechanical damage during turning of the malt, and provides a filtration bed during the extraction of soluble malt components in the mashing process.

The hull, as such or as a result of the high concentration of silica, can slow the attack of storage insects on rice and barley. The palea and lemma in barley are held together by two hook-like structures (Fig. 9.3). In rice the ability of these structures to hold the palea and lemma together without gaps is probably variety dependent. Varieties of rice that had many gaps and separations had greater insect infestations than did varieties with tight husks (Russell 1968; Cohen and Russell 1970; Breese 1960; McGaughey 1970).

Apparently the bran (pericarp, seed coats, nucellus and aleurone) affords little protection against insect infestations, because more insects consistently develop in brown rice than in either rough rice or milled rice. Furthermore, infestation in brown rice is unaffected by variety. Lack of resistance to infestation is probably due to the thinness of the bran, which allows easy insect penetration, and to the large quantities of lipid and protein present in the aleurone which provide nourishment to the insects.

The outer pericarp layers of wheat (epidermis and hypodermis) have no intercellular spaces and are composed of closely adhering, thick-walled cells. The inner layers of the pericarp, on the other hand, consist of thinner-walled cells and often contain intercellular spaces, through which water can move rapidly and in which molds are commonly found (MacMas-

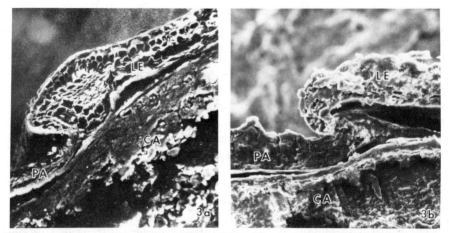

FIG. 9.3. SEM MICROGRAPH TRANSVERSE SECTION, THROUGH THE PALEA (PA), LEMMA (LE), AND CARYOPSIS (CA) OF (a) BARLEY AND (b) RICE (130 ×)

ters 1962). Similarly, molds can enter through the large intercellular spaces at the base of the kernel where the grain was detached from the plant at harvest and where there is no protective epidermis.

The structure of the pericarp, seed coats, and nucellus also explains how the kernel reacts to water. After initial rapid water absorption, the rate decreases significantly (Jones 1949). Hinton (1955) found that the seed coat offered more water resistance than the nucellus. Morrison (1975) studied the development of the seed coat and nucellar cuticles of wheat and found that the cuticle formed by the seed coat was thicker than the one formed by the nucellus. Bechtel and Pomeranz (In press), finding that in mature rice the caryopsis had a thick nucellar cuticle, suggested that the rice seed coat cuticle formed differently from that of wheat. The structure of the caryopsis coat explains its reaction to water. Water is readily absorbed by the sponge like pericarp but is prevented from readily entering the kernel interior by the seed coat and nucellar cuticles. Since the cuticles are deficient in the attachment region, water can enter readily the kernel, mainly via the germ (Krauss 1933). The ability of the germ to absorb and hold considerable amounts of water probably accounts, in part, for the susceptibility of the germ to attack by molds (MacMasters 1962).

It is well known that an intact grain stores much better than a damaged or ground grain and that deteriorative changes (*i.e.*, rancidity, off-flavors, etc.) occur slowly in the whole grain but quite rapidly in ground grain. The hull, apparently, prevents rancidity by protecting the bran layers from mechanical damage during harvesting and subsequent handling.

Once rough rice is dehulled, it rapidly becomes rancid, primarily because of the oxidation of free fatty acids released by the action of lipase (Houston 1972). The lipids and lipase are normally compartmentalized in the aleurone and germ cells. Cell disruption may cause mixing of the cellular constituents. Bechtel and Pomeranz (In press) found that rice aleurone cells were easily disrupted, and that the disruption resulted in the fusion of lipid bodies. Possibly, the dehulling process and subsequent handling of dehulled rice disrupt the aleurone cell and allow the rice to become rancid. The hull, therefore, appears to be necessary to prevent cell disruption during harvesting, storage, and handling.

Mature sorghum may have nucellar material all over the kernel. In some varieties, none commercially important in the U.S.A., the nucellar material carries an objectionable water-soluble pigment, which interferes with processing of sorghum for starch production. When the pigment becomes associated with the starch, it may turn blue-gray with change in acidity (MacMasters 1962).

THE GERM

The site of the germ in the kernel and the extent to which the germ is protected by adjacent layers determine whether it will be retained intact during threshing and, thus, the usefulness of the grain for seeding or malting. The ease with which the germ is removed from the caryopsis during milling depends on several factors.

The germ is a separate structure and generally can be easily separated from the rest of the cereal grain. However, the scutellar epithelium (located next to the endosperm) has fingerlike cells, which in wheat are attached to one another for about one-third of their length (MacMasters 1962). The free ends protrude toward the adjacent starch endosperm cells. The protruding epithelial cells may form an amorphous cementing layer between germ and endosperm. If part of the layer projects into the spaces between the fingerlike cells of the scutellar epithelium and into the folds of the scutellar structure, it may be difficult to separate the germ from the endosperm unless the cementing layer is softened. The softening may be accomplished by steeping, as in corn wet milling, or by conditioning, as in wheat milling. In rice, a layer of crushed cells separating the scutellar epithelium from the starchy endosperm provides a line of easy fracture; hence the germ can be removed intact with minimum effort (Fig. 9.4 and 9.5). Also, cell walls of the starchy endosperm are not well defined, and this structural defect allows the germ to be removed easily.

Germ separation is also facilitated by the fact that the germ takes up

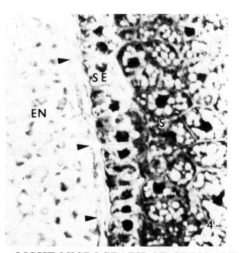

FIG. 9.4.   LIGHT MICROGRAPH OF GERM AND EN-
DOSPERM IN RICE
Scutellar epithellium (SE) and fibrous zone (arrows) sepa-
rate the scutellum (S) from endosperm (EN). (800 ×)

water faster and swells more readily than the endosperm (MacMasters
1962). The strains resulting from differential swelling contribute to easy
separation in milling.

GRINDING AND MILLING

The starchy endosperm is the portion of the caryopsis that is used
primarily as a food source. In the milling of wheat and rice, the bran layers
and the germ are separated, thereby removing much of the lipid and
protein.

Endosperm structure, composition, and associated texture govern both
power requirements in milling and particle size of the products. In wheat
flour milling, the size and shape of the crease affect yield and composition
of flour because the bran in the crease area is difficult to separate from the
starchy endosperm (Fig. 9.6). The thickness and irregularity of the
aleurone around the starchy endosperm and in the crease make it impossible to increase flour yields in wheat milling by processes that successively
"peel" outer layers. The thickness and tenacity with which the pericarp is
bound to the aleurone layer may be responsible for difficulties in the
milling of new European wheat cultivars that yield well in the field but
create problems in processing (Finney et al. 1977).

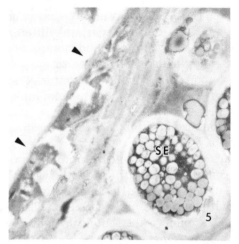

FIG. 9.5.   TRANSMISSION ELECTRON MICROGRAPH
OF RICE GERM DISSECTED FROM WHOLE
CARYOPSIS
Note that germ fractured between endosperm and fibrous
zone (arrows). SE, scutellar epithellium. (6,600 ×)

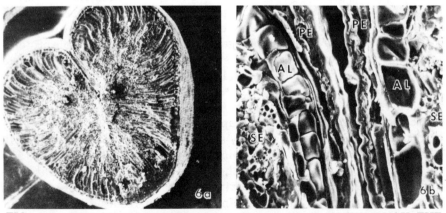

FIG. 9.6.   SEM MICROGRAPH OF (a) CROSS SECTION THROUGH THE
WHEAT KERNEL (× 25) AND (b) THROUGH THE CREASE AREA IN A RYE
KERNEL, PERICARP (PE), ALEURONE (AL), AND STARCHY ENDOSPERM
(SE) (225 ×)

Moss (1973) studied the morphology of three hard and three soft Australian wheat varieties. He found that varietal differences in the rate of penetration of the water used in the conditioning of the wheat prior to milling and the milling yields were related to:

1. The thickness and composition of the outer cuticle and testa.
2. The extent to which the outer epidermal and inner parenchymal cells had been compressed, and
3. The number and size of protein bodies present in the subaleurone endosperm cells.

Moss (1977) used an autoradiographic technique to study the distribution of water in wheat conditioning. Within 1 hr the water had penetrated the aleurone cells and, in many cases, also the starchy endosperm to a depth of 50–60 μm. The embryonic axis and scutellum also absorbed the water rapidly. The cells of the embryo and scutellum bound water more strongly than did those of the aleurone.

The irregular thickness of the aleurone layer in wheat prevents easy separation of all the starchy endosperm from the aleurone. The cells of the starchy endosperm are large and irregularly shaped; in milling the cells are broken. Cell walls in the starchy inner endosperm can be broken with greater ease than those in the protein-rich outer endosperm. The outer cells are smaller and their walls are thicker than those of the inner cells.

Wheat varieties differ in vitreosity and hardness of their mature endosperms. The importance of grain hardness lies in its effect on the milling properties and, in particular, on the amounts of semolina and farina and of damaged starch produced during milling. According to MacMasters and Waggle (1963), particles from the horny endosperm break along the walls of the cells and usually have sharp outlines. Floury endosperm breaks largely through the cell contents (rather than chiefly along the walls of the cells) and yields particles with irregular edges. The water-soluble proteins appear to be important in the genetic determination of wheat hardness (Barlow et al. 1973; Simmonds et al. 1973; Hoseney and Seib 1973). According to Stenvert and Kingswood (1977), however, differences in hardness are related to the continuity of the protein matrix and the strength with which the protein physically entraps starch granules.

Pomeranz (1976B) compared, by scanning electron microscopy (SEM), the physical changes that occurred in a soft white wheat ("Nu Gaines") and in a vitreous durum wheat when they were ground. The soft wheat shattered under the conditions of experimental grinding much more than the vitreous durum wheat did. Grinding soft wheat resulted in a clear-cut separation of the starch granules from the protein matrix in which the granules are imbedded; no clear-cut cleavage line was evident in the vitreous durum wheat when it was ground. The milled particles were a mixture of starch and surrounding protein matrix, and many of the exposed starch granules were damaged. Although some starch granules in the soft wheat did break, the amount was relatively small compared with the starch damage of the hard durum wheat.

This difference in starch damage of the two wheats during grinding is known to have many important technological consequences. Thus, the clean separation of undamaged starch makes soft wheats the preferred material for air classification—a process in which differences in specific gravity are used to separate finely pulverized wheat flours into protein-high and protein-low fractions. Air classification is the basis of numerous patents for producing "tailor-made" flours from a single flour.

Another consequence of the difference in starch damage is that water absorption is greater in the mechanically damaged starch than in the undamaged starch. That is of particular importance to bakers in countries in which the amount of water added to a dough is essentially limited only by the amount required to obtain a dough of adequate consistency for handling and for producing a well-developed bread of adequate shelf life.

Damaged starch granules in flour from hard wheat are more susceptible to enzymatic degradation by amylases than are those from soft wheats; that is important in the production of sugar-free breads in which fermentable sugars are primarily those produced by the action of malt on damaged starch.

The wheats selected to demonstrate the large differences in particle size were the extremes in the spectrum of hardness—soft wheat for producing cakes and cookies and durum wheat for alimentary pastes (Pomeranz 1976B). Differences among wheat cultivars of the same class or even among wheats representing the commonly grown bread wheat classes (hard red winter and hard red spring) would be much smaller. Proper conditioning of wheat prior to milling also may reduce the differences.

GRAIN DRYING

Changes in texture and structure during the drying of corn and rice are important in minimizing breakage during handling. Excessive cracks reduce value of corn for producing foods such as breakfast cereals. Harsh heat treatment during grain drying may reduce starch yields, impair quality of the starch, and create difficulties in corn wet-milling. Used for alcoholic beverages, overheated corn also may cause difficulties in beer brewing and in distillation. The starch granules are imbedded in a proteinaceous matrix that hardens during overheating. The hardened matrix prevents the starch from enzyme attack and conversion to alcohol. In rice milling, harsh drying and accompanying structural cracks substantially reduce yields of head rice and increase the amounts of "brokens," and thus cause economic losses to the miller. The method of rice drying may also affect the texture and color of the milled rice and result in off-color, especially objectionable browning.

NUTRITIONAL IMPLICATIONS

The chemical composition of different cereal grains varies widely, since it is influenced by genetic, soil, cultural and climatic factors. Amounts of proteins, lipids, carbohydrates, pigments, vitamins and total ash vary; mineral elements present also vary widely. Cereals are characterized by relatively low protein and high carbohydrate contents; the carbohydrates consist essentially of starch (90% or more), pentosans, and sugars.

The various components are not uniformly distributed in the kernel. The hulls and bran are high in cellulose, pentosans and ash; the germ is high in lipid and rich in proteins, sugars, and generally, ash. The endosperm, which contains the starch, has a lower protein content than the germ and the bran (in some cereals), and is low in fat and ash.

The cereal grains contain water-soluble proteins (albumins), salt-soluble proteins (globulins), alcohol-soluble proteins (prolamins), and acid- and alkali-soluble proteins (glutelins). The prolamins, together with the glutelins, constitute the bulk of the proteins of cereal grains. In corn, the prolamin called zein is the chief protein; in oats and rice, which are low in prolamins, glutelins make up the bulk of the proteins. The glutelin of oats is called avenin, and the glutelin of rice, oryzenin. Wheat, rye, and barley contain intermediate amounts of prolamins and glutelins.

The various proteins are not distributed uniformly in the kernel. Thus, the proteins fractionated from the inner endosperm of wheat consist chiefly of approximately equal amounts of prolamins (gliadins) and glutelins (glutenins). The embryo proteins consist of nucleoproteins, albumin (leucosin), globulins, and proteoses; in wheat bran prolamins predominate with smaller quantities of albumins and globulins.

The protein of milled rice is unique among the proteins of cereal grains in that it contains at least 80% glutelins. The other protein fractions are 5% albumin, 10% globulin, and less than 5% prolamin. The nutritionally limiting amino acid in cereal grains is lysine. In practice, however, the chief nutritional limitation of rice is its low protein content.

Breeding efforts to improve the nutritive value of cereal grains have been concentrated on increasing protein content without decreasing protein quality (mainly retaining lysine concentration in the protein). The significance of protein distribution in the endosperm depends on the type of product that is likely to be consumed. In the production of highly refined milled products, in which some of the subaleurone layer is removed, a high concentration of protein in the subaleurone layer would not be desirable. However, if the whole kernel is to be consumed, distribution of protein in the kernel is of limited nutritional consequence.

In all cereal grains, the storage protein forms a matrix which surrounds the starch granules. The concentration of protein increases from the inner to the outer starchy endosperm. The increase may be relatively gradual, as in some soft wheats (Kent 1966); or quite steep, as in some high-protein wheat types in which some of the outer subaleurone cells contain few, if any, starch granules. The subaleurone region of the rice endosperm contains much higher concentrations of protein than does the central endosperm. Three types of protein bodies are found in the subaleurone region; only one type in the central endosperm. Since the subaleurone region is only several layers thick and lies directly below the aleurone, the subaleurone layer can be easily removed during milling. It is, therefore, desirable either to mill rice as lightly as possible for a consumer acceptable product, or to breed cultivars with an increased subaleurone layer, or cultivars with a more even distribution of protein throughout the endosperm. The distribution of protein in the endosperm of high-protein rices is more uniform than in low-protein rices. High-protein rices tend to have lower levels of some of the essential amino acids, particularly lysine, than do low-protein rices of the same variety. The drop in lysine, however, is less than proportional to the increase in protein content.

Wolf et al. (1967) found that the protein network in endosperm cells of regular corn was composed of an amorphous matrix in which were imbedded zein-rich granules averaging about 2 μm in diameter. In high-lysine corn, submicroscopic granules (about $1/20$ that in normal corn) had a much lower zein content than did granules in regular corn. The small size of subcellular protein granules in high-lysine maize, as compared with regular corn, correlated with the differences in zein content of the two types of corn.

Sullins and Rooney (1974) used SEM to illustrate differences in corn endosperm structure that account for differences in nutritive value of the grain. High-lysine corn has a reduced amount of protein bodies in the endosperm. SEM of soft endosperms for normal, opaque-2 or modified opaque-2 corn showed loosely packed, nearly round starch granules associated with thin sheets of protein and many intergranular air spaces (Robutti et al. 1974). The hard endosperms had tightly packed, polygonal starch granules associated with a continuous protein matrix, and no intergranular air spaces. Normal hard endosperms had zein bodies embedded in the protein matrix; modified hard endosperms did not. Starch damage was greater in the hard endosperm than in the soft because of a stronger adhesion between starch and protein. The low density and opaqueness of soft endosperm were attributed to the intergranular air spaces. Interaction between protein matrix and starch granules during drying explains the shape of starch granules.

Seckinger and Wolf (1973) studied the structure of grain sorghum endosperm protein of commercial hybrids and experimental lines with the transmission electron microscope (TEM) and SEM. Vitreous endosperm showed a well developed, two-component structure consisting of concentric-ringed protein bodies (2–3 μm in diameter) embedded in an amorphous matrix protein. On the basis of solubility properties of the proteins, they suggested that the protein bodies were the site of prolamin (kafferin) deposition and that the matrix protein was the site of glutelin deposition. Distribution of protein within the sorghum grain was similar to that within other cereal grains in that the peripheral vitreous area of the kernel had the highest protein content. Interior areas had gradually decreasing amounts of protein. Protein bodies accounted for 70–80% of the sorghum protein, as determined by microscopic observations.

The structure of grain sorghum samples representing a wide genetic base was examined by SEM (Hoseney et al. 1973). The soft or opaque endosperm was characterized by relatively large intergranular air spaces. The starch granules were essentially round and covered with a thin sheet of protein. Embedded in the protein sheet were relatively large spherical protein bodies. The hard or translucent endosperm portion was characterized by a tightly packed structure with no air spaces. The starch granules were polygonal and covered with a thick protein matrix. Embedded in the protein matrix were protein bodies. It was suggested that in sorghum, as in wheat, kernel hardness is caused by strong adhesion between protein and starch. When fractured, many starch granules broke, but little breakage occurred at the starch-protein interface. A dwarf variety from Sudan was found to have relatively few protein bodies in the endosperm. Yet because those bodies proved to be kafferin, a prolamin low in lysine, it was surmised and confirmed that the sample was high in lysine.

Sullins and Rooney (1973) conducted light microscopy (LM) and SEM studies of the peripheral endosperm of waxy and nonwaxy endosperm sorghum varieties. Sorghum varieties are known to differ widely in endosperm type (i.e., yellow, sugary, waxy, and nonwaxy). In feeding trials sorghum grains with waxy endosperm tended to have higher feed efficiencies than nonwaxy varieties. Sullins and Rooney found that the subaleurone endosperm area of sorghum was composed of starch granules embedded in an amorphous protein matrix that contained relatively indigestible (alcohol-soluble) protein bodies. The waxy sorghum varieties contained fewer spherical protein bodies and were, therefore, more digestible than the nonwaxy varieties. Because of its low relative proportion of protein bodies, waxy grain, apparently, is more completely broken down than is nonwaxy during processing (*i.e.*, steam-flaking, micronizing, pulverizing, popping, exploding, and reconstituting). The protein differences also may contribute to the difference in feed efficiency between waxy and nonwaxy sorghum grains.

STARCHES

SEM has revealed that starch granules from wheat endosperm vary considerably in shape and surface structure (Evers 1969). Evers demonstrated that an equatorial groove was present in large granules but absent in small granules. That finding provided evidence in support of the different types of starch suggested by results of TEM, and was further evidence for the homology of granules of intermediate size in the subaleurone layer with larger granules in the inner starchy endosperm. In a preliminary study on the structure of wheat starch granules Evers and McDermott (1970) examined in the SEM modifications resulting from alpha-amylolysis. They found that when present in starch granules, the groove was generally the site of amylolytic attack. Complete penetration of alpha-amylase either occurred exclusively at that site or also at other sites randomly located on the major surface.

Baked products and alimentary pastes are affected adversely if made from flour milled from sprouted wheat. Jones and Bean (1972) found that amylases from sprouted wheat attacked wheat starch granules preferentially along the equatorial groove of the granules. When the central part of the granule was penetrated, the starch in that region was digested rapidly. It was suggested that damaged starch granules may easily break up to yield a large number of fragments that would already be primed with amylase. The degraded granules would ultimately yield a dough unsuitable for producing bread or noodles.

Bean *et al*. (1974) found that enzymes from a malted wheat flour extract entered wheat starch granules at certain sites, often preferring an equatorial groove, degrading some of the surface layer, and then following a path toward the interior of the granules. Some granules were attacked at many surface sites. The enzyme then apparently followed a path of least resistance toward the interior layers of the granules. This mode of attack had been postulated previously by Sandstedt (1955) using the LM.

Varietal differences in the surface ultrastructure of rice endosperm cells and in the ultrastructure of native and modified starch granules were examined by SEM (Anon. 1975). Micrographs of the fractured face of rice endosperm verified that opacity of nonwaxy rice and of the endosperm in crumbly rice was caused by the loose arrangement of cell contents. By contrast, only a portion (ventral region) of the waxy rice endosperm was mealy, although the whole endosperm was opaque. Crumbly rice showed predominantly intercellular cleavage planes.

Native starch granules were mostly angular in all samples except crumbly rice. They differed in degree of indentation and signs of attack by α-amylase. Corroding the granules with 2.2N HCl for 4 days at 35° C little changed their surface structure, although 31–47% of their weight was lost. By contrast, α-amylolysis resulted in extensive surface corrosion and in varietal differences in corrosion pattern. Waxy granules showed the fastest and most uniform pitting by α-amylolysis; there were many holes per granule. Nonwaxy granules showed various degrees of corrosion within a sample. Extent of acid corrosion was correlated negatively with gelatinization temperature of starch.

GERMINATION AND MALTING

In malting barley for brewing, the grain is modified into a product that can yield an aqueous extract containing: (a) fermentable products, (b) available substrate for yeast nutrition, and (c) precursors for imparting the desirable organoleptic qualities to the beer (Preece 1954).

The sum total of physical and chemical changes taking place during malting is termed "modification." According to MacLeod (1967), "modification" describes "a rather nebulous but nonetheless real condition which has resulted from the transformation of endospermic constituents to give the best possible material for mashing." In practice, the modification conditions of malting are selected so that yield of extractable solids is maximum, and malting losses and excessive degradation of the high-molecular-weight components of the barley are minimum. Modification transforms tough barley into friable malt. The transformation can be assessed by physical methods ranging from the simple biting test to tests involving elaborate self-recording mechanical devices. Among chemical indices, the increase in soluble proteins is probably the most important single parameter.

Dronzek et al. (1972) used both SEM and LM to study the changes that occur in starch granules during germination of wheat. Enzymatically degraded starch granules were observed near the aleurone layer in grains germinated for 2 days. Most of the degradation was confined to the large starch granules. Differences between the mode of enzymatic attack on the large and small granules were interpreted to indicate differences in physical structure of the granules.

Palmer's studies (1971) indicate that the modification of the endosperm in germinated barley commences at the dorsal (nonfurrowed) surface of the grain. Palmer found that the rate of endosperm modification depended more on the effective dispersal of hydrolytic enzymes than on the total amounts of these enzymes in the grain. Microscopic analyses showed that starch grains and hemicellulosic materials of the cell walls were coated with proteinaceous materials. Proteases were found to play a more active role than carbohydrases in the conversion of hard barley into friable malt.

Changes in the aleurone layer and in the starchy endosperm of steeped, malted, and kilned barley were examined by SEM (Pomeranz 1972). The surfaces of aleurone cells in steeped barley were highly pitted. The walls of aleurone cells were progressively degraded during malting and kilning. Aleurone grains increased in diameter during steeping and were further distorted during kilning. Partial breakdown of cell walls in the center of the starchy endosperm of malted barley was accompanied by extensive dissolution of the protein matrix and the "freeing" of small starch granules that previously were embedded in that matrix; the effect on the appearance of the starch granules was small. In the central endosperm of kilned barley malt, the cell-wall dissolution was extensive and was accompanied by mechanical breakdown of the large starch granules.

The SEM was used to follow the modification in malting of a low-protein barley and a high-protein cultivar (Pomeranz 1974). In the low-protein cultivar, the protein matrix degraded extensively, and some of the degraded protein was deposited in the kilned malt on large starch granules. In the high-protein cultivar, much of the protein matrix was largely intact, and some protein was retained in the form of a modified but coherent and continuous thick film covering the starch granules. It was suggested that the thick film is responsible for difficulties in malting high-protein barleys, for reduction of wort extract, and for persistence of undegraded proteins, which enhance chill haze formation in beer.

Palmer (1974) suggested on the basis of SEM that, during malting, hydrolytic enzymes migrate into the endosperm to disrupt and solubilize mainly the cell walls, complex protein materials, and the small starch granules. Satisfactory modification in malting should result in degradation of cell-wall material throughout the endosperm and release of starch and degraded protein during mashing. However, some areas of the endosperm (especially at the distal end) may contain undegraded endosperm cell walls in which starch extract can be trapped, and the trapped starch gives rise to glucan (gum) materials during mashing.

MICROBIAL DAMAGE

Cereal grains are important as food because of their excellent keeping qualities. Moisture content is the major factor in determining the storage behavior of grain, which is also influenced by temperature, oxygen supply, history and condition of the grain, length of storage, and biological factors (molds and insects). The respiratory rate of dry grain is low. As the moisture content is raised above 14%, the respiration increases gradually until a certain critical moisture is reached above which respiration accelerates rapidly and the grain tends to heat. This sharp increase in respiration is due to the germination and growth of certain molds (predominantly various species of *Aspergillus* and *Penicillium*) commonly found in soil and in previously used storage bins. Molds are invariably found on the grain and within the seed coats, even though the grain is harvested under ideal conditions.

Recent investigations have indicated that the palea is the primary site of infestation by microorganisms in oats (Pomeranz and Sachs 1972). The area between the palea and the crease seems to be favorable for growth of microorganisms, and here they are harbored in the mature and dried grain. Presence of a plaque with microbial growth (probably a slime-producing bacterial colony) beneath the palea of an oat kernel is indicated in Fig. 9.7. Fungi under the hull of rice are shown in Fig. 9.8.

FIG. 9.7.   PLAQUE WITH MICROBIAL GROWTH ON
THE INSIDE OF THE PALEA OF AN OAT KERNEL
(530 ×)

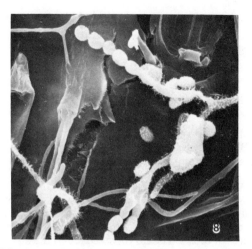

FIG. 9.8.   FUNGI UNDER THE HULL OF RICE
(2,000 ×)

IMPROVING NUTRITIONAL QUALITY BY MODIFYING GRAIN MORPHOLOGY

In recent years many studies have concerned the improvement of the nutritional value of cereal grains. Simple changes in grain morphology could be the basis of improvement. The embryo of cereal seeds is rich in protein (up to 38%), and the protein may contain about 7% lysine. Selection for larger embryos is particularly important if the whole seed (rather than starchy endosperm) is to be consumed. Variations in the number of aleurone cells of the endosperm exist in corn, rice, and barley. The aleurone layer is rich in protein having a good amino acid balance. Selection for a high aleurone cell number could be useful, provided the high number is associated with improved nutritional value. Increasing the relative surface area of the seed could lead to an increase in the number of aleurone cells. In early triticale (crosses between wheat and rye) and in *Avena sterilis* (a wild hexaploid oat), such an increase results from the otherwise undesirable development of long and thin, very small, or shrunken kernels. Both in wheat and in rice, much of the protein is concentrated in the aleurone and outermost subaleurone. Those tissues are diverted to feed during the milling and polishing of rice or during the milling of highly refined wheat flour. "Restructuring" cereal grains for a more even distribution of protein throughout the whole endosperm would increase the protein content of milled products.

The hulls are rich in fibrous materials and low in protein. Many hulless varieties of barley and oats have substantially more protein than the hulled varieties do. However, the low yield of the available hulless varieties discourages their cultivation.

BIBLIOGRAPHY

ANON. 1975. IRRI Annual report for 1975. 83–88.

BARLOW, K. K., BUTTROSE, M. S., SIMMONDS, D. H. and VESK, M. 1973. The nature of the starch-protein interface in wheat endosperm. Cereal Chem. *50*, 443–454.

BEAN, M. M., KEAGY, P. M., FULLINGTON, J. G., JONES, F. T. and MECHAM, D. K. 1974. Dried Japanese noodles. Properties of laboratory-prepared noodle doughs from sound and damaged wheat flours. Cereal Chem. *51*, 416–427.

BECHTEL, D. B. and POMERANZ, Y. (In press.) Ultrastructure of the mature ungerminated rice caryopsis. The caryopsis coat and the aleurone cells. Amer. J. Bot.

BREESE, M. H. 1960. The infestibility of stored paddy by *Sitophilus sasakii* (Tak.) and *Rhyzopertha dominica* (F.). Bull. Entomol. Res. *51*, 599–630.

COHEN, L. M. and RUSSELL, M. P. 1970. Some effects of rice varieties on the biology of the Angoumois grain moth, *Sitotraga cerealella*. Ann. Entomol. Soc. Am. *63*, 930–931.

DRONZEK, B. L., HWANG, P. and BUSHUK, W. 1972. Scanning electron microscopy of starch from sprouted wheat. Cereal Chem. *49*, 232–239.

EVERS, A. D. 1969. Scanning electron microscopy of wheat starch. 1. Entire granules. Die Staerke *21*, 96–99.

EVERS, A. D. and McDERMOTT, E. E. 1970. Scanning electron microscopy of wheat starch. 2. Structure of granules modified by alpha-amylolysis—preliminary report. Die Staerke *22*, 23–26.

FINNEY, K. F., POMERANZ, Y., BOLTE, L. C. and SHOGREN, M. D. 1977. High-yielding European wheats: determination of end-use properties. Baker's Dig. *51*(1), 28–30, 32–34, 36.

HINTON, J. J. C. 1955. Resistance of the testa to entry of water into the wheat kernel. Cereal Chem. *32*, 296–306.

HOSENEY, R. C., DAVIS, A. B. and HARBERS, L. H. 1973. Structure of grain sorghum viewed with a scanning electron microscope. Abstr. No. 150, Proc. 58th Ann. Mtg. Amer. Assoc. Cereal Chem.

HOSENEY, R. C. and SEIB, P. A. 1973. Structural differences in hard and soft wheat. Baker's Dig. *47*(6), 26–28, 56.

HOUSTON, D. F. 1972. Rice bran and polish. *In* Rice Chemistry and Technology, D. F. Houston (Editor). American Association of Cereal Chemists, Inc., St. Paul, MN.

JONES, C. R. 1949. Observations on the rate of penetration of water into the wheat grain. Milling (Liverpool) *113*, 80, 82, 84, 86.

JONES, F. T. and BEAN, M. M. 1972. A light and SEM look at enzyme-damaged wheat starch. Microscope *20*, 333–334.

KENT, N. L. 1966. Subaleurone endosperm cells of high protein content. Cereal Chem. *43*, 585–601.

KRAUSS, L. 1933. Entwicklungsgeschichte der Fruchte von *Hordeum, Triticum,*

Bromus, und *Poa* mit besonderer Berucksichtigung ihrer Samenschalen. Jahrb. Wiss. Bot. *77*, 733–808.

MacLEOD, A. M. 1967. The physiology of malting—a review. J. Inst. Brewing *73*, 146–162.

MacMASTERS, M. M. 1962. Important aspects of kernel structure. Trans. ASAE *5*(2), 247–248.

MacMASTERS, M. M. and WAGGLE, D. H. 1963. The importance of starch in the microscopic identification of cereal grains in feeds. Die Staerke *15*, 7–11.

McGAUGHEY, W. H. 1970. Effect of degree of milling and rice variety on insect development in milled rice. J. Econ. Entomol. *63*, 1375–1376.

MORRISON, I. N. 1975. Ultrastructure of the cuticular membranes of the developing wheat grain. Can. J. Bot. *53*, 2077–2087.

MOSS, R. 1973. Conditioning studies on Australian wheat. 2. Morphology of wheat and its relationship to conditioning. J. Sci. Food Agr. *24*, 1067–1076.

MOSS, R. 1977. An autoradiographic technique for the location of conditioning water in wheat at the cellular level. J. Sci. Food Agr. *28*, 23–33.

PALMER, G. H. 1974. Abrasion and acidulation processes in malting. Brewer's Dig. *49*(2), 40–48.

PALMER, G. H. 1971. Modes of action of gibberellins during malting. Proc. 13th Congr. Europ. Brewery Conv. Estoril., pp. 59–71, Elsevier, Amsterdam.

POMERANZ, Y. 1974. A note on scanning electron microscopy of low- and high-protein barley malts. Cereal Chem. *51*, 545–552.

POMERANZ, Y. 1972. Scanning electron microscopy of the endosperm of malted barley. Cereal Chem. *49*, 5–6, 9–11, 18–19.

POMERANZ, Y. 1976A. Scanning electron microscopy in food science and technology. Adv. Food Res. *22*, 206–307.

POMERANZ, Y. 1976B. Particle size of wheat in infrared analysis. TIS News *2*(1), 4–6.

POMERANZ, Y. and MacMASTERS, M. M. 1968. Structure and composition of the wheat kernel. Baker's Dig. *42*(4), 24–26, 28–29, 32.

POMERANZ, Y. and MacMASTERS, M. M. 1970. Wheat and other cereals. *In* Kirk Othmer's Encyclopedia of Chemical Technology. John Wiley Publ. Co.

POMERANZ, Y. and SACHS, I. B. 1972. Scanning electron microscopy of the oat kernel. Cereal Chem. *49*, 20–22.

PREECE, I. A. 1954. The Biochemistry of Brewing. Oliver Boyd, Edinburgh, Scotland.

ROBUTTI, J. L., HOSENEY, R. C. and WASSOM, C. E. 1974. Modified opaque-2 corn endosperms. 2. Structure viewed with a scanning electron microscope. Cereal Chem. *51*, 173–180.

RUSSELL, M. P. 1968. Influence of rice variety on oviposition and development of the rice weevil, *Sitophilus oryzae*, and the maize weevil, *S. zeamais*. Ann. Entomol. Soc. Am. *61*, 1335–1336.

SANDSTEDT, R. M. 1955. Photomicrographic studies of wheat starch. 3. Enzymatic digestion and granule structure. Cereal Chem. Suppl. *32*, 17.

SECKINGER, H. L. and WOLF, M. J. 1973. Sorghum protein ultrastructure as it relates to composition. Cereal Chem. *50*, 455–465.

SIMMONDS, D. H., BARLOW, K. K. and WRIGLEY, C. W. 1973. The biochemical basis of grain hardness in wheat. Cereal Chem. *50*, 553–562.

STENVERT, N. L. and KINGSWOOD, K. 1977. The influence of the physical structure of the protein matrix on wheat hardness. J. Sci. Food Agr. *28*, 11–19.

SULLINS, R. D. and ROONEY, L. W. 1973. Light and scanning electron micro-
    scopic studies of the peripheral endosperm area of waxy and nonwaxy endosperm
    sorghum varieties. Abstr. No. 149, Proc. 58th Ann. Mtg. Amer. Assn. Cereal
    Chem.
SULLINS, R. D. and ROONEY, L. W. 1974. Effect of endosperm structure on the
    nutritive value of cereals. Paper No. 326. Proc. 34th Ann. Mtg. Inst. Food
    Technol.
WOLF, M. J., KHOO, V. and SECKINGER, H. L. 1967. Subcellular structure of
    endosperm protein in high-lysine and normal corn. Science *157*, 556–557.

CHAPTER 10

CONTRIBUTION OF INDIVIDUAL CHEMICAL CONSTITUENTS TO THE FUNCTIONAL (BREADMAKING) PROPERTIES OF WHEAT

K. F. FINNEY

USDA North Central Region, ARS
U.S. Grain Marketing Research Laboratory
Manhattan, KS 66502

ABSTRACT

Data that relate the contribution of individual chemical constituents to the functional (breadmaking) properties of wheat are reviewed and integrated, after giving background information on the meaning of quality and the effect of formula ingredients on breadmaking potential.

INTRODUCTION

Wheat, like other cereal grains, has many natural advantages as a food. It is nutritious, concentrated, readily stored and transported, and easily processed to give highly refined foods. The products are bland, fit into countless recipes, and suit many tastes (Reitz 1967). Wheat provides about 20% of the total food calories for the people of the world. It is the main staple for about 35% of the world's population (Brown 1963).

Wheat is too often regarded as merely a starchy food crop, but it contains other valuable nutrients, notably proteins, minerals and vitamins. Actually, the amino acid yield per acre from wheat far exceeds that of animal products for every one of the essential amino acids (Hegsted 1965; Mac Gillivray and Bosley 1962). Wheat is an excellent source of nutritious protein when balanced by other foods that supply certain amino acids, such as lysine. The minerals and vitamins are significant nutritionally, especially in foods derived from whole grain products and enriched flour (Hegsted 1965; Reitz 1964). The prominence of wheat is attributable not only to its nutritive value, but also to its unique proteins. Unlike any other plant-derived food, wheat contains gluten protein, which enables a leavened dough to rise by forming a structure of minute cells that retain carbon dioxide produced during fermentation. Dough of a poor quality flour retains carbon dioxide less efficiently than that of a good quality flour. Before reviewing and presenting data that relate the contribution of individual chemical constituents to the functional (breadmaking) properties of wheat, I want to give some background information on the meaning of quality and the effect of formula ingredients on breadmaking potential.

DEFINITION OF QUALITY

Physical and chemical differences vary widely among lots and varieties of wheat. Those differences have far-reaching effects and become the basis for what is loosely referred to as "quality" whether the problem relates to testing and evaluating, cereal chemistry research, processing, or economics. Quality of a wheat is the sum of the effects of soil, climate, and seed stock on the wheat plant and kernel components, particularly gluten protein (Finney and Yamazaki 1967).

The simplest definition of wheat quality is "Wheat that is desired has good quality and wheat that is not desired has poor quality." The basic definition of wheat quality usually varies from one class of wheat to another and depends on the wheat's suitability for a given product. For example, the quality of a soft winter or white wheat variety is defined in terms of suitability for soft wheat milling and production of cakes, cookies, and crackers. The quality of a durum wheat is defined in terms of its suitability for making semolina and macaroni. Whether a hard winter or spring wheat variety has good or poor quality depends on its suitability for hard wheat milling and bread production. Durum wheat with good quality for semolina and macaroni production invariably has undesirable (poor) quality for hard wheat milling and bread production. Thus, quality of any kind of wheat cannot be expressed in a single term, because quality depends on several functional (milling, baking, processing, and physical dough) properties, each important in the production of bread, pastry, or macaroni products.

FUNCTIONAL PROPERTIES AND THE BREADMAKING TEST

When we say a good-quality wheat variety must be suitable for bread production, we are generalizing. Instead, we must consider the functional properties that determine the quality or suitability of wheat for breadmaking. Specifically, a flour of good quality for bread-baking (Table 10.1, left) should have high water absorption, a medium to medium-long mixing requirement, satisfactory mixing tolerance, and good loaf-volume potential (considering protein content), and should yield a loaf with good internal crumb grain and color.

TABLE 10.1

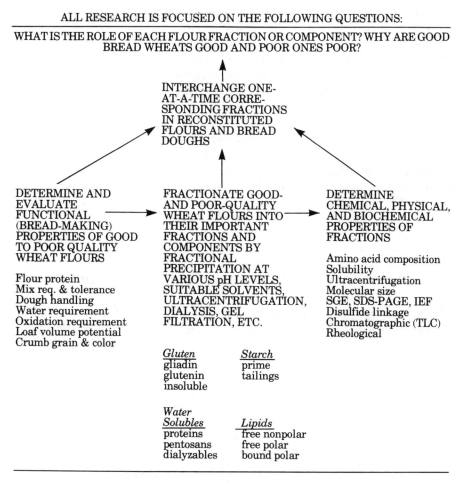

ALL RESEARCH IS FOCUSED ON THE FOLLOWING QUESTIONS:

WHAT IS THE ROLE OF EACH FLOUR FRACTION OR COMPONENT? WHY ARE GOOD BREAD WHEATS GOOD AND POOR ONES POOR?

INTERCHANGE ONE-AT-A-TIME CORRESPONDING FRACTIONS IN RECONSTITUTED FLOURS AND BREAD DOUGHS

DETERMINE AND EVALUATE FUNCTIONAL (BREAD-MAKING) PROPERTIES OF GOOD TO POOR QUALITY WHEAT FLOURS

Flour protein
Mix req. & tolerance
Dough handling
Water requirement
Oxidation requirement
Loaf volume potential
Crumb grain & color

FRACTIONATE GOOD- AND POOR-QUALITY WHEAT FLOURS INTO THEIR IMPORTANT FRACTIONS AND COMPONENTS BY FRACTIONAL PRECIPITATION AT VARIOUS pH LEVELS, SUITABLE SOLVENTS, ULTRACENTRIFUGATION, DIALYSIS, GEL FILTRATION, ETC.

DETERMINE CHEMICAL, PHYSICAL, AND BIOCHEMICAL PROPERTIES OF FRACTIONS

Amino acid composition
Solubility
Ultracentrifugation
Molecular size
SGE, SDS-PAGE, IEF
Disulfide linkage
Chromatographic (TLC)
Rheological

Gluten
gliadin
glutenin
insoluble

Starch
prime
tailings

Water Solubles
proteins
pentosans
dialyzables

Lipids
free nonpolar
free polar
bound polar

For the manifestation of loaf volume and crumb grain potentials and other functional properties, an analytical baking test such as the one developed by Finney and Barmore (1943; 1945A, B; 1948) and Finney (1945) is required. In that test, mixing time, oxidation level, fermentation time, and water absorption are optimized and balanced. Yeast is added at a level that is optimal and not limiting for the fermentation time. Shortening, sugar, and malt are added in excess so that none becomes limiting. Sufficient salt is used to produce an optimal effect, but not enough to impair gas production. Nonfat dry milk or Ardex 550 soy flour buffer the oxidative effects of potassium bromate, so that loaf volume and crumb grain potentials can be measured accurately and at a maximum of differentiation between good- and poor-volume wheat flours. The optimized test can be performed with the analytical precision of most generally accepted biological assays.

Figures 10.1 and 10.2 illustrate the effects and importance of dough ingredients on loaf volume and crumb and on differentiation between good and poor quality samples. Flour [100g, 14% moisture basis (mb)] and water (as needed) formed a dough, but its loaf volume was only 230 cc, equal to that of the dough volume after mixing. Leavening with 2% yeast increased loaf volume to 417 cc. Adding 6% sugar and then both sugar and 1.5% salt to flour, water and yeast further increased loaf volume to 583 and 732 cc, respectively, but all crumb grains were unsatisfactory to varying degrees. Those five bread ingredients are basic. When the four ingredients 4% soy flour, 0.25% malt, 50 ppm ascorbic acid, and 3% shortening were added to the five basic ingredients, a loaf of excellent volume (1006 cc) and crumb grain was produced (Fig. 10.1).

FRACTIONATING AND RECONSTITUTING TECHNIQUES

Breadmaking, which basically involves biochemical and physical-chemical systems, is an essential analytical test for interpreting the

FIG. 10.1.   RELATIVE LOAF VOLUMES AND CRUMB GRAINS OF BREADS MADE FROM FLOUR AND WATER (1) PLUS: YEAST (2), YEAST AND SUGAR (3), YEAST, SUGAR AND SALT (4), YEAST, SUGAR, SALT, MALT, ASCORBIC ACID AND SHORTENING (5), AND YEAST, SUGAR, SALT, MALT, ASCORBIC ACID, SHORTENING AND SOY FLOUR (6)

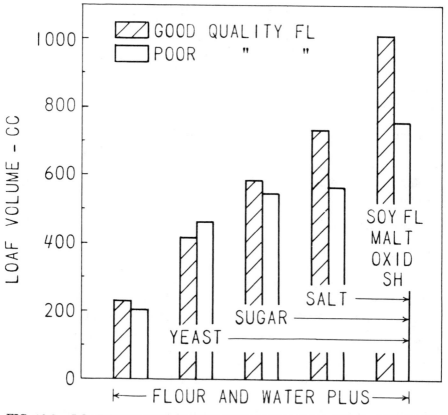

FIG. 10.2.   LOAF VOLUMES OF BREADS MADE FROM GOOD- AND POOR-
QUALITY WHEAT FLOURS (100g) AND VERY LEAN TO OPTIMIZED
FORMULATIONS

biochemical properties of such important wheat-flour components as lipid,
gluten-protein, water-soluble-protein, and starch fractions. Normal functional properties of those fractions can be demonstrated only when each
fraction is chemically and physically unaltered and is allowed to perform
and interact singly and in various combinations with the other essential
ingredients of a fermenting and optimum dough (Finney 1954). Thus,
fractionating and reconstituting techniques have been developed to bridge
the gap between functional baking research and basic biochemical research on wheat and flour. With such techniques, a wheat flour (both
petroleum-ether-extracted and unextracted) can be taken apart by fractionating it first into gluten, starch, and water-solubles, and then into
simpler components or groups of components (Table 10.1, center). The

basic criterion is that any given fractionating technique is not satisfactory until all components or fractions can be reconstituted to give a flour or dough with functional baking properties equal to those of the original flour. Thereafter, a specific flour fraction or component of both good- and poor-quality wheat can be interchanged, one at a time, to establish its contribution to one or more important functional (breadmaking) properties of wheat flour.

When a fractionating technique has proved satisfactory, the flour fractions or components involved can be characterized and evaluated by purely physical, analytical, or biochemical techniques (Table 10.1, right). All research is focused on the following questions: What is the role of each flour fraction or component? Why are good bread wheats good and poor ones poor?

Role of Gluten Protein

By fractionating and reconstituting techniques (Finney 1943), the recognized differences in loaf-volume potentials of Chiefkan (poor) and Kharkof (good) wheats were entirely accounted for by differences in their gluten-protein fractions. The data in Fig. 10.3 illustrate the significance of those techniques for studying, under conditions that eliminate the effect of environment, a given flour fraction or component (such as gluten protein) in relation to a specific property. Below about 8% protein, the relation between protein content and loaf volume is curvilinear, so that all variety curves meet at 0% protein and about 275 cc loaf volume (100g flour).

Volume-protein regression lines for many other varieties of wheat have been added to those in Fig. 10.3 to form a fan-shaped family of lines (Finney 1943; Finney and Barmore 1948). As the loaf volume increases, at a given protein content greater than about 8%, the slope of the regression line increases. The vertical distance between any two regression lines represents the difference in the gas retention of the gluten protein. The loaf volume and protein content of a single sample define its protein quality. Loaf volumes of new wheat flours of different protein contents are evaluated against those of known varieties after correcting or adjusting to an average protein level with the aid of the fan-shaped family of regression lines.

Role of Glutenin and Gliadin Fractions

Gluten that was soluble at pH 4.7 was fractionated by ultracentrifugation at $100,000 \times g$ for 5 hr (Hoseney et al. 1969A). About 15% of the protein was recovered as centrifugate (100-5C; 85% remained in the supernatant (100-5S). Starch-gel electrophoretic patterns 1, 3, and 5 (Fig.

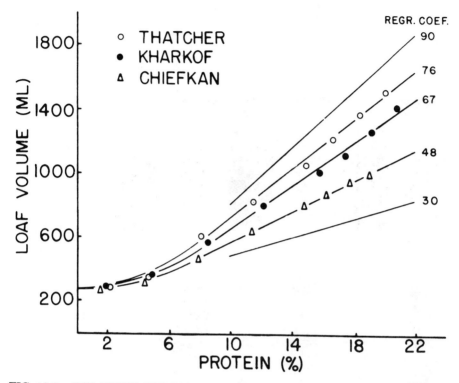

FIG. 10.3. RELATION OF LOAF VOLUME AND PROTEIN CONTENT WHEN APPLYING FRACTIONATING AND RECONSTITUTING TECHNIQUES

10.4) characterize the 100-5C fractions as proteins retained at the origin; patterns 2, 4, and 6 characterize the 100-5S fractions as proteins migrating into the starch gel. The rapidly moving bands in both fractions of each variety (of C.I. 12995, RBS, and K501099) are considered to be water-soluble and salt-soluble proteins trapped during gluten formation; therefore they are impurities. Because a constant amount of protein was used for each electrophoretic pattern, the 100-5C fraction is about seven times as concentrated as it occurs in gluten.

Fig. 10.5 shows cut loaves of bread baked from reconstituted flours containing the 100-5S and 100-5C fractions of C.I. 12995, and from reconstituted flours in which those two fractions of C.I. 12995 and of K501099 were interchanged. When reconstituted, the two fractions of C.I. 12995 interacted and produced a loaf (second row, left) fully equal to the control. When 100-5S of C.I. 12995 was reconstituted with 100-5C of K501099 (third row, left), a loaf fully equal to the control also was produced. How-

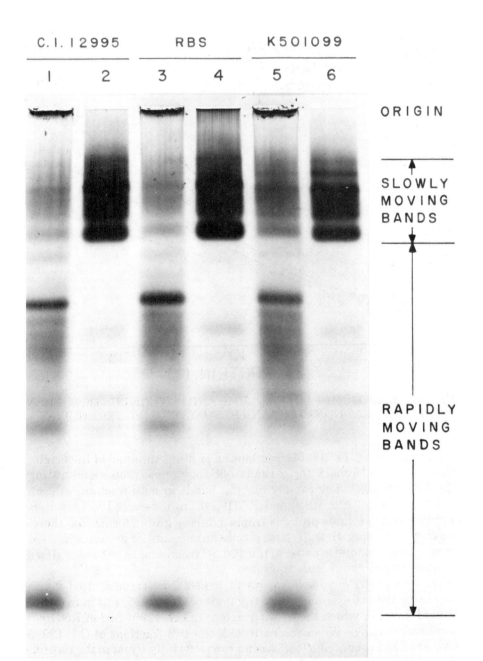

FIG. 10.4.   STARCH-GEL ELECTROPHORETIC PATTERNS OF THE 100-5C
(1, 3, 5) AND 100-5S FRACTIONS (2, 4, 6) OF C.I. 12995, RBS AND K501099

FIG. 10.5.   CUT LOAVES OF BREAD BAKED FROM RECONSTITUTED
FLOURS CONTAINING THE 100-5S AND 100-5C FRACTIONS OF C.I. 12995,
AND RECONSTITUTED FLOURS IN WHICH THOSE TWO FRACTIONS OF
C.I. 12995 AND K501099 WERE INTERCHANGED

All reconstituted flours contained 12.5% protein and C.I. 12995 starch and water-
solubles. Loaves represent, from left to right: top row, original flours of C.I. 12995
(12.5% protein) and K501099 (13.7% protein); middle row, 100-5S plus 100-5C,
100-5S, and 100-5C fractions of C.I. 12995; bottom row, 100-5S of C.I. 12995 plus
100-5C of K501099 and 100-5S of K501099 plus 100-5C of C.I. 12995.

ever, when 100-5S of K501099 was reconstituted with 100-5C of C.I. 12995 (third row, right), a distinctly inferior loaf was obtained. It was poorer than the poor-quality control K501099, at least in one respect, because it was reconstituted to 12.5% protein instead of to 13.7%.

Starch-gel electrophoretic patterns reproduced in Fig. 10.6 show that the 100-5C glutenins, 100-5S glutenins, and total glutenins from pH 4.7-soluble gluten are essentially equal (Hoseney et al. 1969B). All three glutenin fractions (whether from a wheat variety with poor or good loaf-volume potential) were equally functional in maintaining the loaf-volume potential of the 100-5S protein fraction of a good-quality wheat variety.

Because the ratio of gliadin to glutenin apparently does not vary from short- to long-mixing flours, the mixing requirement apparently is governed by differences in one of the fractions or by the mode of interaction of the fractions. Thus, reconstituted flours containing 12.9% protein—50% from the gliadin-rich 100-5S fraction and 50% from the fraction insoluble in 70% ethyl alcohol (total glutenins)—were prepared by interchanging those fractions of long-, medium-, and short-mixing flours (Table 10.2). The gliadin and glutenin composition of the reconstituted flours was constant, so mixing time was obviously a function of differences in the glutenin proteins.

Those data—together with similar data for gluten proteins fractionated by 70% ethyl alcohol and by partial solubilization in 0.002N lactic acid (Hoseney et al. 1969B)—demonstrated conclusively that gliadin proteins are responsible for loaf-volume potential, and that glutenin proteins govern the mixing requirement of a wheat flour.

The complete separation of glutenin from gliadin by ultracentrifugation at 435,000 $\times$ g (Goforth and Finney 1976) and the entirely functional properties of the gliadin fraction (Goforth and Finney, unpublished data) are the ultimate conditions for studying the protein fractions of good- and poor-quality wheats.

Role of Water-Soluble Fraction

When flour is slurried with water, part of it (3–5%) becomes soluble and contains about 8% of the total flour protein. Finney (1943) found that the water-soluble fraction was not responsible for inherent differences in loaf volume (protein quality), but demonstrated that the water-soluble fraction not only behaved as a protein-softening or conditioning material (varying in amount or composition depending on wheat variety), but also shared (probably with the gluten fraction) responsibility for the varied oxidation requirements of different varieties of wheat. The water solubles generally were necessary for normal baking properties. Pence et al. (1951) also found

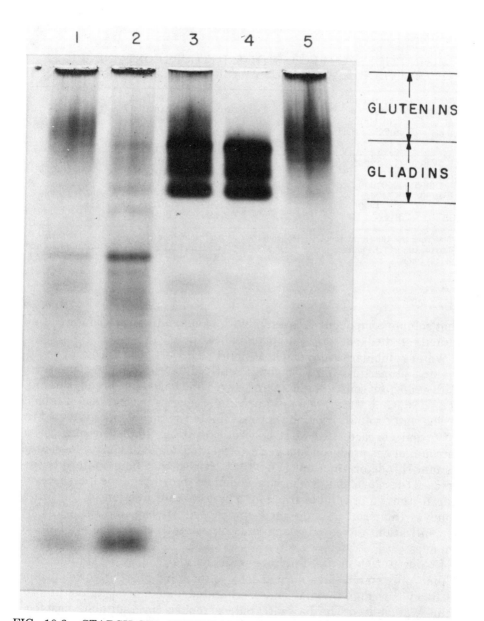

FIG. 10.6.   STARCH-GEL ELECTROPHORETIC PATTERNS OF PROTEIN
FRACTIONS OBTAINED BY ULTRACENTRIFUGING AND SOLUBILIZING
IN 70% ETHYL ALCOHOL THE pH 4.7-SOLUBLE GLUTEN
The patterns represent the following fractions: 1, insoluble in 70% ethyl alcohol
(total glutenins); 2, 100-5C (glutenins); 3, 100-5S; 4, soluble in 70% ethyl alcohol
(gliadins); 5, insoluble in 70% ethyl alcohol from 100-5S (100-5S glutenins).

TABLE 10.2

MIXING TIMES FOR RECONSTITUTED FLOURS CONTAINING 12.9% PROTEIN[1] AND STARCH PLUS WATER-SOLUBLES FROM C.I. 12995

Source of 100-5S	Source of Fraction Insoluble in 70% EtOH	Mixing Time[2] (min)	Source of 100-5S	Source of Fraction Insoluble in 70% EtOH	Mixing Time[2] (min)
C.I. 12995	C.I. 12995	9	RBS	K501099	4
C.I. 12995	RBS	4⅞	RBS	K14042	1¾
C.I. 12995	K501099	5⅝	K501099	C.I. 12995	9
C.I. 12995	K14042	1⅞	K501099	RBS	5¼
RBS	C.I. 12995	8⅞	K501099	K501099	4
RBS	RBS	5⅜	K501099	K14042	1

[1]50% from the 100-5S fraction and 50% from the fraction insoluble in 70% ethyl alcohol (EtOH).
[2]Mixing times of the unfractionated flours: C.I. 12995, 6½ min; RBS, 3⅞ min; K501099, 1⅞ min; and K14042, 1 min.

that soluble components were required for maximum performance of all glutens studied, except for a durum wheat flour.

Water-soluble fractions of wheat flours that were good to poor in bread-making properties were not significantly different when reconstituted with a constant source of gluten and starch and baked into bread (Hoseney *et al.* 1969C), but they were required to produce a normal loaf of bread.

The water-soluble fraction was fractionated by the scheme given in Fig. 10.7, and characterized electrophoretically (Fig. 10.8). Gassing power (amount of gas produced in a yeast-fermented dough or slurry, measured as mm Hg) determinations (Fig. 10.9) showed that flour materially increased the rate of gas production, and more than 50% of the increase was due to the water-soluble fraction. The dialyzate fraction of the water solubles (material passing through the dialysis bag and recovered by lyophilization) increased gas production as much as did the total water solubles.

Reconstitution of the dialyzate with the dialyzed water solubles and gluten plus starch gave normal loaf volume and gassing power (Table 10.3), even though the material precipitated upon dialysis (globulin proteins) was omitted. The dialyzate could be replaced by a suitable yeast food (Bacto yeast nitrogen base or ammonium chloride). The dialyzed water solubles were boiled, cooled, and centrifuged to denature and remove the albumin proteins. The resulting fraction (water solubles, dialyzed and boiled), when reconstituted with gluten plus starch and a suitable yeast food, gave the entire water-solubles response. Thus, the albumin and

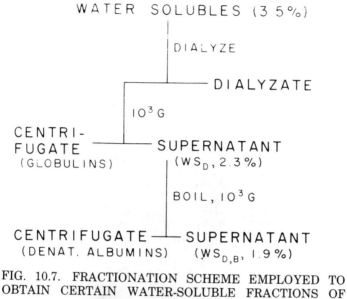

FIG. 10.7. FRACTIONATION SCHEME EMPLOYED TO OBTAIN CERTAIN WATER-SOLUBLE FRACTIONS OF FLOUR

Percentages are based on total flour weight. WS and subscripts D and B are abbreviations for water-solubles, dialyzed, and boiled, respectively.

globulin protein fractions were not involved in breadmaking performance. The dialyzed and boiled water-soluble fraction contained 26% protein and probably consisted of water-soluble pentosans and the glycoproteins described by Kuendig *et al.* (1961A, B) and studied by Fausch *et al.* (1963), Tracey (1964) and Wrench (1965).

Microbaking tests showed that water-soluble pentosans (Patil *et al.* 1976) were required to obtain normal loaf volume from reconstituted gluten and starch doughs, and that pentosans and bromate, in the absence of other water-soluble components, had an additive effect of overoxidation, which caused dough rigidity and reduced loaf volume. Diethylaminoethyl (DEAE) cellulose fraction II, a high-molecular-weight glycoprotein, greatly improved loaf volume of gluten-starch loaves in the absence of bromate. Water-soluble pentosans (no bromate), in place of the total water solubles, produced a loaf-volume-improving effect equal to that of the water solubles plus bromate. The rigidity of reconstituted doughs containing pentosans and bromate (usually characterized by reduced loaf volume) possibly results from a combination of two factors: (a) removal of water-

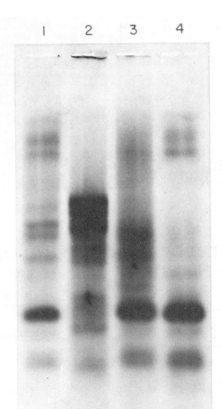

FIG. 10.8. STARCH-GEL ELECTROPHORETIC PAT-
TERNS OF THE FOLLOWING PROTEIN FRACTIONS
OBTAINED FROM THE WATER-SOLUBLES: 1,
WATER-SOLUBLE PROTEINS; 2, GLOBULINS; 3, AL-
BUMINS AND GLYCOPROTEINS; AND 4, GLYCO-
PROTEINS

soluble components responsible for gluten-protein extensibility and/or
for oxidation requirement (for suppressing the detrimental effect of
overoxidation, and (b) oxidation of the pentosan-glycoprotein interaction
product. Excessive rigidity of dough impairs oven spring and dependent
loaf volume. When dough extensibility is decreased before starch gela-

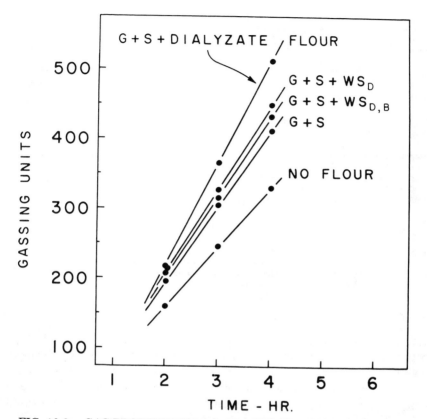

FIG. 10.9.   GAS PRODUCTION (mm Hg) VS FERMENTATION TIME
OF RECONSTITUTED RBS (REGIONAL BAKE STANDARD) FLOURS
IN WHICH THE WATER-SOLUBLE FRACTION AND ONE OR MORE
OF ITS COMPONENTS WERE OMITTED OR REPLACED BY YEAST
FOOD
G, S, and WS and subscripts D and B are abbreviations for gluten, starch,
water-solubles, dialyzed, and boiled, respectively. Other reconstituted
flours (not shown) that had gas production equal to the original flour
included the following: G + S + WS, G + S + 17.5 mg Difco nitrogen base,
and G + S + 5 mg ammonium chloride.

tinizes, gas cells cannot expand before gluten is denatured by the baking
temperature.

The role of water solubles in baking appears to be related to two fractions: (a) the dialyzable fraction contributes to gas production and can be
replaced by a suitable yeast food, and (b) the fraction containing soluble
pentosans and glycoproteins contributes to gas retention and/or gluten
extensibility.

TABLE 10.3

BAKING RESULTS FOR THREE WATER-SOLUBLE FRACTIONS
RECONSTITUTED WITH GLUTEN AND STARCH FROM RBS FLOUR

Original or Reconstituted Flour	Mixing Time (min)	Loaf Volume (cc)
RBS flour	3¾	82
Gluten + starch	3	60
Gluten + starch + WS[1]	2⅝	83
Gluten + starch + dialyzate	2⅝	72
Gluten + starch + WS D	2⅝	73
Gluten + starch + WS D + dialyzate	2⅝	82
Gluten + starch + WS D & B	2⅝	70
Gluten + starch + WS D & B + dialyzate	2¾	81
RBS flour + YF	3¾	81
Gluten + starch + YF	3¾	73
Gluten + starch + WS D + YF	3½	84
Gluten + starch + WS D & B + YF	3	80

[1]Abbreviations used: WS = water solubles; WS D = dialyzed water solubles; WS D & B = water solubles dialzyed and boiled; YF = 17.5 mg of yeast food.

Role of Lipids

The lipids in flour have been divided into "free" and "bound" groups (Daftary and Pomeranz 1965). The free (about 60% of the total lipids) have been defined as those extractable by petroleum ether or similar nonpolar solvents. The remaining bound lipids (practically all polar) have been extracted by more highly polar solvents such as water-saturated butanol (WSB). When flour is mixed into a dough, all the free polar and about 50% of the free nonpolar lipids become bound and are no longer extractable with petroleum ether (Olcott and Mecham 1947).

Free lipids have been fractionated further by silicic-acid column chromatography into polar (25% of total free) and nonpolar (remaining 75%). The role of free lipids and their fractions has been studied extensively (Pomeranz et al. 1968; Daftary et al. 1968). Free polar lipids increased loaf volume substantially; the increase was smaller when bound polar lipids were added instead of free polar lipids.

In shortening-free doughs (Daftary et al. 1968) nonpolar lipids decreased loaf volume (Fig. 10.10) and impaired crumb grain of bread baked from petroleum-ether-extracted flour; the deleterious effects were counteracted by polar lipids. The effects on bread depended on the levels and ratios of polar to nonpolar lipids. Lipid fractions isolated from a poor-quality flour were equal to the corresponding fractions of a good-quality flour.

Evidence obtained by unique fractionating techniques (Hoseney et al. 1970) indicated that free polar lipids (principally glycolipids) are bound to

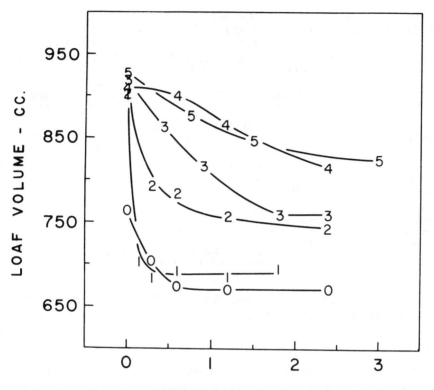

FIG. 10.10.   RELATION BETWEEN LOAF VOLUME OF BREAD BAKED
(WITHOUT SHORTENING) FROM EXTRACTED MARQUIS FLOUR
AND VARIOUS COMBINATIONS OF NONPOLAR AND POLAR LIPIDS
Numbers 1 to 5 denote levels of 0.1–0.5g polar lipids per 100g flour,
respectively.

gliadin protein by hydrophilic bonds and to glutenin protein by hydrophobic bonds. In unfractionated gluten, the lipid apparently is bound to
both protein groups at the same time. The simultaneous binding of polar
lipids to gliadin and glutenin may contribute structurally to gas-retaining
complexes in gluten in a manner not previously proposed.

Total free lipids from each of six flours that varied greatly in loaf volume
and crumb grain were reconstituted with the petroleum-ether-extracted
hard red wheat composite flour (Table 10.4). The source of the free lipids
had no effect on loaf volume or crumb grain. Free lipids from durum were
as effective as free lipids from hard winter or spring wheat flour in restor-

TABLE 10.4
EFFECTS OF FREE FLOUR LIPIDS FROM VARIOUS FLOURS ON LOAF
VOLUME AND CRUMB GRAIN OF BREAD BAKED FROM THE
PETROLEUM-ETHER-EXTRACTED HARD WINTER WHEAT COMPOSITE
FLOUR (3% SHORTENING)

Flour Description	Source of Added Free Lipids (0.8 g)	Loaf Volume (cc)	Crumb Grain
Original	—	920	S
Extracted	—	780	Q
Extracted	Composite (HRW)[1]	920	S
Extracted	C.I. 12995 (HRW)	913	S
Extracted	K501099 (HRW)	930	S
Extracted	Marquis (HRS)	913	S
Extracted	Seneca (SRW)	918	S
Extracted	Wells (Durum)	918	S
Extracted	Omar (Club)	915	S

[1]Class of wheat is in parentheses. S = satisfactory; Q = questionable.

ing the shortening response or loaf-volume potential to the extracted hard winter composite flour (Pomeranz *et al.* 1968).

Role of Starch

Fractionation and reconstitution studies have shown that the starch-tailings fraction (Table 10.5), although contributing to water absorption and dough-feel, is not essential for optimum loaf volume (Hoseney *et al.* 1971). In those studies, reconstituted prime starches (Table 10.6) from three hard red winter wheats, three hard red spring wheats, one soft white wheat, and one soft red winter wheat had essentially equal loaf volume when all were baked with constant gluten and water-soluble fractions. Prime starch of durum wheat was significantly lower and prime starches of four club wheats were significantly higher in loaf volume than that of the control starch from hard red winter wheat.

Technological and Nutritional Importance of Wheat Flour Proteins

Techniques for fractionating and reconstituting have established that the gluten proteins are responsible for most of the functional (breadmaking) properties of wheat flours, namely, gliadin for loaf volume potential (gas retention) and glutenin for mixing requirement, which is closely related to mixing tolerance and other important physical properties of dough. Also, glutenin apparently controls water absorption required for optimum dough consistency, because flours that have medium-long to long mixing times almost invariably have high water absorptions at a given

TABLE 10.5

BAKING DATA FOR RECONSTITUTED FLOURS (12.8% PROTEIN) CONTAINING GLUTEN AND WATER SOLUBLES FROM RBS AND VARIOUS RBS STARCH FRACTIONS

Starch Fraction	Absorption (%)	Loaf Volume (cc)
Unfractionated flour	66	80
Total	65	80
Prime and tailings (80/20)	65	80
Prime	61	80
Tailings	93	43

TABLE 10.6

BAKING DATA FOR RECONSTITUTED FLOURS (12.8% PROTEIN) CONTAINING GLUTEN AND WATER SOLUBLES FROM RBS AND PRIME STARCH FROM VARIOUS SOURCES

Source of Prime Starch	Wheat Type	Absorption (%)	Loaf Volume (cc)
RBS	HRW	61	80
Thatcher	HRS	61	79
Selkirk	HRS	61	81
Lee	HRS	61	79
Seneca	SRW	61	79
Wells	Durum	63	74
Omar	Club	61	85
Moro	Club	61	83
Elgin	Club	61	84
Elmar	Club	61	83
Kharkof	HRW	61	80
Wanser	HRW	60	78
Nugaines	Soft white	60	79

protein level. Although the water solubles of a poor quality flour generally are as functional as those of a good one, certain of the water-soluble proteins, together with the corresponding gluten proteins, control oxidation requirement. Wheat-flour lipids and the water-soluble pentosans and accompanying glycoprotein did not account for differences between good- and poor-quality bread wheat flours, but they contributed to gas retention and/or gluten extensibility and were necessary to obtain normal loaf volume. The starch of a poor-quality bread-wheat flour was as functional as that from a good one. However, most of the breadmaking properties are functions of protein content and the variety of wheat (Finney and Yamazaki 1967; Finney and Barmore 1948). For example, loaf volume, flour absorption, and oxidation requirement generally increase with protein content within a variety.

In consideration of commercially grown wheats released because of their good breadmaking and agronomic properties, protein content is extremely important. For many years, however, wheat yields have gradually increased at the expense of protein content, which now is a limiting factor for both home and export requirements. Today, a flour containing 11.25% protein does the job because the gradual decrease in average wheat protein content during the past 25 years has been offset by improved protein quality for breadmaking.

Plant breeders from commercial companies and from the Kansas, Nebraska, and South Dakota Agricultural Experiment Stations have developed genetically high-protein hard winter wheats that are agronomically superior and have other good milling and breadmaking properties. Varieties that have at least two percentage points more protein than comparably grown and leading commercial varieties already appear to have halted the commercial decline in protein content of hard winter wheat (Finney 1977).

Bread wheat gluten proteins are excellent carriers of materials such as rye flour, whole wheat products, and highly nutritious and concentrated foreign proteins such as soy flour. The higher the quantity and/or the quality of wheat flour protein, the greater the carrying power. Wheat protein is an excellent source of nutritious protein when balanced by other foods that supply certain amino acids, such as lysine. When 14% of wheat flour is replaced by defatted soy flour (Finney 1975), lysine content of the wheat flour is nearly tripled and a good amino acid balance is established. High-protein, nutritious breads and related products made from that blend of wheat flour and soy flour would be one way to feed millions of human beings who need more protein in their diets.

BIBLIOGRAPHY

BROWN, L. R. 1963. Man, land, and food; Looking ahead at world food needs. Foreign Agr. Econ. Rept. No. 11, Economic Research Service, U.S. Dept. of Agr. 153 pp., illus.

DAFTARY, R. D. and POMERANZ, Y. 1965. Changes in lipid composition in maturing wheat. J. Food Sci. 30, 577–582.

DAFTARY, R. D., POMERANZ, Y., SHOGREN, M. D. and FINNEY, K. F. 1968. Functional bread-making properties of wheat flour lipids. 2. The role of flour lipid fractions in bread-making. Food Technol. 22, 79–82.

FAUSCH, H., KUENDIG, W. and NEUKOM, H. 1963. Ferulic acid as a component of a glycoprotein from wheat flour. Nature 199, 287.

FINNEY, K. F. 1943. Fractionating and reconstituting techniques as tools in wheat flour research. Cereal Chem. 20, 381–396.

FINNEY, K. F. 1945. Methods of estimating and the effect of variety and protein level on the baking absorption of flour. Cereal Chem. 22, 149–158.

FINNEY, K. F. 1954. Contributions of the hard winter wheat quality laboratory to wheat quality research. Trans. Amer. Assoc. Cereal Chem. 12, 127–142.

FINNEY, K. F. 1975. A sugar-free formula for regular and high-protein breads. Baker's Digest 49, 18–22.

FINNEY, K. F. 1977. Quality of Kansas wheat varieties. Proc. of the 5th Annual Wheat Marketing Field Day for Kansas Wheat Producers. Report of Progress 295, Kansas Wheat Commission and Kansas Agri. Exp. Station.

FINNEY, K. F. and BARMORE, M. A. 1943. Yeast variability in wheat variety test baking. Cereal Chem. 20, 194–200.

FINNEY, K. F. and BARMORE, M. A. 1945A. Varietal responses to certain baking ingredients essential in evaluating the protein quality of hard winter wheats. Cereal Chem. 22, 225–243.

FINNEY, K. F. and BARMORE, M. A. 1945B. Optimum vs. fixed mixing time at various potassium bromate levels in experimental bread making. Cereal Chem. 22, 244–254.

FINNEY, K. F. and BARMORE, M. A. 1948. Loaf volume and protein content of hard winter and spring wheats. Cereal Chem. 25, 291–312.

FINNEY, K. F. and YAMAZAKI, W. T. 1967. Quality of hard, soft, and durum wheats. In Wheat and Wheat Improvement K. S. Quisenberry and L. P. Reitz (Editors). Amer. Soc. Agron., Madison, WI.

GOFORTH, D. R. and FINNEY, K. F. 1976. Communication to the editor: Separation of glutenin from gliadin by ultracentrifugation. Cereal Chem. 53, 608–612.

GOFORTH, D. R. and FINNEY, K. F. (Unpublished data), U.S. Grain Marketing Research Lab., USDA, ARS, Manhattan, KS.

HEGSTED, D. M. 1965. Wheat: challenge to nutritionists. Cereal Sci. Today 10, 257–259, 360.

HOSENEY, R. C., FINNEY, K. F. and POMERANZ, Y. 1970. Functional (bread-making) and biochemical properties of wheat flour components. 6. Gliadin-lipid-glutenin interaction in wheat gluten. Cereal Chem. 47, 135–140.

HOSENEY, R. C., FINNEY, K. F., POMERANZ, Y. and SHOGREN, M. D. 1969B. Functional (breadmaking) and biochemical properties of wheat flour components. 4. Gluten protein fractionation by solubilizing in 70% ethyl alcohol and in dilute lactic acid. Cereal Chem. 46, 495–502.

HOSENEY, R. C., FINNEY, K. F., POMERANZ, Y. and SHOGREN, M. D. 1971. Functional (breadmaking) and biochemical properties of wheat flour components. 8. Starch. Cereal Chem. 48, 191–201.

HOSENEY, R. C., FINNEY, K. F., SHOGREN, M. D. and POMERANZ, Y. 1969A. Functional (breadmaking) and biochemical properties of wheat flour components. 3. Characterization of gluten protein fractions obtained by ultracentrifugation. Cereal Chem. 46, 126–135.

HOSENEY, R. C., FINNEY, K. F., SHOGREN, M. D. and POMERANZ, Y. 1969C. Functional (breadmaking) and biochemical properties of wheat flour components. 2. Role of water-solubles. Cereal Chem. 46, 117–125.

KUENDIG, W., NEUKOM, H. and DEUEL, H. 1961A. Investigations of the cereal pentosans. 1. Chromatographic fractionation of water-soluble wheat flour pentosans on diethylaminoethyl cellulose. Helv. Chim. Acta 44, 823–829.

KUENDIG, W., NEUKOM, H. and DEUEL, H. 1961B. Investigations of the cereal pentosans. 2. The gelling properties of water-soluble solutions of wheat flour pentosans caused by oxidative materials. Helv. Chim. Acta *44*, 969–976.

MAC GILLIVRAY, J. H. and BOSLEY, J. B. 1962. Amino acid production per acre by plants and animals. Econ. Bot. *16*, 25–30.

OLCOTT, H. S. and MECHAM, D. K. 1947. Characterization of wheat gluten. 1. Protein-lipid complex formation during doughing of flours. Lipoprotein nature of the glutenin fraction. Cereal Chem. *24*, 407–414.

PATIL, S. K., FINNEY, K. F., SHOGREN, M. D. and TSEN, C. C. 1976. Water-soluble pentosans of wheat flour. 3. Effect of water-soluble pentosans of loaf volume of reconstituted gluten and starch doughs. Cereal Chem. *53*, 347–354.

PENCE, J. W., ELDER, A. H. and MECHAM, D. K. 1951. Some effects of soluble flour components on baking behavior. Cereal Chem. *28*, 94–104.

POMERANZ, Y., SHOGREN, M. D. and FINNEY, K. F. 1968. Functional bread-making properties of wheat flour lipids. 1. Reconstitution studies and properties of defatted flour. Food Technol. *22*, 76–79.

REITZ, L. P. 1964. Wheat quality components responsive to genic control. Qual. Plant Mater. Veg. *11*, 1–16.

REITZ, L. P. 1967. World distribution and importance of wheat. *In* Wheat and Wheat Improvement. K. S. Quisenberry and L. P. Reitz (Editors). Amer. Soc. Agron., Madison, WI.

TRACEY, M. V. 1964. The role of wheat flour pentosans in baking. 1. Enzymatic destruction of pentosan in vitro. J. Sci. Food Agr. *15*, 607–611.

WRENCH, P. M. 1965. The role of wheat flour pentosans in baking. 3. Enzymatic degradation of pentosan fractions. J. Sci. Food Agr. *16*, 51–54.

CHAPTER 11

RELATIONSHIP BETWEEN FINE STRUCTURE AND COMPOSITION AND DEVELOPMENT OF NEW FOOD PRODUCTS FROM LEGUMES

LOUIS B. ROCKLAND

Western Regional Research Center
Agricultural Research Service
United States Department of Agriculture
Berkeley, CA 94710

ABSTRACT

Basic research on the fine structure and composition of dry beans has been applied to the development and optimization of new technology for the production of more convenient and acceptable products from dry beans and other legume seeds. Processing procedures have been developed and adapted to the production, from most commercial types of dry beans, of a broad spectrum of new products including: redried, frozen and hydrastable (refrigerator-stable) quick-cooking beans; precooked frozen as well as canned mixed bean salads and other products; and combinations of complementary proportions of animal proteins and cooked beans which enhance nutritive value, digestibility and overall acceptability of products produced from the dry legume seeds.

INTRODUCTION

The use of legumes for human food probably evolved concurrent with the appearance of modern man about 10,000 years ago following the recession of the last great glacier in the Mediterranean section of the Near East (Zohary and Hopf 1973). Raw legumes are not acceptable foods because they contain several mildly toxic ingredients and antinutritional factors (Liener 1969) which interfere with digestive processes and must be removed by a rudimentary type of processing, *i.e.* soaking and cooking. Therefore, the use of legume seeds for food was precluded until two interrelated tools were developed: the controlled use of fire; and the manufacture of nondestructible cooking vessels (Leopold and Ardrey 1972), *i.e.* fired ceramics or ductile copper (Wertime 1973). The procedure advocated for the preparation of dry beans in some of the most modern cook books, is essentially identical with that which has been used for several thousand years. The washed bean seeds are soaked in water, generally overnight, or until they reach approximately twice their initial weight. The rehydrated beans are heated in boiling water until tender. Cooking time may vary from slightly less than 1 hr to up to about 4 hr depending upon the legume variety, moisture content, age and storage history (Morris and Wood 1956). Up until the 19th Century, when cooking stoves became popular, the cast iron kettle, suspended over an open fire, was the principle cooking method used for controlled heating at a constant temperature, *i.e.* boiling water.

Cooking may be defined as the art and science of heating food to make it more palatable, nutritious and free from natural toxic constituents. Although other legume processing technologies, *i.e.,* fermentation, sprouting, etc. are used as alternative procedures for utilizing legumes in many areas of the world, the soaking and boiling procedure is probably used most widely.

On a world-wide basis, legumes are important sources of food, rivaling many of the cereal grains in respect to their contribution to protein nutrition (Table 11.1). Based on the 1975, *FAO Production Yearbook,* world production of legumes including dry beans, peas, soybeans and peanuts totaled about 133,000 metric tons, almost equivalent in protein content to the 355,000 metric tons of wheat and over 50% more than the protein production from either rice or corn.

There has been an obvious need for the development of improved and more convenient procedures for preparing more nutritious and acceptable food products from dry beans and other legumes. This need has been met,

TABLE 11.1

TOTAL WORLD ACREAGE, PRODUCTION AND PROTEIN YIELD OF THE MAJOR FOOD LEGUME CROPS COMPARED WITH WHEAT, RICE AND CORN[1]

Product (Protein, %)	Area (1000 hectares)	Crop Production (1000 metric tons)	Protein Production (1000 metric tons)
Pulses (25%)	69,500	46,000	11,500
Soybeans (34%)	46,500	68,000	23,100
Peanuts (48%)	19,400	19,000 (in shell)	4,100
Total legumes	135,400	133,000	38,700
Wheat (12%)	228,000	355,000	42,600
Rice (%)	141,000	344,000	24,000
Corn (8%)	115,000	323,000	25,800

[1]Area and production data taken from FAO Production Yearbook (1975).

in part, by the development of new, simple and economical processing technology which reduces the cooking times for dry beans by 80% or more, while improving their overall flavor, texture and general acceptability, and without decreasing their nutrient content or nutritional qualities Rockland 1972; Rockland *et al.* 1967, 1971, 1977A). A flow sheet, outlining the new technology , is presented in Fig. 11.1.

The present contribution is designed to illustrate how basic research investigations on fine structure and composition have been used to provide a rational basis for the development of improved processing technology for preparing a variety of economical quick-cooking products from dry beans and other legume seeds.

FINE STRUCTURE AND PHYSICAL PROPERTIES

Shear Press Measurement of Cooking Rates

Insight concerning the problems attending the development of quick-cooking beans was obtained from a study of changes, during cooking, in the physical and rheological properties of large dry Lima beans (Binder and Rockland 1964). A L.E.E.-Kramer shear press equippped with a flat multibladed shear head, was used to measure the resistance to shear by whole, rehydrated beans as function of cooking time in boiling water. Recorder tracing of only partially cooked beans revealed two distinctive peaks, a broad area, followed by a high slender peak (Fig. 11.2). At further intermediate stages of cooking, both peaks regressed continuously until the beans were cooked completely, leaving a more diminutive, single broad peak. It was shown that the two peaks observed in the under-cooked beans were due to the independent effects of seed coats and cotyledons. Under pressure from the shear head, the seed coats of the partially tenderized beans had become detached from the cotyledons and stratified at the top, just under the descending shear head and the softer cotyledon mass was extruded through the stationary slotted base. When the beans were completely cooked, both tissues were soft enough to be extruded without disproportionation of tissues. These studies demonstrated that the cooking rates and total cooking time for seed coats and cotyledons varied independently. In boiling water, rehydrated large Lima bean cotyledons softened more rapidly than the seed coat. The latter encapsulated the cotyledon preventing dispersion in the cooking water and thereby helped to maintain the shape of the original, whole bean. Distinctive, differential rates of softening during cooking of seed coats and cotyledons have also been observed for other legume seeds. For example, Garbanzo bean seed coats are softer relative to the cotyledon, so that they may "slip" during cooking. However, the firm texture of the garbanzo cotyledon generally prevents

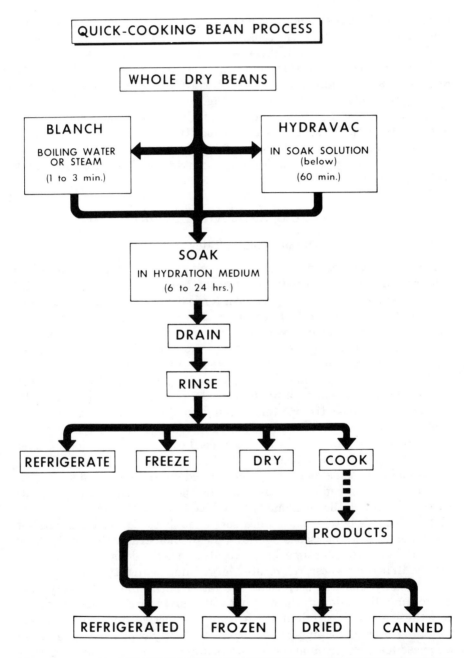

FIG. 11.1. FLOW SHEET FOR PREPARATION OF QUICK-COOKING BEANS

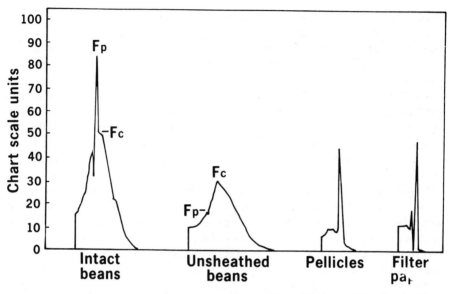

From Binder and Rockland (1964)

FIG. 11.2.   SHEAR-PRESS TRACINGS OF COOKED LARGE LIMA BEAN
TISSUES

dispersion. Processing technology designed to produce satisfactory pre-cooked legume products must accommodate differential cooking times of seed coat and cotyledons for each specific type of legume seed.

Because of differences in the cooking characteristics of legume seeds, it appeared that there were intrinsic limitations to the development of a general process of producing precooked legume products which would retain their physical integrity and normal appearance. Precooked products, produced previously, all contained high proportions of cracked, split or "butterfly-shaped" beans and had a generally unaesthetic appearance. Therefore, emphasis was placed upon the elaboration of a general pre-treatment which would condition legume seeds for quick-cooking properties rather than to develop technology for partial or total pre-cooking. This concept was consistent with the premise that the final product would have an improved, more natural appearance, if the product required only a single heat treatment to prepare it for table use.

Influence of Inorganic Salts on Cooking Rates of Rehydrated Beans

Based on assumptions concerning the effects of specific inorganic salts on major constituents (*i.e.* protein, starch, hemicellulose, pectin) present in seed coats and cotyledons, a variety of ingredients were added to the rehydration medium for evaluation of their specific influences on the cooking characteristics of legume seeds (Rockland 1964). Final formulations for inducing quick-cooking properties in dry beans were based on a large number of trials in which individual and subsequently combinations of neutral salts, acid, alkali, oxidizing and reducing agents, surface active materials, enzyme preparations and other miscellaneous additives to the hydration medium were evaluated. For example, it was observed that after rehydration in a 2% solution of sodium chloride, the cooking rate of large Lima bean cotyledons in fresh boiling water decreased slightly while the seed coat cooked significantly faster, bringing the relative cooking times for both tissues into closer alignment. Rehydration of beans in dilute alkaline solutions decreased cooking time for the whole bean, but often imparted undesirable flavors and odors to the cooked products. In a buffered solution at pH 9, in the presence of other ingredients, the unfavorable influence of the alkali was eliminated. Various phosphates were tested primarily because of their recognized metal chelating properties and the premise that the cotyledon cells may be held together by a calcium or other bivalent cation bridge in the protein-carbohydrate matrix of the middle lamella. Ingredients which appeared to have greatest utility were combined in various proportions and evaluated for their integrated effects.

During these studies a casual observation helped to orient the direction of the project. It was noted that after standing for several hours, a white (semi)crystalline precipitate formed after large Lima beans had been decanted from a rehydration medium which contained sodium tripolyphosphate. Chemical microscopy studies and elemental analyses indicated that the precipitate was largely trisodium magnesium tripolyphosphate, hydrate, probably contaminated with a small amount of isomorphous crystals of trisodium calcium tripolyphosphate. Fig. 11.3 compares crystals of synthetic $Na_3MgP_3O_{10} \cdot 14\ H_2O$ and the recrystallized preparation isolated from the Lima bean soak solution. The two products had the same elemental composition and identical refractive indices along the principle optical directions of the crystalline compounds. Subsequent experiments using tripolyphosphate solutions indicated a reasonable consistency between the amount of precipitate formed in the hydration medium and reduction in cooking time of beans from which the solution had been decanted (Rockland et al., unpublished data). These observations were consistent with the premise that: (1) divalent cations, such as magnesium and calcium, may form an insoluble complex between polymeric carbohydrates and proteins which compose the intracellular cement of the middle lamella of cells within the cotyledon; and (2) that the softening of the tissue is due, in part, to extraction of calcium and/or magnesium and dissociation of the complexes during cooking.

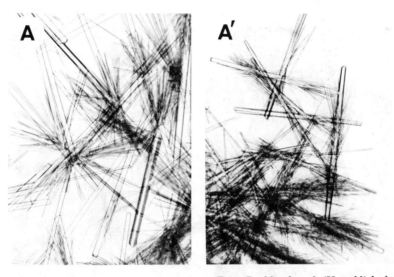

From Rockland et al. (Unpublished data)

FIG. 11.3. COMPARISON OF CRYSTALS OF SYNTHETIC TRISODIUM MAGNESIUM TRIPOLYPHOSPHATE HYDRATE WITH RECRYSTALIZED ISOLATE FROM HYDRATION MEDIUM OF LARGE DRY LIMA BEANS
A. Synthetic $Na_3MgP_3O_{10} \cdot 14\ H_2O$, A'. Recrystalized isolate from Lima bean hydration medium

On the basis of numerous trials with different inorganic salt combinations and concentrations, optimized hydration media were developed for preparing raw, quick-cooking products from most domestic commercial dry beans, soybeans, and several South American and African legume varieties. A National Academy of Sciences report (1975) has suggested that the Winged Bean *(Psophocarpus tetragonolobus)* may have exceptional merit as a crop plant of the future particularly if the normal cooking time of 3–4 hr for the dry bean seed could be lowered to within a useful limit. Quick-cooking Winged Beans, which require about 20 min to cook, may fulfill the specifications detailed in the National Academy of Sciences report.

Effects of Cooking on Cell Structure

Light and scanning electron microscopy (SEM) were employed to study the effects of cooking on the cellular structure of Lima beans (Rockland and Jones 1974). It was also of interest to determine if there were any differences between standard and quick-cooking beans in respect to physical changes that occurred during cooking in boiling water.

A SEM photomicrograph of whole cells in raw, dry Lima bean cotyledon is shown in Fig. 11.4. Individual cells resembled oblate plastic bags par-

FIG. 11.4. SCANNING ELECTRON PHOTOMICROGRAPH OF WHOLE
CELLS IN RAW, DRY, LARGE LIMA BEAN COTYLEDON
Unit cell is about 100 microns.

tially collapsed on nearly spherical (25–50 mμ) starch granules, and appeared to constitute the complete cell contents. There were no apparent differences between cells in redried, salt soaked or standard large Lima bean cotyledons. A cross section of a wet-sliced and redried cotyledon (Fig. 11.5) indicated apparent uniformity of the cell wall-middle lamella matrix. Although protein constitutes about 20% of the bean solids, its presence was not obvious. Raw, dry-fractured cotyledon sections frequently contained small, amorphous particles presumably protein, but there was no evidence of protein bodies analogous to the material present in soybeans. An apparent thin film, bridging some of the intracellular starch granules (Fig. 11.5C), may be cytoplasmic protein which coated the granules during drying.

Major changes in cell structure occurred during cooking, but no major differences were observed between standard and quick-cooking beans. However, analogous changes in standard beans required four to five times as long to occur. After heating in boiling water for 2 min, the quick-cooking bean cells cleaved at the cell wall interface without cell rupture, and the intracellular starch granules were changed significantly (Fig. 11.6). In contrast, raw, hydrated cell walls were severed during cotyledon fracture,

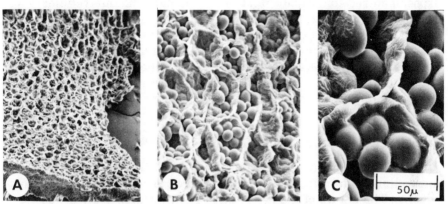

From Rockland and Jones (1974)

FIG. 11.5.   SCANNING ELECTRON PHOTOMICROGRAPHS OF A SECTION
THROUGH RAW, DRY LIMA BEAN COTYLEDONS
Unit cell width about 100 microns.

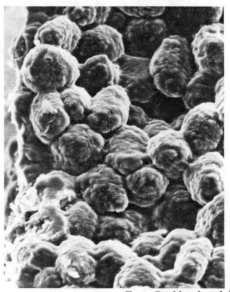

From Rockland and Jones (1974)

FIG. 11.6.   SCANNING ELECTRON PHOTOMICROGRAPH
SHOWING THE EFFECTS OF COOKING FOR 2 MINUTES
ON CELL STRUCTURE IN COTYLEDONS FROM QUICK-
COOKING LARGE LIMA BEANS

exposing intracellular material. Further changes and intercellular deformations occurred (Fig. 11.7) progressively for 10 min until the beans were cooked completely. These cells appeared to be identical to those in standard beans which required 45 min to cook and reach the same cotyledon tenderness (Fig. 11.8). In both cooked tissues, the intercellular adhesive within the middle lamella had either dissolved or plasticized to the extent that the cells could be displaced by mild pressure and easily dispersed in water (Fig. 11.9).

Extracellular and Intracellular Gelatinization of Starch

Cell deformations during cooking were related directly to changes in the intracellular starch granules which appeared to be the primary cell components. The nature of the bean starch gelatinization process was elucidated by SEM and light microscopy studies on intracellular (Hahn *et al.* 1977) and extracellular (Rockland *et al.* 1977B) gelatinization of Lima

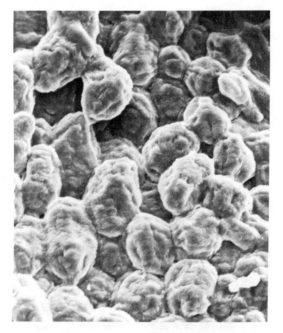

From Rockland and Jones (1974)

FIG. 11.7. SCANNING ELECTRON PHOTOMICROGRAPH SHOWING THE EFFECTS OF COOKING FOR 3 MIN ON CELL STRUCTURE IN COTYLEDONS FROM QUICK-COOKING LARGE LIMA BEANS

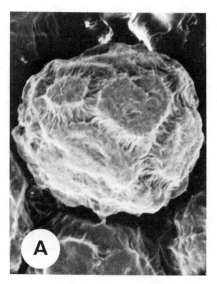

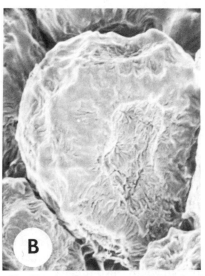

From Rockland and Jones (unpublished data)

FIG. 11.8. SCANNING ELECTRON PHOTOMICROGRAPH OF COM-
PLETELY COOKED CELLS FROM STANDARD AND QUICK-COOKING
LARGE LIMA BEANS
Unit cell about 100 microns. A. Standard; B. Quick-cooking.

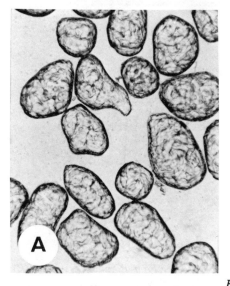

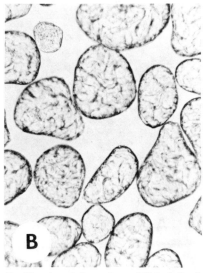

From Rockland and Jones (unpublished data)

FIG. 11.9. LIGHT PHOTOMICROGRAPHS OF WHOLE CELLS FROM
COOKED STANDARD AND QUICK-COOKING BEANS
Unit cell about 100 microns. A. Standard; B. Quick-cooking.

bean starch in water and also in a salt solution used to process quick-cooking beans.

Raw Lima bean starch granules are normally smooth, spherical or kidney-shaped particles (Fig. 11.11A) which exhibit double refraction under polarized light. Between crossed polarizers, the starch is birefringent, each granule showing the characteristic "maltese cross" (Fig. 11.11B).

In heated water, the gelatinization process was initiated independently by individual granules as a function of temperature. The process was initiated by apparent "melting" at the center of each granule (Fig. 11.10), disappearance of concentric rings and loss of birefringence in the melted portion. Progressive melting, granule deformation, and the number of granules initiating gelatinization, increased with increasing temperature. Gelatinization was initiated by a small proportion of dispersed granules in water at 71° C and in the salt solution at 79° C. The number of granules initiating gelatinization continued until the process was completed when the temperature approached 79° C in water or 85° C in the salt solution. Sequential configurational changes in starch granules (Fig. 11.12) were characterized arbitrarily as: (1) swollen; (2) dimpled; (3) doughnut; (4) rubber-raft; (5) pancake; and (6) dispersed.

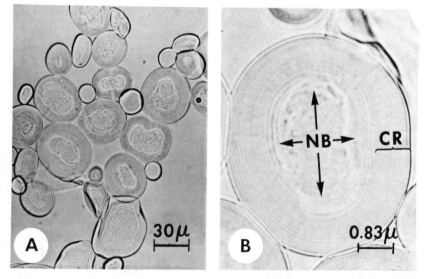

From Rockland et al. (1977B)

FIG. 11.10. PHOTOMICROGRAPHS OF PARTIALLY GELATINIZED LIMA BEAN STARCH GRANULES SHOWING CONCENTRIC RINGS IN OUTER UNGELATINIZED PORTIONS
(A) 500 ×; (B) 1800 ×.

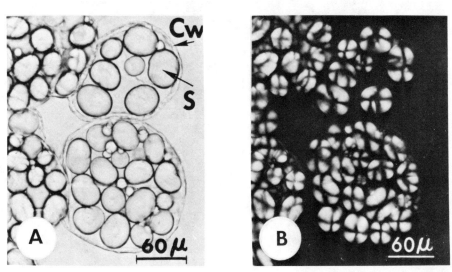

From Hahn et al. (1977)

FIG. 11.11.   LIGHT PHOTOMICROGRAPHS OF CELLS IN RAW LIMA BEAN
COTYLEDONS AFTER SOAKING IN SALT SOLUTION AT 22° C
(A) Cells containing raw, ungelatinized starch. (B) Same field as (A) under
polarized light showing birefringent starch granules. (250 ×)

Intracellular gelatinization appeared to follow the same general sequence, except that expansion and configurational changes were restricted by adjacent granules and the constraints of the intact cell wall. However, intracellular gelatinization was initiated 5–10° C higher than corresponding extracellular gelatinization temperatures, and gelatinization was not completed until the temperature approached 100° C (Fig. 11.13).

In boiling water, both intracellular and extracellular starch gelatinization was complete within 3 min. In whole beans, plasticization of the middle lamella and cell separation is a much slower and probably limiting step in cooking both standard and quick-cooking beans.

COMPOSITION

Influence of Cooking on Legume Proteins

The cooking process may also be limited by chemical and physical changes in legume proteins. Although Lima bean seeds contain about 20% protein, specific protein structure was not apparent in either light or SEM

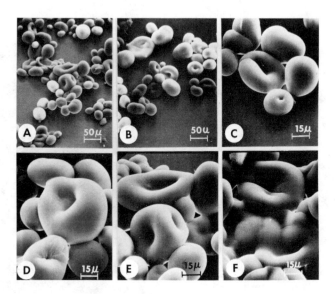

From Rockland et al. 1977B

FIG. 11.12. SCANNING ELECTRON PHOTOMICROGRAPHS OF CHANGES IN THE STRUCTURE OF LARGE LIMA BEAN STARCH GRANULES DURING EXTRACELLULAR GELATINIZATION IN AQUEOUS MEDIA

(A) 73° C, in water (300 ×); (B) 82° C, in dilute salt solution (300 ×); (C) Another field analogous to 1B (1000 ×); (D) Another field analogous to 1A (1000 ×); (E) Enlargement of a field in 1A (100 ×); and (F) Another field analogous to 1A (1000 ×).

photomicrographs. Amorphous, dense particles were observed occasionally in SEM photomicrographs of Lima *(P. lunatus)* or pink *(P. vulgaris)* beans. However, no particles were observed that could be identified unequivocally as proteins similar to the protein present in soybeans and some of the cereal grains. It is probable that in the rehydrated, wet tissue, proteins are dispersed in the cell cytoplasm as well as in the cell wall-middle lamella matrix. SEM photomicrographs of redried, fractured raw cells suggest that a thin, perhaps protein film, encapsulates the starch granules (Fig. 11.5).

Intracellular proteins and other constituents may be expected to influence starch gelatinization and may be responsible, in part, for the higher temperatures required for intracellular starch gelatinization.

Additional information concerning the role of proteins in the cooking process was obtained from polyacrylamide gel electrophoresis studies of both raw and cooked beans. The distinctive genetic constitution of each bean variety was reflected in an array of protein bands on gel elec-

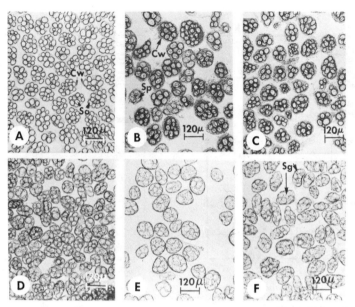

From Hahn et al. (1977)

FIG. 11.13. LIGHT PHOTOMICROGRAPHS OF INTACT CELLS IN LIMA
BEAN COTYLEDON AFTER SOAKING IN QUICK-COOKING SALT SOLU-
TION AT VARIOUS TEMPERATURES
(A) 70° C; (B) 85° C for 6 hr; (C) 85° C for 24 hr; (D) 90° C; (E) 96° C; and (F) 100° C.

trophoretograms (Rockland *et al.* 1974A). Definitive differences were observed between closely related members of the same sub-species of *P.
vulgaris, i.e.,* light and dark red kidney beans.

Water extracts of standard and quick-cooking beans produced slightly
different protein patterns, suggesting that the processing salts may have
facilitated dissociation of protein complexes (Fig. 11.14). After cooking,
the electrophoretograms of both standard and quick-cooking beans were
identical to each other and related to, but distinctly different from, those
obtained from raw bean extracts. In addition, about five times as much
extract was required to obtain a satisfactory electrophoretogram, indicating that the proteins had been denatured and their extractability reduced
about 80% during cooking. A separate study was conducted on the influence of cooking time, in boiling water, on the electrophoretic patterns of
proteins in standard and quick-cooking Lima beans. Changes in electrophoretic patterns were relatively slow, requiring up to 10 min to reach
the stable pattern which characterized the cooked bean.

On the basis of the chemical, light microscope, SEM and gel electrophoresis studies of proteins, it was concluded that the cooking process

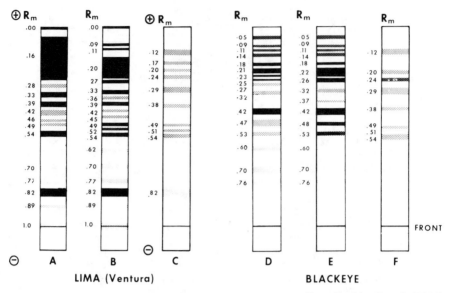

From Rockland et al. (1974)

FIG. 11.14. POLYACRYLAMIDE GEL DISC ELECTROPHORETOGRAM
TRACINGS OF MINOR PROTEINS IN DRY BEANS
(A) Raw, water-soaked large Lima beans; (B) Raw, salt-soaked (quick-cooking)
large Lima beans; (C) Cooked, water-soaked or salt-soaked large Lima beans; (D)
Raw, water-soaked Blackeye beans; (E) Raw, salt-soaked (quick-cooking) Blackeye
beans; and (F) Cooked, water-soaked or salt-soaked Blackeye beans.

involves at least four distinct chemical and/or physical changes including:
1. Partial release of calcium and magnesium into the cooking water.
2. Rapid intracellular starch gelatinization.
3. Gradual plasticization or partial solubilization of components of the
   middle lamella and separation of bean cells along the planes of the
   cell wall without cell rupture.
4. Progressive slow denaturation of protein.

Biochemical Properties

Early modern man recognized that most legume seeds must be rehydrated and cooked in boiling water for extended periods in order to convert
them to acceptable food products. In addition to improving their texture
and flavor, cooking significantly improved their nutritional qualities by
removing or inactivating most of the more undesirable, non-adventitious
factors that are present in raw, dry legumes (Table 11.2). Selection and
breeding of the more attractive and digestible varieties of legumes further
reduced some levels of the other undesirable factors in the dried seeds.

TABLE 11.2
NONADVENTITIOUS FACTORS AFFECTING UTILIZATION OF
COMMERCIAL EDIBLE DRY BEANS[1]

Factor	Active Principle	Major Effects	Heat Lability[2]
Antitrypsin	protein	digestibility, pancreas hypertrophy	+
Antiamylase	protein	uncertain	+[2]
Hemagglutinins	protein	growth inhibition	+
Essential-metal binding	protein, phytate	inhibits metal absorption	+
Cyanogenetic glycoside	linimarin	HCN toxicity	?
Protein quality	methionine deficiency	growth inhibition	−[3]
Nitrogen utilization	unknown	growth inhibition	−[3]
Food digestibility	unknown	inefficient food utilization	−[3]
Flatulence	unknown	gastrointestinal discomfort	−

[1]Adams et al. (1978).
[2]Cooked in boiling water or autoclaved.
[3]Heat activation required.

However, some of the most successful and satisfactory present day legume varieties contain a number of recognized, although not chemically identified antinutritional factors. Several antienzymes, such as amylase and trypsin inhibitors, are heat labile. However, it has been reported that other antienzymes may not be inactivated or are inactivated incompletely during normal soaking and cooking. Antienzyme inactivation may not be complete if the legume or legume meal is heated in a dry or semimoist condition (Rackis et al. 1974). The influences of soaking and cooking on several of the undesirable factors is still unknown. However, commercial varieties of dry beans have been selected to have minimal, nominal levels of these factors. For example, the cyanogens found in many plant food and feed products (Table 11.3) are present at innocuous levels in commercial dry bean varieties, or are eliminated during soaking and/or cooking. Haemagglutinins and antienzymes in raw legumes (Table 11.4), being proteinaceous, are normally inactivated during cooking.

Quick-cooking bean products require 20% or less of the normal cooking time for analogous standard products. Therefore it was necessary to establish that the brief cooking period required for these products was sufficient to inactivate the heat labile antinutrient factors without impairing other qualities. It was shown that haemagglutinating activity in raw pink beans was eliminated within the normal 10 min and 60 min cooking times for quick-cooking and standard beans, respectively (Table 11.5). Similarly it

TABLE 11.3

CYANOGENS IN DRY BEANS AND OTHER PLANT MATERIALS[1]

Material	HCN yield (mg/10g)
Sorghum	200
Cassava	113
Linseed meal	53
Large Lima bean *(Phaseolus lunatus)*	17
Blackeye bean *(Vigna unguiculata)*	2
Garden pea *(Pisum sativum)*	2
Kidney bean *(Phaseolus vulgaris)*	2
Garbanzo or Bengal gram *(Cicer arietinum)*	1
Pigeon pea or Red gram *(Cajanus cajans)*	0.5

[1]Data taken from Montgomery (1969).

TABLE 11.4

HEMAGGLUTINATING AND ANTITRYPTIC ACTIVITIES
OF CRUDE EXTRACTS OF RAW LEGUMES[1]

Legume	Hemagglutinating Activity (units/ml)	Antitryptic Activity (units/ml)
Phaseolus vulgaris		
Black bean	2450	2050
Kidney bean	3560	1552
Cicer arietinum (Garbanzo)	0	220
Cajanus cajan (Pigeon Pea)	0	418
Phaseolus aureus (Mung bean)	0	260

[1]Data taken from Honavar and Liener (1962).

TABLE 11.5

HAEMAGGLUTINATING ACTIVITY OF RAW AND
COOKED SUTTER PINK BEANS[1]

Condition	Activity (Activated Cow Erythrocytes)	
	Raw	Cooked
Dry	12	—
Water-soaked	14	0 (60 min)
Quick-cooking	14	0 (10 min)

1Rockland and Jaffe (unpublished data).

was shown that trypsin inhibitor activities in pink and large Lima beans were eliminated completely during cooking (Table 11.6).

Some unfavorable anomalous effects have been observed for completely cooked legumes, including growth inhibition (Bressani 1975), low food and nitrogen digestibility in rats, increased pancreas weights in rats, and flatulence (formation and rectal elimination of intestinal gases by many higher animals including humans). For the most part, these generally nominal deficiencies are compensated by the general overall acceptability, high nutritional value and low cost of legumes. For example, the Protein Efficiency Ratios (PER) of cooked standard and quick-cooking large Lima beans were both about 60% of standard casein (Table 11.7). The whole diet digestibilities of the beans were both about 3% lower and pancreas weights about 10% higher than the casein standard. The raw Lima beans were consumed at very low levels and produced no growth of weanling rats. However, pancreas weights were increased significantly.

Since quick-cooking beans require less than 20% of the normal cooking time, it was of interest to determine the influence of cooking time on PER values of quick-cooking large Lima beans. The PER value found for quick-cooking Limas, cooked for 5 min, was slightly but not significantly higher than a comparable sample of standard beans cooked ten times longer (Table 11.8). Extending the cooking time to six times normal, reduced the observed PER about 10%. No significant differences were observed between PER values for paired samples of standard and quick-cooking beans (Table 11.9) prepared from pink and pinto *(P. vulgaris)* blackeye *(V. unguiculata)* and another lot of large Lima *(P. lunatus)* beans. The colored *P. vulgaris* types had significantly lower PER values than the two light-colored varieties. The whole diet and nitrogen digestibilities of all four

TABLE 11.6

TRYPSIN INHIBITOR ACTIVITY OF RAW AND COOKED DRY BEANS[1]

(Large Lima and Sutter Pink Beans)

| Condition | Trypsin Inhibitor Activity, U/mg | | | |
| | Raw | | Cooked[4] | |
	Pink[2]	Lima[3]	Pink	Lima
Dry	18	62	—	—
Water-soaked	16	56	0 (60)	0 (50)
Quick-cooking[5]	16	72	0 (10)	0 (8)

[1]Rockland and Jaffe (unpublished data).
[2]Commercial, 1970 crop.
[3]Ventura, UCD, 1970 crop.
[4]Cooking time within parenthesis.
[5]Soaked in quick-cooking salt solution.

TABLE 11.7

NUTRITIONAL AND BIOLOGICAL PROPERTIES OF CALIFORNIA LARGE DRY LIMA BEANS[1]

Treatment	Protein Efficiency Ratio		Digestibility, Whole diet, %		Pancreas weight g/100g body wt	
	Raw	Cooked	Raw	Cooked	Raw	Cooked
Water-soaked	0.0	1.5	83	93	0.90	0.55
Quick-cooking	0.0	1.4	84	91	0.67	0.54
Casein (control)	2.5		95		0.49	

[1]Reported by Rockland et al. (1974B).

TABLE 11.8

EFFECT OF COOKING TIME ON PROTEIN QUALITY OF CALIFORNIA LARGE DRY LIMA BEANS[1] (Ventura, SCFS)

Condition	Cooking time (min)	Protein Efficiency Ratio[2]
Water-soaked	50	1.92
Quick-cooking	5	1.96
Quick-cooking	15	1.84
Quick-cooking	30	1.71

[1]Reported by Rockland et al. (1974B).
[2]Relative to casein (2.50).

TABLE 11.9

PROTEIN QUALITY OF WATER-SOAKED AND QUICK-COOKING CALIFORNIA DRY BEAN VARIETIES[1]

Variety	Protein Efficiency Ratio[2]	
	Water-soaked	Quick-cooking
Pinto	1.2	1.3
Pink, Sutter	1.2	1.2
Blackeye	1.5	1.5
Large Lima, Ventura	1.5	1.5

[1]Rockland et al. (1974A).
[2]Cooked beans relative to casein (2.50).

types of beans were lower than for the casein standard (Table 11.10). The nitrogen digestibilities of the light-colored Lima and blackeye beans were also higher than the pink and pinto beans. In general, light-colored beans appear to have greater overall nutritional quality than more pigmented types. The colored, *P. vulgaris* types appear to contain larger amounts of lectins and antitrypsin than light-colored beans (Table 11.4). However, the known lectins and antienzymes are heat labile and therefore cannot be directly responsible for lower apparent nutrient qualities. Elias *et al.* (1976) have suggested that tannins in the seed coat of colored beans may influence protein quality evaluations, and Becker *et al.* (1976) have reported a heat stable trypsin inhibitor in California small white beans.

Supplementation of large Lima beans with small amounts of complementary meat protein improved the amino acid balance, and increased the PER of the mixture by 50% (Table 11.11). Further supplementation of this mixture with milk protein raised the PER from 1.4 to 2.5, an im-

TABLE 11.10
PROTEIN QUALITY OF CALIFORNIA DRY BEANS[1,2]

| Variety | Protein Efficiency Ratio | Digestibility, % | |
		Whole Diet	Nitrogen
Pinto	1.2	90	75
Pink, Sutter	1.2	89	73
Blackeye	1.5	93	79
Large Lima, Ventura	1.5	90	79
Casein	2.5	95	94

[1]Reported by Rockland *et al.* (1974B).
[2]Cooked, water-soaked beans.

TABLE 11.11
EFFECTS OF NATURAL SUPPLEMENTS ON PROTEIN QUALITY
OF QUICK-COOKING DRY LIMA BEANS[1] (Ventura variety, UCD, 1971)

| Condition | Material | Protein Efficiency Ratio | Digestibility, % | |
			Whole Diet	Nitrogen
Raw	Lima Beans		83	
Cooked	Lima Beans	1.4	90	76
Cooked	Lima-meat casserole	2.1	96	92
Cooked	Lima-meat casserole + milk protein	2.5	96	92
Control	Casein	2.5	96	92

[1]Rockland *et al.* (1974A).

provement of almost 80%, making the blend equivalent to the standard protein, casein. Concurrently the whole diet digestibility of the mixture increased by 13% to the same level as the casein standard. Similarly nitrogen digestibility of the complete mixture was 21% higher than Lima beans alone and also equivalent to pure casein. Similar experiments with other varieties of dry beans confirmed the unanticipated parallel increases in PER, diet and nitrogen digestibilities when appropriate levels of complementary proteins were used as supplements to cooked beans. Further research is needed on interrelationships between chemical and biological properties of the legumes and their overall nutritional qualities.

Composition and Nutrients

Evidence for essentially nutrient and compositional equivalence of standard and quick-cooking beans was obtained from comparisons of proximate composition (Table 11.12), mineral and trace element analyses (Table 11.13) and several B-complex vitamins including thiamine, pyridoxine, niacin and folacin (Table 11.14).

Authentic, isogenic bean varieties were examined. Analogous, paired samples of cooked standard and quick-cooking beans contained essentially identical protein content, although individual varieties each had a characteristic protein level. For example, pink beans contained about 27%, blackeye beans 25%, and large Lima 21% crude protein (N × 6.25). Cooked, quick-cooking beans contained slightly higher levels of nonvolatile ash, most of which may be related to an increase in sodium chloride which was irreversibly absorbed from the original processing solution. Quick-cooking beans also contained slightly higher levels of phosphorous (Table 11.13) for the same reason. Lower levels of calcium and/or magnesium in quick-

TABLE 11.12

PROXIMATE COMPOSITION OF STANDARD AND QUICK-COOKING BEANS[1,2]

Variety	Type	Protein[3] (%)	Lipid (%)	Fiber (%)	Ash (%)	Carbohydrate (%)
Large Lima	Standard	20.8	1.9	7.4	4.9	65.0
	Quick-cooking	20.4	1.5	5.9	5.3	64.9
Blackeye	Standard	25.4	2.2	3.5	2.8	66.1
	Quick-cooking	24.7	2.4	3.6	4.4	64.9
Pink	Standard	26.6	1.7	4.7	3.8	63.2
	Quick-cooking	26.5	1.6	4.9	4.0	63.0

[1]Rockland *et al.* (1974A).
[2]Cooked, drained beans. Calculated to a dry weight basis.
[3]Nitrogen × 6.25.
[4]By difference.

TABLE 11.13

MINERALS IN STANDARD AND QUICK-COOKING BEANS[1,2]

Variety	Type	Phosphorous (mg %)	Calcium (mg %)	Magnesium (mg %)	Iron (mg %)
Large Lima	Standard	370	40	150	9
	Quick-cooking	450	30	120	8
Blackeye	Standard	510	60	170	6
	Quick-cooking	640	40	170	6
Pink	Standard	530	100	150	8
	Quick-cooking	530	100	120	10

[1]Rockland *et al.* (1974A).
[2]Cooked, drained beans. Calculated to a dry weight basis.

TABLE 11.14

VITAMINS IN STANDARD AND QUICK-COOKING BEANS[1,2]

Variety	Type	Thiamine (B[1]) (mg %)	Pyridoxine (B[6]) (mg %)	Niacin (mg %)	Folacin (μg %)
Large	Standard	0.55	0.55	1.1	100
	Quick-cooking	0.60	0.60	1.7	100
Blackeye	Standard	0.55	0.35	1.2	400
	Quick-cooking	0.75	0.40	1.1	600
Pink	Standard	0.55	0.40	0.90	150
	Quick-cooking	0.35	0.30	0.65	100
Pinto	Standard	0.55	0.35	0.75	100
	Quick-cooking	0.20	0.25	0.60	150

[1]Taken from Rockland *et al.* (1977C).
[2]Cooked, drained beans. Calculated to a dry weight basis.

cooking beans may be related to the extraction of these bivalent cations as complex tripolyphosphates in the original processing solution. However, standard beans also have lower levels of these elements than the raw beans.

B-complex vitamin levels in standard and quick-cooking beans varied slightly but probably not significantly with the bean type (Table 11.14). In general, cooked legumes are good sources of several B-complex vitamins and some of the dry beans compare favorably with some of the best natural sources of these vitamins (Rockland *et al.* 1977C).

Biological Properties

Gastrointestinal distress and flatulence, which often result from the ingestion of legume products, may be the prime deterrent to greater utilization of legume products for human food. Spontaneous release of intestinal gas is uncomfortable and distressing. Characterization of the flatulence factor(s) has been one of the most intriguing and elusive prob-

TABLE 11.15
RELATIVE FLATULENCE ACTIVITY OF DRY BEANS[1,2]

| Variety | Relative Activity | |
	Human Assay	Microbiological Assay
Large Lima	1.0	1.0
Small white (Calif.)	1.9	1.9
Red Kidney	2.1	2.2

[1]Adapted from Rockland et al. (1969).
[2]Cooked beans with cook water.

lems of the past two decades. During the past few years it has become generally recognized that the primary flatulence gases are predominantly hydrogen and carbon dioxide. These gases are believed to be produced by intestinal bacteria stimulated by unknown factors in legume and other foods as the bolus progresses through the alimentary canal. The presence of hydrogen in the flatus gases precludes its endogenous origin since there is no known biochemical mechanism which permits the formation of hydrogen gas in primate tissues (Rockland *etal.* 1969). Hydrogen and carbon dioxide are recognized end products of phosphoclastic cleavage of pyruvate by gram positive, obligate anaerobes, such as *Clostridium perfringens* and related bacteria which dominate the anaerobic flora in higher animals.

Using a pure culture of *C. perfringens* (Type A, No. 3624) it was demonstrated that legume extracts elicit rapid proliferation of hydrogen and carbon dioxide (Fig. 11.15). It was suggested that the response of microorganisms might be employed as the basis for a presumptive test for flatulence activity in legumes and serve as an assay method for following extraction and isolation of the stimulating factor(s) in legume seeds. The reliability of a presumptive microbiological assay procedure is contingent upon obtaining analogous results from human assays on the same test materials. Available human assay procedures for estimating flatulence activity depend upon volumetric measurement of rectal gases. Biochemical individuality and normal physiological variability are predisposed toward observing large variations in responses from random test subjects. However, in comparisons between results of human and microbiological assays of similar, but not identical samples of three varieties of beans, the estimates of flatulence activities agreed remarkably well (Table 11.15). Concurrence between the relative activities of identical samples of standard and quick-cooking large Lima beans observed in precision human and microbiological studies (Table 11.16) increased the credibility of the microbiological procedure. It was of special significance that both assay

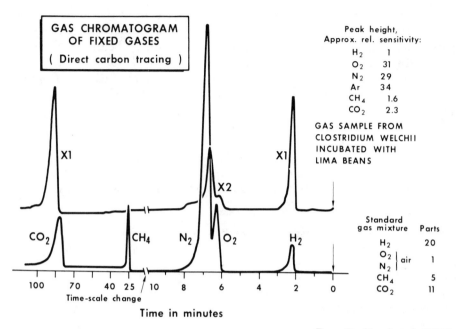

From Rockland et al. (1969)

FIG. 11.15. QUALITATIVE, GAS CHROMATOGRAPHIC ANALYSES OF GASES GENERATED OVER CULTURE OF *C. PERFRINGENS* TYPE A (NO. 3624) INCUBATED WITH A HOMOGENATE OF LARGE LIMA BEANS

TABLE 11.16

ESTIMATION OF RELATIVE GAS STIMULATING ACTIVITY OF VARIOUS COOKED DRY BEAN PRODUCTS[1,2]

Bean Variety		Relative Activity		
		Human Assay (I)	Microbiological Assay (I)	(II)
Blackeye	quick-cooking, drained			0.26
	standard + cookwater			0.67
Large Lima	quick-cooking, drained	0.42	0.43	0.37
	standard + cookwater	1.00	1.00	1.00
Pinto	quick-cooking, drained			1.2
	standard + cookwater			2.2

[1]Adapted from Rockland *et al.* (1971).
[2]Large Lima plus cookwater taken as unity.

procedures confirmed subjective judgments concerning the lower flatulence activity of quick cooking large Lima beans. Microbiological assays of paired samples of cooked standard and quick-cooking beans, evaluated both with and without added cook-water, confirmed the proposal that quick-cooking beans have lower apparent flatulence activities than samples of the same lots of beans prepared in the conventional manner (Table 11.17).

TABLE 11.17

PRESUMPTIVE MICROBIOLOGICAL ASSAY FOR RELATIVE FLATULENCE ACTIVITIES OF COOKED STANDARD AND QUICK-COOKING BEANS[1,2]

Variety	With cook-water		Drained	
	Standard	Quick-cooking	Standard	Quick-cooking
Blackeye	1.8	1.5	1.0	0.7
Large Lima	2.7	2.1	1.4	1.0
Pinto	6.0	4.0	4.5	3.1

[1]Rockland et al. (1971).
[2]Relative to drained, quick-cooking Lima beans = 1.0.

BIBLIOGRAPHY

ADAMS, M. W., ROCKLAND, L. B. and MONFORT, G. W. 1978. Food legumes as protein sources. In Protein Resources and Technology: Status and Research Needs. M. Milner, N. S. Scrimshaw and I. C. Wang (Editors). AVI Publishing Co., Inc., Westport, CT.

BECKER, R., OLSON, A. C., MEIRS, J. C., GUMBMANN, M.R. and WAGNER, J. R. 1976. Effects of legume trypsin inhibitor on hydrogen production by the rat. Abstract of paper (No. 392) presented at the 36th Annual Meeting of the Institute of Food Technologists, Anaheim, CA, June 6–9.

BINDER, L. J. and ROCKLAND, L. B. 1964. Use of the automatic recording shear press in cooking studies of large dry Lima beans (Phaseolus lunatus). Food Technol. 18(7), 127–130.

BRESSANI, R. 1975. Legumes in human diets and how they might be improved. In Nutritional Improvement of Food Legumes by Breeding. M. Milner (Editor). Wiley & Sons, New York.

ELIAS, L. G., deFERNANDEZ, D. G. and BRESSANI, R. 1976. Studies on the possible effects of seed coat pigments of beans on the nutritional value of its protein. Abstract of paper (No. 391) presented at the 36th Annual Meeting of the Institute of Food Technologists, Anaheim, CA, June 6–9.

FAO PRODUCTION YEARBOOK. 1975. FAO Statistics Series. Food and Agriculture Organization of the United Nations, Rome. 1976.

HAHN, D. M., JONES, F. T., AKHAVAN, I. and ROCKLAND, L. B. 1977. Intracellular gelatinization of starch in cotyledons of large dry Lima beans. J. Food Sci. 42, 1208–1212.

HONAVAR, P. M. and LIENER, I. E. 1962. The inhibition of growth of rats by purified hemagglutenin isolated from *Phaseolus vulgaris*. J. Nutr. *17*, 109.

LIENER, I. E. 1969. Toxic Constituents of Plant Foodstuff. Academic Press, New York.

LEOPOLD, A. C. and ARDREY, R. 1972. Toxic substances in plants and food habits of early man. Science *176*, 512.

MONTGOMERY, R. D. 1969. Cyanogens. *In* Toxic Constituents of Plant Foodstuffs. I. E. Liener (Editor). Academic Press, New York.

MORRIS, H. J. and WOOD, E. R. 1956. Influence of moisture content on keeping quality of dry beans. Food Technol. *10*(5), 1–5.

NATIONAL ACADEMY OF SCIENCES. 1975. The winged bean, a high-protein crop for the tropics. Report of Ad Hoc Panel of the Advisory Committee on Technology Evaluation. National Academy of Sciences, Washington, D.C.

RACKIS, J. J., McGHEE, J. E., LIENER, I. E., KAKADA, M. L. and PUSKI, G. 1974. Problems encountered in measuring trypsin inhibitor activity of soy flour. Cereal Science Today *19*(11), 513–516.

ROCKLAND, L. B. 1964. Effects of hydration in salt solutions on cooking rates of Lima beans. Proc. Seventh Research Conference on Dry Beans. Ithaca and Geneva, New York, Dec. 2–4, 1964. ARS 74–32, U.S. Dept. of Agriculture.

ROCKLAND, L. B. 1969. Search for a convenient assay method for the flatulence factor in dry beans. Proc. Ninth Research Conference on Dry Beans. Fort Collins, Colorado, August 13–15, 1968. ARS 74–50, U.S. Dept. of Agriculture, January, 1969.

ROCKLAND, L. B. 1972. Quick-cooking soybean products. U.S. Pat. 3,635,728. Jan. 18.

ROCKLAND, L. B., GARDINER, B. L. and PIECZARKA, D. 1969. Stimulation of gas production and growth of *Clostridium perfringens* Type A (No. 3624) by legumes. J. Food Sci. *34*, 411–414.

ROCKLAND, L. B., HAHN, D. M. and ZARAGOSA, E. M. 1977A. Frozen, quick-cooking beans prepared from dry beans. Food Prod. Develop. *11*(3), 34.

ROCKLAND, L. B., HAYES, R. J., METZLER, E. and BINDER, L. 1967. Process for producing quick-cooking legumes. U.S. Pat. 3,318,708. May 9.

ROCKLAND, L. B., HEINRICH, J. D. and DORNBACK, K. J. 1971. Recent progress on the development of new and improved quick-cooking products from Lima and other dry beans. Proc. Tenth Dry Bean Research Conf. Davis, California, August 12–14, 1970. ARS-74–56, U.S. Dept. Agriculture, Feb. 1971.

ROCKLAND, L. B. and JAFFE, W. G. unpublished data.

ROCKLAND, L. B. and JONES, F. T. 1974. Scanning electron microscope studies on dry beans; Effects of cooking on the cellular structure of cotyledons in rehydrated large Lima beans. J. Food Sci. *39*, 342–346.

ROCKLAND, L. B., JONES, F. T. and HAHN, D. M. 1977B. Light and scanning electron microscope studies on dry beans: Extracellular gelatinization of Lima bean starch in water and a mixed salt solution. J. Food Sci. *42*, 1204–1207.

ROCKLAND, L. B. and JONES, F. T. and McCREADY, R. M. unpublished data.

ROCKLAND, L. B., METZLER, E. 1967A. Preparation of frozen quick-cooking legumes. U.S. Pat. 3,352,687. Nov. 14.

ROCKLAND, L. B., METZLER, E. A. 1967B. Quick-cooking Lima and other dry beans. Food Technol. *21*(3A) 26A–30A.

ROCKLAND, L. B., MILLER, C. F. and HAHN, D. M. 1977C. Thiamine, pyridoxine, niacin and folacin in quick-cooking beans. J. Food Sci. *42*, 25–28.

ROCKLAND, L. B., ORACCA-TETTEH, R. and ZARAGOSA, E. M. manuscript in preparation.

ROCKLAND, L. B., ZARAGOSA, E. M. and HAHN, D. M. 1974A. New information on the chemical, physical and biological properties of dry beans. Proc. Bean Improvement Cooperative and National Dry Bean Research Association Conference, M. H. Dickson (Editor). New York State Agric. Expt. Station, N.Y., Feb. 1974. pp. 93–107.

ROCKLAND, L. B., ZARAGOSA, E. M., HAHN, D. M., BOOTH, A.N. and ROBBINS, D. J. 1974B. Summarized Text. Proc. Fourth International Congress of Food Science and Technology, Madrid, Spain. Sept. 22–27.

WERTIME, T. A. 1973. The beginnings of metallurgy: A new look. Science 182(4115), 875–887.

ZOHARY, D. and HOPF, M. 1973. Domestication of pulses in the old world. Science 182(4115), 887–894.

CHAPTER 12

ENZYME ACTION AND MODIFICATIONS OF CELLULAR INTEGRITY IN FRUITS AND VEGETABLES: CONSEQUENCES FOR FOOD QUALITY DURING RIPENING, SENESCENCE AND PROCESSING

SIGMUND SCHWIMMER

Western Regional Research Center
Agricultural Research Service
U.S. Department of Agriculture
Berkeley, CA 94710

ABSTRACT

Modification of cellular integrity of fruits and vegetables leading to cellular disruption can create identity, be deleterious or be beneficial with respect to food quality. Frequently these consequences of cell disruption arise as the result of the potentiation of the action of the food's endogenous enzymes, even when, as in the case of a recent study of the keeping quality of refrigerated coleslaw, the deterioration is suspected as being due to microbial growth. Desirable potentiation of endogenous enzyme action may occur preharvest in vivo *or postharvest* in situ. *Examples of beneficial effects of water-deprivation stress and insect damage illustrate desirable preharvest potentiation. Among the many approaches which lead to desirable postharvest potentiation, we zero in on the application of a transient temperature increase, 40–80°, as well as suggest that regeneration of blanch-inactivated enzymes may be beneficial. A model is presented of how this temperature-induced potentiation occurs. It is based on the development of a cascade of physical, chemical and enzymatic events leading to increasing disarray of the structural elements of cellular envelopes, especially the temperature-sensitive organelle membranes, via disorientation, desensitization, deregulation, disfunction, and finally disruption. The action of many of these enzymes gives rise to a wide diversity of end products which confer upon foodstuffs their altered quality attributes. While the nature and the level of the substrates and isozymes are of importance in generating this diversity, the outstanding feature of the action of many cellular-disruption-induced potentiated enzymes is the production of unstable, highly reactive primary products which become intermediates in an ever-widening network of interrelated secondary reactions, as well as the development of alternative pathways leading to alternative food quality attributes. Three pathways leading from the primary product (enzymatic or nonenzymatic) encompass: rearrangements; decomposition, degradation, or polymerization; and interaction with food constituents. The origin of diversity of both food quality and of these end products and their interrelations are illustrated by an account of new findings and reinterpretations of old findings on established, food-related enzymes: lipoxygenase, polyphenol oxidase, alliinase and glucosinolase; e.g., the latter includes "epithio specifier" protein, syn-anti isomerism of substrates, and its hitherto unexpected intracellular localization within the dilated cysternae of the site of protein synthesis, the rough endoplasmic reticulum.*

We also undertake a discussion of the effect on food quality of nonlipoxygenase-induced formation and removal—via superoxide dismutase—of active forms of oxygen in fruits and vegetables, including the nonhydrolase depolymerization of polysaccharides. We shall touch on the impact and promise of these new insights for the future of food quality creation and maintenance, as well as their implications for efficacious utilization of diminishing agricultural and energy resources.

INTRODUCTION

A giant stride in the evolution of pre-agricultural human cultures took place with adaptation of cooking and leaching of the edible plants laboriously gathered in the course of the day's quest for food. According to Leopold and Ardrey (1972) these innovations, by removing a variety of toxic substances, made available to man an entirely new class of foods and thus had a profound influence on the subsequent development of social characteristics. We are now in the midst of new cultural adaptation processes in which technology is attempting to make available to man highly nutritious and inviting foodstuffs which, for a variety of reasons, including the presence of toxic constituents, make these sources unavailable as food. Among the approaches employed and proposed to achieve this goal (cooking, leaching and their sophisticated modern engineering refinements and variants) is the harnessing and controlling of the action of enzymes. Control of the action of enzymes already present in the food (endogenous enzymes) *in vivo* or *in situ* prior to ingestion may manifest itself either as attempts to completely suppress such action where it is undesirable or to potentiate such action in the presence of endogenous substrate when the result of such action is beneficial. In a previous publication (Schwimmer 1972) we pointed out that this overt manifestation of enzyme action is frequently dependent on cellular disruption, or more generally, on the modification of cellular integrity.

While cell disruption may occur by brute mechanical shear forces during handling and processing of fruits and vegetables, it may also occur during the normal life cycle of the plant. For instance, cell disruption accompanying senescence is essential for harvesting of high quality dates (Coggins *et al*. 1968). As shown with the aid of light microscopy (Fig. 12.1), the cells of both immature and of low quality dates are intact, as evidenced by the well-defined cell walls. On the other hand, in tree-ripened dates or in poor-quality dates upgraded by vacuum hydration or by vacuum infiltration at elevated temperature as shown in Fig. 12.1, the boundaries between cells are abolished or become indistinct. This cell disruption is accompanied by the activation of pectinases, polyphenol oxidase, and invertase as well as cellulase (Hasegawa and Smolensky 1971). The activation of these autolytic enzymes converts a light colored, tasteless and turgid fruit into a dark, moist, soft product typical of a high quality date.

To observe more subtle intracellular modification of intracellular integrity it is frequently necessary to go from the light to the electron microscope, as illustrated in Fig. 12.2 (Schwimmer 1972). Potatoes respond to stresses such as temperature extremes and ionizing radiations by converting part of their starch to sugar, with reducing sugars predominating in the case of low temperature storage. Upon subsequent processing, these reducing sugars, as is well-known, give rise to brown products. The question arises as to what triggers the conversion of starch to sugars. Most of the enzymes involved appear to maintain their normal activity. Paez and Hultin (1970) showed that potato mitochondria display a temperature response typical of enzyme reactions. [An exception, which accounts for the appearance of reducing sugar, is the sucrose-hydrolyzing enzyme in-

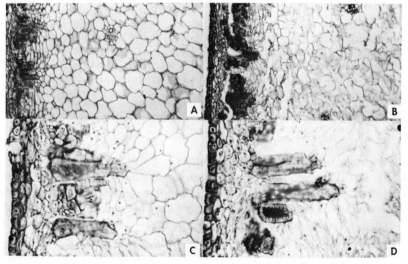

From Schwimmer (1972)

FIG. 12.1.   CELL DISRUPTION IMPROVES DATE QUALITY

Photomicrographs (21×) of cross section of date tissue located about midway from stem to stylar end of fruit: (A) Kimri stage of development, turgid and green; (B) Tree-ripe stage of development but not as ripe as a completely ripe date; (C) Number 2 dry grade of date; (D) Number 2 grade of date after filtration and incubation for 24 hr at 50° C. These photographs were supplied through the courtesy of Dr. C. W. Coggins, University of California, Riverside.

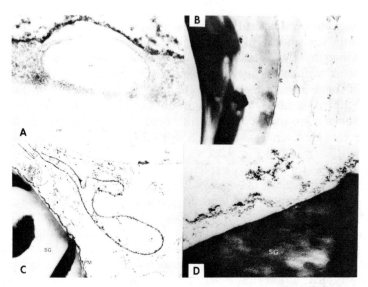

From Schwimmer (1972)

FIG. 12.2. DEVELOPMENT AND DISINTEGRATION OF STARCH-GRANULE-ASSOCIATED MEMBRANES

(A) Section through the cell of an immature potato tuber. Within the cytoplasm is a protoplast (pp) surrounded by a double membrane (27,800×); (B) Section for a mature tuber immediately after harvest. A starch granule is seen surrounded by the intact plastic membrane, pm (6,400×); (C) Section from a tuber stored at 25° C for 31 days (13,900×); (D) Section through a tuber stored at 5° C for 12 days. The membrane has disintegrated and moved away from the starch granule, whereas the vacuolar and cytoplasmic membranes remain intact (25,200×) (Ohad *et al.* 1971).

vertase whose activity rises in cold-storage potatoes (Schwimmer 1962). Pressey (1966) showed that expressed invertase activity is the resultant of the action among a specific invertase inhibitor, de novo enzyme synthesis and enzyme turnover.] The actual trigger may be, according to Ohad *et al.* (1971), the disruption and/or modification at low temperatures, of a membrane surrounding starch granules of freshly harvested or room temperature-stored potatoes. This membrane disruption putatively renders the granule susceptible to starch-degrading enzymes which can give rise to reducing sugars.

Figure 12.2 displays electron micrographs of potato sections showing the protoplastid in immature tubers which becomes a membrane-encased starch granule in the mature and room temperature-stored tubers. Thus, we have a "physiological" internal cell disruption functioning to protect the cell against the external stress of low temperature.

While the question of the trigger responsible for the starch-sugar transformation in potatoes is still far from solved (Isherwood 1976) the foregoing does serve to illustrate the prominent role of membranes in the thinking of investigators about biological phenomena in general and more recently about the modification of foodstuffs in particular. Thus, although Isherwood could not confirm the presence of disintegrated amyloplast membranes in potatoes stored at 2° C, he did find them in association with sugar accumulation during sensecence.

Microbial *vs* Enzymatically Induced Food Quality Deterioration—Coleslaw

While development of off-odors in frozen vegetables is commonly and correctly attributed to endogenous attrite enzyme action, off-odor in fresh salad vegetables such as sliced cabbage held at refrigerator temperatures has usually been attributed to microbial degradation of the tissue. In response to requests for information concerning keeping qualities of delicatessen coleslaw in a sour cream dressing, microbiologically oriented investigators at the Western Regional Research Center conducted an inquiry as to the cause of the development of an off-odor (King *et al.* 1976). Instead of pinpointing a particular microorganism as the cause of the spoilage they found, as shown in Fig. 12.3, that the total microbial count actually decreased with time at the typical display temperature of 7° C. Furthermore they observed that the off-odor was no worse at 14° C, where there was evidence for microbial infestation, than it was at 7° C. The authors suggested that the cabbage portion of the coleslaw at first continued to respire and thus used up the oxygen whose access to the tissue was impeded by the plastic bags used and by the oily emulsion (mayonnaise and sour cream) surrounding the cabbage strips. This lack of oxygen resulted in tissue death, disruption of the cells, and enzyme degradation. They considered the latter to be the probable main cause of deterioration.

BENEFICIAL EFFECTS OF STRESS AND CELLULAR MODIFICATION

While low temperatures and other stresses result in the undesirable accumulation of sugars in potatoes, stresses to which other food plants may be subject, both in the pre- and postharvest period, can lead to improved food quality. Sugar build-up is but one specific example of accumulation of low molecular weight metabolites in such stressful situations as extremes of temperatures, nutrient deficiency and water deficit. In each case, membrane modification and its entrainment of altered enzyme action is implicated.

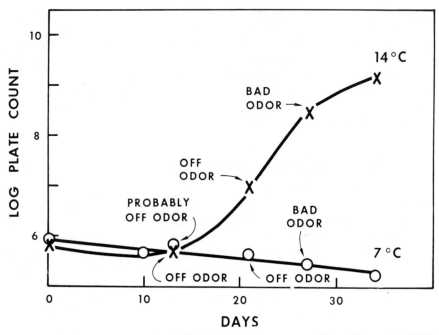

From King et al. (1976)

FIG. 12.3.   TOTAL AEROBIC BACTERIAL COUNT AND ODOR JUDGMENT
ON COLESLAW
At 7° C the cold-susceptible organisms gradually died and off-odor was probably
due to enzyme action ensuing after cell death and disruption.

Quality Improvement *via* Water Stress

Freeman and Mossadegh (1973) demonstrated that onions grown under
water-stress conditions accumulate S-propenyl-L-cysteine sulfoxide, the
precuror-substrate of the enzyme responsible for the development of onion
odor and that watercress accumulated glucosinolates (mustard oil
glucosides) under similar conditions. Prolonged overwinter postharvest
storage and sprouting result in further increases in onion strength
(Freeman and Whenham 1976A) due to the release of the flavor precursor
from gamma-glutamyl peptide bondage by the action of a gamma-
glutamyl transpeptidase (Schwimmer 1973; Schwimmer and Austin
1971).

Enzyme action during the growth of the tea leaf and the manufacture of
the beverage illustrate the beneficial effects, both pre- and postharvest, of
water deprivation in relation to cellular organelle damage. According to

Wickremasinghe (1974), "flavory" weather with respect to the growth of tea leaves in Sri Lanka corresponds to hot, dry, cloudless days followed by cold nights. The resulting improved flavor of the tea, according to the author, may be traced back to a partial disorganization and injury of the chloroplasts imposed by the lack of adequate water. This in turn results in a shift of some metabolic pathways from the chloroplast to the cytoplasm, so that the aroma-bearing volatile terpenoids, which contribute to the tea flavor, are synthesized at an enhanced rate outside the chloroplasts, apparently from the amino acid leucine (instead of acetate) as the ultimate precursor. Further along in the manufacture of tea, the leaf is subject to a withering process which further promotes or selectively potentiates those enzymes which improve the ultimate quality of the tea. Conditions for operation of home dehydrators afford an example of the beneficial effects of the dehydration regime at relatively low temperatures advocated by Gee *et al.* (1977).

In a reversal of their usual role as undesirable agents of cell disruption, certain insects such as the Ceylon mite or the greenfly improve tea flavor, also by damaging the chloroplasts.

MEMBRANE DISRUPTION AND THERMAL POTENTIATION OF DESIRABLE ENZYME ACTION

Some approaches to the potentiation of desirable enzyme activity are shown in Table 12.1. A special category in which cellular membrane modification probably plays a prominent role, is that elicited by simply exposing the food for a relatively short period, from seconds to a few hours, to temperatures in the range of about 40–80° C. Some of the beneficial results derived from such treatments and proposals for incorporation into food processing lines are listed in Table 12.2. Of course, such treatment does not invariably result in quality improvement, and not all of those

TABLE 12.1

METHODS OF ENZYME POTENTIATION

Chemicals	Preprocessing	Process variables
Hormones	Ripening	Temperature
Salts	Storage	Time
Coenzymes	Curing	pH
Vitamins	Germination	Water
Activators	Genetic Engineering(?)	Mechanical

listed in Table 12.2 have been adopted for large-scale use in the food industry. They are, however, indicative of the versatility and potential of controlled potentiation of endogenous food enzymes, especially when the intricacies of the consequences of the secondary reactions some of which are herein adumbrated, are more thoroughly explored, understood and applied.

Perhaps the most successful example is the interposition of the "pre-blanch" in the canning of vegetables where a certain degree of firmness is desirable and sloughing is to be avoided. These foods include snap or string beans, cauliflower, potatoes and tomatoes as well as canned apples, cherries and apricots.

We propose that regeneration of heat inactivated enzymes, although involving a somewhat higher temperature range than that in the other

TABLE 12.2
IMPROVEMENT OF FOOD QUALITY BY ENZYME POTENTIATION,
$60 \pm 20°$ C

Food/Process	Improvement	Enzymes	References[1]
Fruits and vegetables	Firmness	Pectin esterase (+ Calcium salt)	1
Fruits and vegetables	Flavor Appearance	Regenerated enzymes	2
Beans, dry Wheat	Nutrition Acceptance	Phytase, Alpha-galactosidases	3
Soybean flour	Flavor	Lipoxgenase(?)	4
Vegetables, green	Color, Health(?)	Chlorophyllase(?)	5
Sweet Potato, dehydrated	Color, Appearance	Alpha-Amylase	6
Mushrooms, Potatoes	Flavor (5'-nucleotides)	Ribonucleases	7
Single cell protein (SCP)	Removal of gout-causing purines via nucleotides	Ribonucleases	8
Tea	Flavor, Color	Phenolase	9
Dates	Color, Texture, Flavor	Phenolase, Invertase Cellulase, PG	10
Oranges	Flavor, removes bitterness	Limonin-catabolizing enzymes	11
Unconventional protein sources— seeds, SCP, FPC forages	Nutrition, Texture Flavor	Proteinases	12
Meat	Texture	Collagenase	13
Meat, aging at low temperature	Texture	Calcium-activated proteinase, weakens Z-line	14

[1] References: 1. Appleman and Conrad (1927), Labelle (1971), Bartolme and Hoff (1972); 2. Winter (1976); 3. Becker et al. (1974), Chang and Schwimmer (1977); 4. Schwimmer (1975); 5. Clydesdale and Francis (1968), Holden (1965); 6. Debold et al. (1968); 7. Shimazono (1964), Buri and Solms (1971); 8. Ohta et al. (1971); 9. Sanderson (1975); 10. Coggins et al. (1968); 11. Rockland et al. (1957); 12. Schwimmer (1975); 13. Laakonen (1973); 14. Olson et al. (1977).

examples cited in Table 12.2, may constitute another example of enzyme potentiation. Some processors have noticed that packs in which regeneration has occurred, although posing problems of meeting perhaps arbitrary specifications, have superior storage stability in comparison with packs in which no regeneration has occurred. It would appear that regeneration of desirable enzymes such as superoxide dismutase, discussed below, and perhaps catalase or even peroxidase is fortuitously favored over undesirable enzymes. Another possibility is that the pathways of secondary reactions of the same enzyme are shunted to yield final products which confer improved food quality. Indeed, some potentiations, such as that of soy treatment (Schwimmer 1975), may be due to such pathway shifts.

Included in Table 12.3 are two examples of potentiation of meat tenderizing proteinases to show that this phenomenon is not confined to fruits and vegetables, nor to temperatures above ambience.

Potentiation and Membrane Modification

With the possible exception of the action of α-amylase in sweet potatoes (where the potentiation may be due, at least in part, to the accessibility of the enzyme to the substrate, starch, occasioned by the hydration of the latter by heating thus making it amenable to enzyme attack) the invocation of enzyme action in the examples of Table 12.3 may be interpreted in terms of some modification of cellular envelopes including walls but especially membranes (Fig. 12.4). A clue to the initial events is afforded by the observations of investigators on the altered physical state of the membrane phospholipids as detected, for instance, by spin-labelling of cold-hardy plants as compared to that of cold-sensitive plants (Raison *et al.* 1971) and the association of lowered phospholipid level to increased tendency of potatoes to undergo enzymatic browning (Mondy and Mueller 1977). In the latter case the deficiency of phospholipid may be indicative of thinner membranes which would be more susceptible to disruptive forces.

TABLE 12.3
PRIMARY PRODUCTS OF SOME PLANT FOOD RELATED ENZYMES

Polyphenol oxidase (Phenolase)	Quinones
Lipoxygenases	Lipohydroperoxides
Flavoprotein oxidases and others	Superoxide anion, $O_2^{\bar{}}$, Singlet oxygen, O^*_2
Fatty Acid α-Oxidation	Long chain aldehydes
Alliinase (L-Cysteine sulfoxide lyase)	Alkan(en)e sulfenic acids
Thioglucosidase (Glucosinolase, myrosin)	Thiohyroxamic-O-sulfonates
Anthocyanases (β-glucosidases)	Anthocyanidins

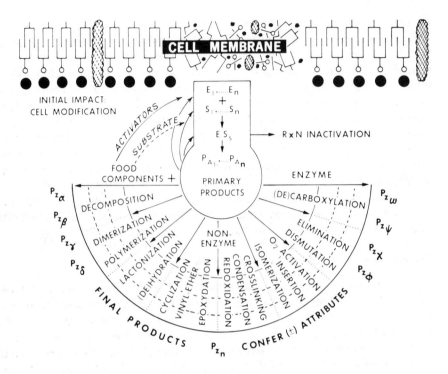

FIG. 12.4. DEPICTION OF ENZYME POTENTIATION FOLLOWING MOD-
IFICATION OF CELLULAR INTEGRITY (MEMBRANE DISRUPTION) AND
THE FORMATION OF UNSTABLE PRIMARY PRODUCTS OF ENZYME AC-
TION WHICH UNDERGO SUBSEQUENT REACTION LEADING TO PROD-
UCTS WHICH CONFER CHANGED QUALITY ATTRIBUTES ON THE FOOD

It is not too far fetched to assume that in the temperature range under
consideration some of the membrane lipids "melt" as indicated by the
nonaligned phospholipid symbols in Fig. 12.4. This disorientation of the
aligned phospholipids would undoubtedly injure the membrane with respect to its normal physiological functions in active transport and hormone
transmission but may not, perhaps, render it particularly adapted to
enzyme and substrate interaction. It may, however, produce conditions
under which lysosomal acylhydrolases (lipases) now can have access to the
disoriented polyunsaturated phospholipids and thus produce unsaturated
free fatty acids as suggested by Galliard and Matthew (1977). The cell is
thus now susceptible to further degradative modification via several
mechanisms. In the presence of oxygen in vegetables the ubiquitous
lipoxgenases (Wardale and Galliard 1975) apparently localized in distinct,

exceptionally fragile organelles may be considered as mediators for the production of active oxygen in the form of lipohydroperoxides (see below). Other means of producing activated oxygen are available at the site of the membrane disintegration or less drastic perturbation. The role of hydroperoxides in the onset of senescence processes is discussed in detail elsewhere in this book by Frenkel (1978). As indicated in Fig. 12.4, further membrane disintegration may be promoted by the action of proteolytic enzymes on membrane proteins especially in senescent tissue. Endopeptidase activity (in contrast to exopeptidase which remains at a low but constant level) is induced at the onset of senescence (Feller *et al.* 1977). Such proteinase activity also elicits mitochondrial ATPase whose action would further interfere with the membrane associated action of synthetases (E.C.6.) and with other membrane-associated functions such as active transport (Jung and Laties 1975).

The above sequence of events thus sets the stage for interaction of the enzyme of interest, by allowing for passage of substrate and activators, if the enzyme is immobilized, or for passage of the enzyme to the site of the substrate. Of course, in the intact cell the membrane is not an inert gossamer web but actively participates in the selective access and egress of activators, inhibitors and enzymes. It is also the site of attachment of hormones which exert their effect *via* the "second messenger," cyclic AMP. This, as is now well known, initiates a cascade of sequential activation of a series of enzymes, winding up finally in the "potentiation" of a specific metabolic enzyme such as phosphorylase or lipase and the accomplishment of a particular physiological function. In some ways the events taking place during the thermal potentiation of enzymes in foods may be considered as a quasi-cascade of disorganization, with death mimicing life. Frequently, this charade of a cascade results in the improvement of food quality.

PRIMARY ENZYME PRODUCTS AS INTERMEDIATES
IN SECONDARY REACTIONS

As previously alluded to, the ultimate consequence of cellular modifications in terms of food quality will depend, in part, on the course of the subsequent transformations of the primary products of the enzyme action. Frequently enzyme activity and subsequent train of events as a consequence of cellular disruption is initiated by the formation of relatively unstable primary products as depicted in Fig. 12.4. Some of these primary products formed from the action of enzymes associated with fruit and vegetables are shown in Table 12.3. These unstable products can then usually undergo a plethora of reactions leading to a diversity of products. The primary source of this diversity for a given enzyme reaction type is, as indicated, the presence of numerous substrates and secondarily, the presence in a given food of distinct isozymes. Furthermore, the primary reaction product can then: undergo rearrangements and relatively minor reactions; revert back to the original substrate; react with the enzyme so as to inactivate the latter (reaction inactivation); undergo gross decomposition; or react with other stable food constituents in ever widening ripples of complex interrelated reactions, winding up in stable products which confer quality, desirable or undesirable, on the food. These secondary reactions may be enzymatic or nonenzymatic. It is difficult to tell where to draw the line when considering consequences of enzyme action, in contrast to strictly nonenzyme action, because initially all freshly harvested produce in which the cells are intact may be considered as primary products of enzyme reaction. Thus one might consider the Maillard type of nonenzymatic browning as the result of secondary reactions arising from the action of invertase to produce the primary product, glucose. However, it is only when they are heated that such "primary" products become unstable and yield such an abundance of quality-affecting constituents which confer color, flavor and antioxidant properties on the food. Hence we usually include only those primary enzyme products which undergo transformations readily at temperatures at which enzymes usually act. For a more complete collation and discussion of such reactions see Schwimmer (1979). We shall forthwith examine the consequences of these transformations both in terms of food quality and in terms of the chemistry involved for a selected list of more or less familiar food enzymes.

Phenolase Action

Factors influencing browning.—Traditionally, the action of phenolases (polyphenol oxidases) has been associated with the development of melanin pigments, usually undesirable, in most fruits and vegetables. The ultimate gross causes of discoloration are: mechanical (cutting, crushing, insects, birds); climatic (wind, sun, frost); pathological (virus, bacteria, molds); and physiological (cold storage, internal bruising, senescence). At the cellular level we mentioned the role of the amount of phospholipid of potatoes as an index of membrane fragility and hence tendency to brown. Control of browning in strictly chemical or enzymatic terms may manifest itself in several different ways. Thus at least two fruits, the kiwi and the Sunbeam peach, do not turn brown because of the absence of substrate but not of enzyme. On the other hand cantaloupes and the oranges do not ordinarily brown because they have potential substrates for phenolase but no or very low levels of activity at the pH of the crushed fruit.

When a phenolase-containing fruit or vegetable browns, it is often not clear, but it may be important to know, whether the substrate or the enzyme level limits the rate and extent of browning. In the case of the avocado, it appears to be the overall phenolase activity—as measured, for instance, in crude extracts or by reflectance from slices—and not the level of phenols (although phenol level frequently exhibits some correlation with phenolase activity) nor presence of inhibitor nor isóenzyme pattern nor shift in properties of the enzyme from variety to variety (Kahn 1977).

Diversity of End-Products due to Substrate Variability.—In most fruits and vegetables the substrates of phenolases are tyrosine and/or the chlorogenic acids. The resulting change is usually an undesirable brown or black color. But in some fruits, especially berries and grapes, phenolase action results in the *loss* of desirable pigmentation (anthocyanidins) rather than in the development of undesirable color (Cash *et al*. 1976). When the substrates are flavanol tannins, as in tea and cocoa, the final products are desirable colors such as the thearubigens in teas (Sanderson *et al*. 1972). On the pathway from these tannins to the final products, a rearrangement takes place which results in the degallation of one of the epimerized oxidized intermediates. Thus phenolase may be considered to possess tannase activity with certain flavanols as substrates.

When the substrates are phenols attached to pentosans present in wheat flour, the consequences of phenolase action may be manifested as alteration in dough properties and bread texture (Neukom 1976; Patil *et al*. 1976). However, perhaps a more important concern of the consequences of phenolase action in wheat flour is the problem of acceptance of the off-color bread made from dwarf wheat varieties of the Green Revolution (Singh and Sheoran 1972). Added phenolase (i.e. from mushrooms) may also affect bread textures (Kuninori *et al*. 1976).

Diversity of End-Products due to Transformations of Primary Products.—Classically the primary product of phenolase action (aside from the abberation of the tyrosine-to-DOPA transformation) are the quinones, usually *ortho*. DOPAquinone undergoes a rearrangement to DOPAchrome which then polymerizes to melanins (Lerner 1953). In the presence of reducing substances the quinone will revert to the original substrate. The action of ATP in creating a mitochondrial-dependent reduction of quinone *via* reversal of oxidative phosphorylation (Table 12.4) constitutes an interesting instance of cell disruption in which at least one

TABLE 12.4
INHIBITION OF COLOR FORMATION BY MITCHONDRIA AND ATP
(POTATOES)[1]

	Color	
	−ATP	+ATP
A. Potato Extract	100	66
B. Mitochondria from A.	66	5
C. B ++ "Uncoupler"	43	42
(Dinitrophenol)		

[1]From Makower and Schwimmer (1954).

of the subcellular organelles survives and potentially can participate in the control of food quality (Schwimmer 1972).

Another pathway for phenolase-induced "primary" product is that in which an intermediate (in this case not the quinone but an earlier one) reacts with the enzyme and inactivates it (reaction inactivation, suicide substrate) as indicated in Fig. 12.4. Of course, it is theoretically possible for the intermediate to create an activator or more enzyme if the substrate is a proenzyme. A specific example is the autocatalytic transformation of trypsinogen into trypsin.

Quinones can also interact with amino acids—lysine and cysteine—both in the free form and when they are part of protein molecules and thus alter the functional and health-associated properties of the proteins by covalent interaction (Schwimmer 1975, 1979). In the manufacture of tea and cocoa, and perhaps coffee, phenolase-produced quinone can reduce other constituents of these foods, especially during the latter firing stages of their manufacture and thus influence or contribute to the typical aroma of these beverages.

Lipoxygenase Action

Even more than that of phenolase the lipoxygenases, after cell disruption, manifest their protean action in a diversity of products which confer a wide variety of quality attributes on vegetables, cereals and even some fruits such as bananas. We previously alluded to possible involvement of lipoxygenases in cell envelope damage. Its use as an external source of enzyme to improve the color quality of wheat flour dough by bleaching carotene (Haas and Bohn 1934) and its effect on textural properties of dough by shifting SS-SH ratios (Frazier et al. 1973) are perhaps its principle beneficial commercial benefits. However, as an endogenous enzyme it also contributes to the desirable aroma of fresh salad vegetables, especially cucumbers (Hatanaka et al. 1975; Galliard and Phillips 1976) although at least one investigator believes that cucumber odor arises as the result of the direct attack of fatty acid by singlet oxygen (Grosch *et al*. 1974) through dioxetane intermediates (Cilento 1975). Its action contributes to the flavor of string beans (Stone *et al*. 1975), tomatoes (Galliard and Matthew 1977) and undoubtedly others. One vegetable in which it is not involved and in which the aroma is preformed, *i.e.*, does not rise as the result of cellular disruption, is parsley (Freeman *et al*. 1975).

On the other side of the ledger, the end-products of lipoxygenase action have been implicated in the unwanted removal of color from semolina products, again via co-oxidation of carotenes (Walsh *et al*. 1970). It may also remove carotene from leafy vegetables (Blain 1970) and has been implicated in the loss of green color in peas by promoting the oxidation of chlorophyll (Buckle and Edwards 1970; Holden 1965, 1974). It could bleach added dyes. It is generally agreed its action is largely responsible for the objectionable odor of soybean flour and ground beans when these are mixed with water. It also contributes to the reduced shelf life of cereal products with relatively high levels of unsaturated fats such as those containing wheat germ. Its ability to undergo reaction inactivation via potentiation by temporarily increasing the water content of such products, has been used to eliminate its long-range deleterious effects (Wallace and Wheeler 1972).

Origin of Diversity of End Products.—Unlike phenolase, diversity does not arise from a variety of available substrates. Only two widely occurring natural substrates, linoleic and linolenic acids, possess the rigid specificity requirement of this class of enzyme. As is well known, the substrates must contain the *cis, cis*-penta-1,4-diene moiety. It also acts on the naturally occurring arachidonic acid, but this is more important in animals where a lipoxygenase-like enzyme (prostaglandin synthase) converts this acid to prostaglandins (Skinner 1975; Samuelsson *et al*. 1975). Action on its two substrates results in more than two primary products because of the formation of both positional (9- and 13-) and geometric (*cis-trans*) isomers of these primary products, the lipohydroperoxides (LOOHs) of linoleic and linolenic acids. Thus, substances largely responsible for cucumber flavor may be derived from 9-*D*-hydroperoxy-octadeca-*trans*-10, *cis*-12-dienoic acid whereas the ubiquitous hexanal is a descendant of 13-*D*-hydroperoxy- octadeca-*cis*-9,*trans*-11-dienoic acid (Galliard *et al*. 1976). Whereas before the advent of high-pressure liquid chromatography, it was believed that lipoxgenase action produced only *cis-trans* LOOHs, Pattee and Singleton (1977) isolated *trans-trans* isomers of the 13- and 9-linoleic acid hydroperoxides as well as the *cis-trans* isomer of the 13-hydroperoxide from lipoxygenase- rich peanut homogenates. The relative amounts of the LOOH isomers can vary over a wide range of values and are influenced not only by the source of the lipoxygenases and the isozyme pattern but also by pH, the presence of calcium and even by temperature (Axelrod 1974).

Transformation of the Primary Lipohydroperoxides.—These can undergo a variety of reactions (Gardner 1975; Axelrod 1974; Boldinghe, 1976). Gardner (1975) distinguishes among the following pathways as indicated in Fig. 12.5: (1) reduction or nucleophilic reactions involving peroxidase systems including glutathione (Omaye *et al.* 1975); (2) enzymatic isomerization to polyhydroxy derivatives and ketols; (3) epoxidations by flour-water suspensions; (4) vinyl ether formation by a potato enzyme; (5) anaerobic lipoxygenase catalyzed reactions including dimerization due to a reaction between linoleic acid and its hydroperoxide and production of pentane; peanut enzymes also produce pentane aerobically (Singleton *et*

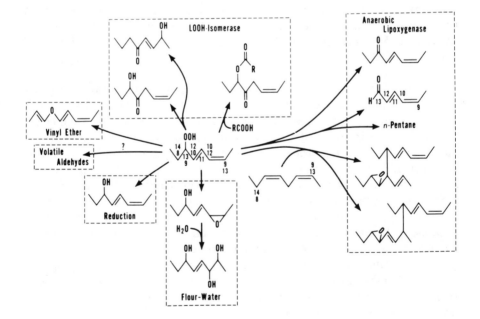

From Gardner (1975)

FIG. 12.5. FURTHER ENZYMATIC TRANSFORMATIONS OF LINOLEIC ACID HYDROPEROXIDES (LOOH), PRIMARY PRODUCTS OF THE ACTION OF LIPOXYGENASE ON LINOLEIC ACID

Carbon numbers of linoleic acid, tridecanaoic acid, and the 9- and 13-LOOH isomers are indicated as a guide to the formation of the transformed products.

al. 1975); (6) production of volatile aldehydes and ketone. There has been some controversy as to whether a special "lyase" is needed to accomplish the scission of the LOOH. The latest consensus appears to favor the participation of lipoxygenase itself (Gardner and Sessa 1977; Galliard and Phillips 1976). More definitive evidence for the presence of such an enzyme is now available (Phillips and Galliard 1978).

These various modes of transformation yield literally hundreds of different products so that it is not surprising that the same enzyme can influence food quality in both desirable and undesirable ways. In many instances, however, our knowledge of the chemical transformations outpaces our understanding of just how these transformations affect the relevant food properties and how they can be applied practically. For instance, little (Kalbrenner *et al*. 1974) known data are available on the organoleptic properties of the C-18 transforms. If they turn out to be inocuous, *i.e*. the vinyl-ether colneleic acid (Galliard and Matthew 1975) their formation might be used to prevent *via* competition some of the undesirable reactions, such as formation of some of the unsaturated aldehydes and ketones. These may, in the proper proportions, contribute to the unacceptable flavor of bean products and unblanched frozen foods (Murray *et al*. 1976).

While lipoxygenase is largely responsible for the formation of the lower aldehydes, attention is now starting to be focused on the presence of intermediate aldehydes (C12–C17). These arise as the result of the action of the fatty acid α-oxidation system present in a variety of plant tissues including vegetables (Shine and Stumpf 1974; Galliard and Matthew 1975). Plants also contain an oxygen-dependent system which decarboxylates long chain fatty acids to alkanes (Khan and Kolattukudy 1974). Although not yet observed in plant tissue, animal microsomes have a powerful NADPH-dependent enzyme system for lipid peroxidation which may also be involved in lipid oxidation in biological membranes in general (Kaschnitz and Hafeti 1975).

Superoxide Dismutase and Non-Lipoxygenase Generators of Active Oxygen

The latter oxygen-dependent enzyme systems constitute examples of oxygen activation without the intervention of lipoxygenase. As indicated in Fig. 12.6, active forms of oxygen may be generated in organisms which utilize oxygen for their normal functioning. These active forms of oxygen which may, as we suggested, participate in postharvest modification of cellular integrity, may also be involved not only in the transformation of lipids, but in the uncontrolled oxidation and decomposition of other major food components as well. Thus, superoxide anion, O_2^-, like quinones and LOOHs, may also be considered as an unstable primary product of enzyme action. The difference between it and some of the other primary products is

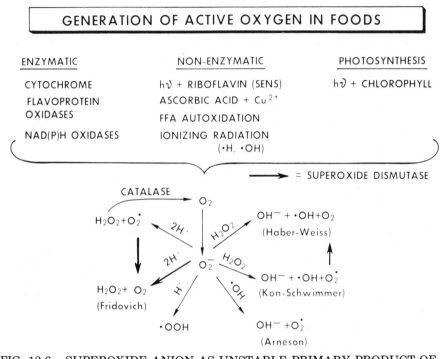

FIG. 12.6. SUPEROXIDE ANION AS UNSTABLE PRIMARY PRODUCT OF ENZYME ACTION LEADING TO OTHER ACTIVE OXYGEN SPECIES PUTA-TIVELY RESPONSIBLE FOR ALTERED FOOD QUALITY
References to the indicated reactions are: Haber and Weiss (1934); Kon and Schwimmer (1977); Arneson (1970); Fridovich (1974).

that it is more unstable and can be transformed into even more active reactants (Fig. 12.6). These, in turn, can interact with food constituents *via* a wide diversity of secondary reactions to yield end-products and food with altered quality attributes.

Among these secondary reactions is the degradation of numerous polysaccharides including that of pectin (Kon and Schwimmer 1976, 1977). An analysis of data such as that shown in Fig. 12.7 and Table 12.5 indicated that both $\cdot$ OH and O^*_2 and, to a lesser extent, perhaps $\cdot$ OOH participate directly in the depolymerization and that the former two arise as a result of a reaction between H_2O_2 and O_2. While there is usually an adequate supply of polygalacturonase and cellulase in fruit and vegetables to account for the postharvest degradation and resulting softening, one possible function of these active forms of oxygen might be to participate in the derangement of the intact supramolecular structure of pectin, such

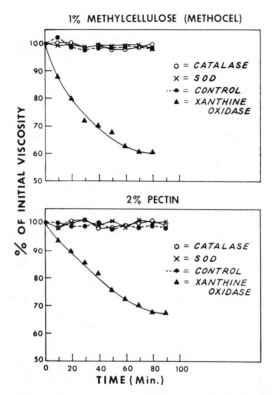

From Kon and Schwimmer (1977)

FIG. 12.7. DEPOLYMERIZATION OF PECTIN AND METHYL CELLULOSE AND ITS PREVENTION BY SUPEROXIDE DISMUTASE OR CATALASE

TABLE 12.5
SUPPRESSION OF XANTHINE OXIDASE-INDUCED DEPOLYMERIZATION OF PECTIN[1]

Depolymerization suppressed by:[2]					
CAT	SOD	EtOH	Mann	DABCO	Mann + DABCO
Which quench, remove or trap:[3]					
H_2O_2	O_2^- (O^*_2?)	$\cdot OH$	$\cdot OH$	O^*_2	$O^*_2 + \cdot OH$
Suppress depolymerization to the following extent (%):					
100	100	59	64	62	86

[1]From Kon and Schwimmer (1977).
[2]CAT = Catalase; SOD = superoxide dismutase; EtOH = ethanol; DABCO = 1,4-diazabicyclo(2,2,2) octane; Mann = mannitol.
[3] H_2O_2, hydrogen peroxide; O_2, superoxide anion; $\cdot$ OH, hydroxy free radical; O^*_2, singlet oxygen.

as the fascicles and elementary fibrils (Dull and Leeper 1975). This action would facilitate cell separation and expose the cell wall-associated cellulose and pectin to attack by the endogenous cellulase and polygalacturonase.

Superoxide Dismutase.—All organisms which utilize oxygen contain the enzyme superoxide dismutase (SOD). At first the sole function of this enzyme was, as indicated in Fig. 12.7, considered to be the acceleration of the spontaneous dismutation of the superoxide anion (McCord and Fridovich 1969). However, as also indicated in the same figure, its primary purpose may be to prevent the formation of singlet oxygen (O^*_2) in this dismutation or to catalyze the conversion of singlet to triplet or ordinary oxygen. For instance it inhibits lipoxygenase which energizes triplet oxygen to the singlet form on the way to producing lipohydroperoxide (Richter *et al.* 1975). Studies on superoxide dismutase are appearing with ever-increasing frequency in the food science literature. It will be important both as an enzyme in fruits and vegetables where its action will have to be taken into account not only in assessing the overall effect of enzyme on food quality but, potentially, as a commercial enzyme to be added to foods to protect them from the deleterious effects of active oxygen. Michelson and Monod (1974) were issued a patent for such use some three years ago. Some of the claims in the patent are set forth in Table 12.6. It is perhaps seemingly surprising that SOD should prevent what would appear to be phenolase-induced melanin formation in potatoes and mushrooms. This suggests that more needs to be known about the enzymatic origin of off-color in foods.

TABLE 12.6
SUPEROXIDE DISMUTASE (SOD) AS FOOD ANTIOXIDANT[1]

Food/Constituent	Effect of SOD
Ascorbic acid	Suppresses oxidation (one week)
Antioxidant	Prolongs effectiveness
Anchovy fat	Inhibits autoxidation (78%) for 44 hr
Mushroom slice in H_2O	Suppresses color in supernatant
Apple: slice, dry	Suppresses color
Potato: slice, dry	Color, odor improved
Potato, carrot: cook, freeze	Flavor, odor improved
Reduce ribonuclease (SH)	Prevents reactivation of enzyme by preventing SH→SS
Salmonella medium	Prolongs effectiveness of medium by inhibiting oxidation of sodium tetrathionate

[1]From Michelson and Monod (1974).

Cysteine Sulfoxide Lyase (Allinase) in Onion

A vegetable in which the action of one enzyme on essentially one substrate results in the production of a diversity of organoleptic sensations is the onion. When the structural integrity of the onion cell is destroyed by mastication or other means of comminution, the tongue burns, the eyes water, and one senses the typical onion aroma. Freshly prepared onion juice develops a bitter alkaloid-like taste within a minute after comminution and, after several hours, will turn pink. All these sensory perceptions can be elicited by adding a suitable lyase preparation to a solution of the most abundant flavor precursor *trans*-(+)-*S*-prop-l-enyl-*L*-cystein S-oxide or, more simply, propenylcysteine sulfoxide (Schwimmer 1969; Schwimmer and Friedman 1972; Whitaker 1976; Freeman and Whenham 1976B; Banyopadhyay and Tewari 1976). In this case the primary product may be propenylsulfenic acid but the first really detectable but still unstable product is the lachrymator, thiopropanal *S*-oxide (Brodnitz and Pascale 1971). All of the other sensations perceived appear to arise from the reactions of this intermediate with itself or with other food constituents (Fig. 12.8).

Glucosinolase and Intracellular Enzyme Localization

As a final example of the origin of a diversity of products as the result of the action/reactions of unstable primary-enzyme products we have selected thioglucosidases or glucosinolases, widely distributed in cruciferous vegetables and in some oilseeds. The traditional appellation "mustard oils" usually refers to distillates of isothiocyanate derivatives. Some of these isothiocyanates, especially the allyl derivative sinigrin, $CH_2 = CH-CH_2-N=C=S$, are highly pungent and are mainly responsible for the odor of brown mustard and horseradish and contribute to the fresh flavor of other crucifers such as cabbage. They are formed from mustard oil thioglucosides (glucosinolates) by an enzyme that catalyzes the following over-all reaction.

$$\begin{array}{c} S-glucose \\ | \\ R - C = NOSO_3^- + H_2O \xrightarrow{\quad Glucosinolase \quad} R - N = C = S + HSO_4^- + Glucose \end{array}$$

It will be noted that there is a formal similarity between the above reaction and that which produces *Allium* flavors. Thus three main products are formed, only one of which is a sulphur volatile. The analogy does not end here. Both plant enzymes coexist in the intact cell without interacting with their respective substrates; they interact only after cellular

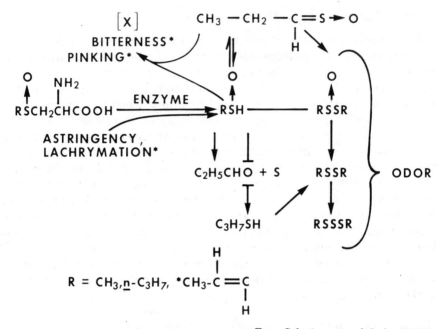

From Schwimmer and Joslyn (1978)

FIG. 12.8. PRODUCTION THROUGH ALLIINASE ACTION OF THE SUB-
STANCES RESPONSIBLE FOR SENSORY PERCEPTIONS PERCEIVED
WHEN ONION TISSUE IS DISRUPTED

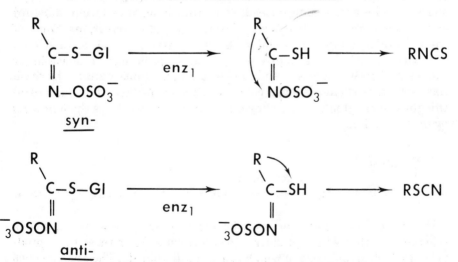

FIG. 12.9. SUGGESTED *syn-anti* ISOMERISM OF THE GLUCOSINOLATES
TO ACCOUNT FOR OCCURRENCE OF ORGANIC THIOCYANATE AND
ISOTHIOCYANATE IN CRUSHED CRUCIFERS

disruption. Neither enzyme is inhibited, at moderate concentrations, by the final stable products. In both cases the primary enzymic event appears to produce a relatively unstable intermediate. In the case of glucosinolase action this is a thiohydroxamic O-sulphonate (Schwimmer 1961).

$$\underset{\substack{| \\ R - C = NOSO_3^- + H_2O}}{S-glucose} \longrightarrow R - \overset{S}{\overset{\|}{C}NH - OSO_3^- + Glucose}$$

Most investigators now look upon this action as due to one enzyme of the general class of thioglucosidases. The release of sulphate is considered to be spontaneous, accompanying the subsequent Lossen rearrangement (Ettlinger et al. 1961).

$$R- \overset{S}{\overset{\|}{C}}NH - OSO_3^- \longrightarrow RNCS + HSO_4^-$$

As in the case of lyase action, the unstable primary intermediate is transformed to one stable major product, but also may be transformed to other products (including free sulphur), depending on the conditions and the substrate. Thus at low pH values the allyl derivative decomposes to vinylacetonitrile and sulphur or hydrogen sulphide (Schwimmer 1960).

$$CH_2 = CH - CH_2 - \overset{S}{\overset{\|}{C}} - NH - OSO_3^- \rightarrow CH_2 = CH - CH_2 - C \equiv N + S + HSO_4^-$$

Under proper conditions indolyl glucosinolate (glucobrassicin) will also yield nitrile as well as indole ethyl alcohol and isothiocyanate, form thiocyanate ion and react with ascorbic acid to form ascorbigen.

Twelve nitriles derived from the action of glucosinolase have been identified in autolyzing cabbage (Daxenbichler et al. 1977; Tookey et al. 1977). Conditions which favor formation of thiocyanate esters parallel those for the formation of epithionitiles in crambe seed. Also, the latter requires both FE^{++} and an "epithio specifier protein" (Tookey et al. 1977). Crambe seed yields unusual products from progoitrin (Van Etten et al. 1969) including the health-associated goitrin, (S)-5-vinyl-oxazoladine-2-thione, which is formed by cyclization of 1-hydroxy-2-butenyl isothiocyanate.

Thiocyanate Formation.—A rather novel source of diversity of products resulting from glucosinolases action arises from a rather subtle difference in the chirality of the substrate. Gmelin and Virtanen (1959) found that glucosinolase acting on benzyl glucosinolate in plants such as garlic, mustard and pennycress resulted in the formation of benzyl-thiocyanate, which was a garlic-like odor, instead of the isothiocyanate derivative. They suggested that either the specificity of the pennycress glucosinolase is different or that a second enzyme catalyzes the isomerization of benzyl isothiocyanate to thiocyanate. This hypothetical enzyme has been raised to the dignity of an E.C. number (Table 12.7). More recently Schlueter and Gmelin (1972) suggested that such glucosinolates (i.e. the 4-methylthio-derivative) exist in the plant in a slightly different geometric form from that obtained when the glucosinolate is isolated. They tentatively suggest that *syn-anti* isomerism may be involved, as depicted in Fig. 12.9 The corresponding products possess a "scorched-pork" odor.

Localization of the Enzyme in the Endoplasmic Reticulum.—Most enzymes of the type discussed are typically found or presumed to be associated with such conventional intracellular structures as the mitochondrion (Paez and Hultin 1970) or within the lysosome, *i.e.*, α-amylase of gibberellic acid-treated wheat (Gibson and Paleg 1975). The glucosinolases of the crucifers appear to be unique in this respect. Iversen and coworkers found that the glucosinolases seem to be localized in structures unique to crucifers, namely, dilated cysternae of the rough endo-

TABLE 12.7
ENZYME COMMISSION (E.C.) NUMBERS, ENZYMES MENTIONED IN THIS PAPER

Enzyme	E.C. No.	Enzyme	E.C. No.
Allinase, L-Cysteine sulfoxide lyase	4.4.1.4	Lipase	3.1.1.3
α-Amylase	3.2.1.1	Lipoxygenase	1.13.11.7
β-Amylase	3.2.1.2	Pectin esterase	3.1.1.11
Anthocyanase; β-Glucosidase	3.2.1.21	Peroxidase	1.11.17
Alliinase	4.4.1.4	Phenolase, polyphenol oxidase catecholase	3.2.1.15
Carboxypeptidase	3.4.12.	Phospholipase	3.1.1.4
Catalase	1.11.1.16	Phytase	3.2.1.21
Cellulase	3.2.1.4	Polygalacturonase(PG)	3.2.1.15
Chlorophyllase	3.1.1.14	Prostaglandin synthase	1.14.99.1
Collagenase	3.4.24.3	Proteinase, endoproteinase	3.4.x.y
α-Galactosidase	3.2.1.22	Ribonuclease	3.1.4.2x
Glucosinolase	3.2.3.1	Superoxide dismutase	1.15.1.1
γ-Glutamyl transpeptidase	3.1.3.26	Tannase	3.1.1.20
Glutathione peroxidase	1.1.1.19	Thiocyanate isomerase	5.99.1.1
Invertase	3.2.1.26	Xanthine oxidase	1.2.1.37

plasmic reticulum (Iversen 1973; Pihakaski and Iversen 1976). The latter is the site of synthesis of all enzymes and other proteins in all eukaryotes. Unlike the conventional cell arrangement where some enzymes are packaged in Golgi bodies which evolve into enzyme-bearing packets, it appears that the glucosinolases remain associated with the site of their synthesis. Even more surprising is the finding of Hoefert (1975) that these dilated cisternae contain tubules as shown in Fig. 12.10. These tubules are about 30 nm wide and the cisternae that contain them are about 10 nm long. Hoefert suggests these tubular inclusions may actually be built up of globular glucosinolase isozymes. Just why they are present in such a complicated and unusual array and what their function might be must await further investigation. It would also be interesting to ascertain the localization of the substrate. Perhaps the elucidation of this problem will yield a more intimate understanding of how and why enzyme action is initiated and potentiated when the living cell is subject to the kinds of perturbations discussed and will lead to rational and purposeful control of the modification of the cellular integrity in fruits and vegetables.

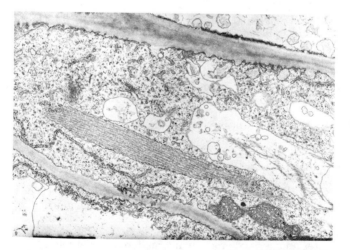

From Hoefert (1975)

FIG. 12.10.   ELECTRON MICROGRAPHS OF VASCU-
LAR PARENCHYMA CELLS OF PENNYCRESS,
SHOWING TUBULES WITHIN DILATED CISTERNAE
OF THE ROUGH ENDOPLASMIC RETICULUM
Glucosinolases are localized within these cisternae.

BIBLIOGRAPHY

APPLEMAN, O. and CONRAD, C. M. 1927. The pectic constituents of tomatoes and their relation to the canned product. U. Maryland Agr. Exp. Sta. Bull. No. 291, p. 17.

ARNESON, R. M. 1970. Substrate induced chemiluminescence of xanthine oxidase and aldehyde oxidase. Arch. Biochem. Biophys. *136*, 352–360.

AXELROD, B. 1974. Lipoxygenases. *In* Food Related Enzymes, J. R. Whitaker (Editor). American Chemical Society, Washington, D.C.

BANDYOPADHYAY, C. and TEWARI, M. 1976. Lachrymatory factor in sprouting onion (*Allium cepa*). J. Sci. Food Agric. *27*, 733–735.

BARTOLME, L. G. and HOFF, J. E. 1972. Firming of potatoes: biochemical effects of preheating. J. Agric. Food Chem. *20*, 266–270.

BECKER, R., OLSON, A. C., FREDERICK, D. P., KON, S. and GUMBMANN, M. R. 1974. Conditions for the autolysis of alphagalactosides and phytic acid in California small white beans. J. Food Sci. *39*, 766–769.

BLAIN, J. A. 1970. Carotene-bleaching activity in plant tissue extracts. J. Sci. Food Agric. *21*, 35–38.

BOLDINGHE, J. 1976. Reaction mechanisms and functions of plant lipoxygenases. *In* International Congress for Fat Research, Milan, 1974. Raven Press, New York.

BRODNITZ, M. H. and PASCALE, J. V. 1971. Thiopropanal S-oxide: a lachrymatory factor in onions. J. Agric. Food Chem. *21*, 269–274.

BUCKLE, K. A. and EDWARDS, R. A. 1970. Chlorophyll degradation in frozen unblanched peas. J. Sci. Food Agric. *21*, 307–312.

BURI, R. and SOLMS, J. 1971. Ribonucleic acid—a flavor precursor in potatoes. Naturwiss. *58*, 56–57.

CASH, J. N., SISTRUNK, C. A. and STUTTE, C. A. 1976. Characteristics of concord grape polyphenoloxidase involved in juice color loss. J. Food Sci. *41*, 1398–1402.

CHANG, R. and SCHWIMMER, S. 1977. Characterization of phytase of beans (*Phaseolus vulgaris*). J. Food Biochem. *1*, 45–56.

CHANG, R., SCHWIMMER, S. and BURR, H. K. 1977. Phytate: removal from whole dry beans by enzymatic hydrolysis and diffusion. J. Food Sci. *42*, 1098–1101.

CILENTO, G. 1975. Dioxetanes as intermediates in biological processes. J. Theor. Biol. *55*, 471–479.

CLYDESDALE, F. M. and FRANCIS, F. J. 1968. Chlorophyll changes in thermally processed spinach as influenced by enzyme conversion and pH adjustment. Food Technol. *22*, 792–796.

COGGINS, C. W. JR., KNAPP, J. C. F. and RICKER, A. L. 1968. Postharvest softening studies of Deglet Noor dates: Physical, chemical and historical changes. Date Growers' Institute Report 45, p. 3.

DAXENBICHLER, M. E., VAN ETTEN, C. H. and SPENCER, G. F. 1977. Glucosinolates and derived products in cruciferous vegetables. Identification of organic nitriles from cabbage. J. Agric. Food Chem. *25*, 121–124.

DEOBOLD, H. J., MCLEMORE, T. A., HASLING, V. C. and CATALANO, E. A. 1968. Control of sweet potato α-amylase for producing optimum quality precooked dehydrated flakes. Food Technol. *22*, 627–630.

DULL, G. G. and LEEPER, G. F. 1975. Ultrastructure of polysaccharides in relation to texture. *In* Symposium: Postharvest Biology and Handling of Fruits and Vegetables, N. F. Haard and D. K. Salunkhe (Editors). Avi Publishing Co., Westport, CT.

ETTLINGER, M. G., DATEO, G. P., HARRISON, B. W., MABRY, T. M. and THOMPSON, C. P. 1961. Vitamin C as a coenzyme: the hydrolysis of mustard oil glucosides. Proc. Nat. Acad. Sci. U.S.A. *12*, 1875–1880.

FELLER, U. K., HAGEMAN, R. H. and SOONG, T. S. 1977. Leaf proteolytic activities and senescence during grain development of field-grown corn. (Zea mays L.) Plant Physiol. *59*, 290–294.

FRAZIER, P. J., LEIGH-DUGMORE, F. A., DANIELS, N. W. R., EGGIT, P. W. and COPPOCK, J. 1973. The effect of lipoxygenase on the mechanical development of wheat flour doughs. J. Sci. Food Agric. *24*, 421–436.

FREEMAN, G. G. and MOSSADEGHI, N. 1973. Studies on the relationship between water regime and flavor strength in water cress, cabbage and onion. J. Hort. Sci. *48*, 365–378.

FREEMAN, G. G. and WHENHAM, R. J. 1976A. Effect of overwinter storage at three temperatures on the flavor intensity of dry bulb onions. J. Sci. Food Agric. *27*, 37–42.

FREEMAN, G. G. and WHENHAM, R. J. 1976B. Nature and origin of volatile components of onion and related species. International Flavors *7*, 222–229.

FREEMAN, G. G., WHENHAM, R. J., SELF, R. and EAGLES, J. 1975. Volatile flavor components of parsley leaves. J. Sci. Food Agric. *26*, 465–470.

FRENKEL, C. 1977. Role of hydroperoxides in the onset of senescence processes in plant tissue. *In* Postharvest Biology and Biotechnology, H. O. Hultin and M. Milner (Editors). Food & Nutrition Press, Westport, CT.

FRIDOVICH, I. 1974. Superoxide dismutases. Adv. Enzymol. *41*, 35–97.

GALLIARD, T. and MATTHEW, J. A. 1975. Enzymic reactions of fatty acid hydroperoxides in extracts of potato tuber. Biochem. Biophys. *398*, 1–9.

GALLIARD, T. and MATTHEW, U. K. 1977. Lipoxygenase-mediated cleavage of fatty acids to carbonyl fragments in tomato fruits. Phytochem. *16*, 339–343.

GALLIARD, T. and PHILLIPS, D. R. 1976. The enzymic cleavage of linoleic acid to C_9 carbonyl fragments in extracts of cucumber fruit and the possible role of lipoxygenase. Biochim. Biophys. Acta *431*, 278–287.

GALLIARD, T., PHILLIPS, D. R. and REYNOLDS, J. 1976. The formation of *cis*-3-nonenal, *trans*-2-nonenal and hexanal from linoleic acid hydroperoxide isomers by a hydroperoxides cleavage enzyme system in cucumber *(Cucumis sativus)* fruits. Biochim. Biophys. Acta *441*, 181–192.

GARDNER, H. W. 1975. Decomposition of linoleic acid hydroperoxides. Enzymic compared with nonenzymic. J. Agr. Food Chem. *23*, 129–136.

GARDNER, H.W. and SESSA, D. J. 1977. Degradation of fatty acid hydroperoxide in cereals and a legume. Annales de Technologie de I.N.R.A. *26*, 151–159.

GEE, M., FARKAS, D. and RAHMAN, A. R. 1977. Some concepts for the development of intermediate moisture foods. Food Technol. *31*, No. 4, 58, 60, 62, 64.

GIBSON, R. A. and PALEG, L. G. 1975. Further experiments on the α-amylase of wheat aleurone cells. Aust. J. Plant Physiol. *2*, 41–49.

GMELIN, R. and VIRTANEN, A. I. 1959. A new type of enzymatic cleavage of mustard oil glucosides. Formation of allyl thiocyanate in *Thlaspi arvense* L. and benzylthiocyanate in *Leipdium ruderale* L. and *Lepidium sativum* L. Acta Chem. Scand. *13*, 1474–1475.

GROSCH, W., LASKAWY, G. and FISCHER, K. H. 1974. Oxidation of linoleic acid in the presence of heamoglobin, lipoxygenase or by singlet oxygen. Identification of the volatile carbonyl compounds. Lebensm. Wiss. Technol. 7, 335–338.

HAAS, L. W. and BOHN, R. M. 1934. Bleaching bread dough. U.S. Pat. 1,957,333.

HABER, F. and WEISS, J. 1934. The catalytic decomposition of hydrogen peroxide, by iron salts. Proc. Roy. Soc. London, Ser. A, 147, 332–351.

HASEGAWA, S. and SMOLENSKY, D. C. 1972. Cellulase in dates and its role in softening. J. Food Sci. 36, 966.

HATANAKA, A., KAJIWARA, T. and HARADA, T. 1975. Biosynthetic pathway of cucumber alcohol: trans-2, cis-6-nonadienol via cis-3, cis-6-nonadienal. Phytochem. 14, 2589–2582.

HOEFERT, L. L. 1975. Tubules in dilated cysternae of the endoplasmic reticulum of Thlaspi arvense (Cruciferae). Amer. J. Bot. 62, 756–760.

HOLDEN, M. 1965. Chlorophyll bleaching by legume seeds. J. Sci. Food Agric. 16, 312–325.

HOLDEN, M. 1974. Chlorophyll degradation in leaf protein preparations. J. Sci. Food Agric. 25, 1427–1432.

ISHERWOOD, F. A. 1976. Mechanism of starch-sugar interconversion in Solanum tuberosum. Phytochem. 15, 33–41.

IVERSEN, T.-H. 1973. Myrosinase in cruciferous plants. In Electron Microscopy of Enzymes, Principles and Methods. Vol. 1, M. A. Hyat (Editor). Van Nostrand Reinhold, New York.

JUNG, D. W. and LATIES, G. G. 1975. Trypsin-induced ATP-ase activity in potato mitochondria. Plant Physiol. 36, No. 2, Supp., p. 72.

KAHN, V. 1977. Some biochemical properties of polyphenoloxidase from two avocado varieties differing in their browning rates. J. Food Sci. 42, 38–43.

KALBRENNER, J. E., WARNER, K. and ELDRIDGE, A. C. 1974. Flavors derived from linoleic and linolenic acid peroxides. Cereal Chem. 51, 406–415.

KASCHNITZ, R. M. and HAFETI, Y. 1975. Lipid oxidation in biological membranes. Arch. Biochem. Biophys. 171, 292–304.

KHAN, A. A. and KOLATTUKUDY, P. E. 1974. Decarboxylation of long chain fatty acids by cell-free preparation of pea leaves (Pisum sativum). Biochem. Biophys. Res. Comm. 61, 1379–1386.

KING, A. D. JR., MICHENER, H. D., BAYNE, H. G. and MIHARA, K. L. 1976. Microbial studies on shelf life of cabbage and coleslaw. Appl. Environ. Microbiol. 31, 404–407.

KON, S. and SCHWIMMER, S. 1976. Degradation of plant and other polysaccharides and its inhibition by catalase and superoxide dismutase. Fed. Proc. 35, 1553.

KON, S. and SCHWIMMER, S. 1977. Depolymerization polysaccharides by active oxygen species derived from a xanthine oxidase system. J. Food Biochem. 1, 141–152.

KUNINORI, T., NISHIYAMA, J. and MATSUMOTO, H. 1976. Effect of mushroom extract on the physical properties of dough. Cereal Chem. 53, 420–428.

LAAKONEN, E. 1973. Factors affecting tenderness during heating of meat. Adv. Food Res. 20, 257–324.

LABELLE, R. L. 1971. Heat and calcium treatment for firming red tart cherries in a hot fill process. J. Food Sci. 36, 323–326.

LEOPOLD, A. and ARDREY, R. 1972. Toxic substances in plants and food habits of early man. Science 176, 512–513.

LERNER, A. B. 1953. Metabolism of phenylalanine and tyrosine. Adv. Enzymol. 14, 73–128.

MCCORD, J. M. and FRIDOVICH, I. Superoxide dismutase, an enzymatic function for erythrocuprein (hemocuprein). J. Biol. Chem. 244, 6049–6055.

MAKOWER, R. U. and SCHWIMMER, S. 1954. Inhibition of enzymic color formation in potato by adenosine triphosphate. Biochim. Biophys. Acta 14, 156–157.

MICHELSON, A. M. and MONOD, J. 1974. Superoxide dismutase from sea bacteria as antioxidant for foods. Ger. Offen. 2, 417, 508. Nov. 7.

MONDY, N. I. and MUELLER, T. O. 1977. Potato discoloration in relation to anatomy and composition. J. Food Sci. 42, 14–18.

MURRAY, K. E., SHIPTON, J., WHITFIELD, F. B. and LAST, H. H. 1976. The volatiles of off-flavored unblanched peas (Pisum sativum). J. Sci. Food Agric. 27, 1093–1107.

NEUKOM, H. 1976. Chemistry and properties of the non-starch polysaccharides (NSP) of wheat flour. Lebensm. Wiss. Technol. 9, 143–148.

OHAD, I., FRIEDBERG, I., NEEMAN, Z. and SCHRAMM, M. 1971. Biogenesis and degradation of starch. The fate of the amyloplast membranes during maturation and storage of potato tubers. Plant Physiol. 47, 465–477.

OHTA, S., MAUL, S., SINSKEY, A. J. and TANNENBAUM, S. R. 1971. Characterization of a heat shock process for reduction of the nucleic acid content of Candida utilis. Appl. Microbiol. 22, 415–421.

OLSON, D. G., PARRISH, F. C. JR., DAYTON, W. R. and GOLL, D. E. 1977. Effect of postmortem storage and calcium activated factor on the myofibrillar proteins of bovine skeletal muscle. J. Food Sci. 42, 117–124.

OMAYE, S. T., TAYLOR, S. L. and TAPPEL, A. L. 1975. Lipid peroxidation and reactions of glutathione peroxidase. Fed. Proc. 34, 538.

PAEZ, L. E. and HULTIN, H. O. 1970. Respiration of potato mitochondria and whole tubers and relation to sugar accumulation. J. Food Sci. 35, 46–51.

PATTEE, H. E. and SINGLETON, J. A. 1977. Isolation of isomeric hydroperoxides from the peanut lipoxygenase-linoleic reaction. J. Am. Oil Chem. Soc. 54, 183–185.

PATIL, S. K., FINNEY, K. F., SHOGREN, M. D. and TSEN, C. C. 1976. Water-soluble pentosans of wheat flour. 3. Effect of water-soluble pentosans on loaf volume of reconstituted gluten and starch doughs. Cereal Chem. 53, 347–354.

PHILLIPS, D. R. and GALLIARD, T. C. 1978. Flavour biogenesis. Partial purification and properties of a fatty acid hydroperoxide cleaving enzyme from fruits of cucumbers. Phytochem. 17, 355–358.

PIHAKASKI, K. and IVERSEN, T.-H. 1976. Myrosinase in Brassicaceae. 1. Localization of myrosinase in cell fraction of roots of Sinapis alba L. J. Exp. Bot. 97, 242–258.

PRESSEY, R. 1966. Separation and properties of potato invertase and invertase inhibitor. Arch. Biochem. Biophys. 113, 667–674.

RAISON, J. K., LYONS, J. M., MEHLHORN, R. J. and KEITH, A. D. 1971. Temperature-induced phase change in mitochondrial membrane detected by spin labeling. J. Biol. Chem. 246, 4036–4040.

RICHTER, C., WENDEL, A., WESER, U. and AZZI, H. 1975. Inhibition by superoxide dismutase of linoleic peroxidation by lipoxidase. FEBS Lett. 51, 300–303.

ROCKLAND, L. B., BEAVENS, E. A. and UNDERWOOD, J. C. 1957. Debittering of citrus fruit. U.S. Pat. 2,816,835. Dec. 17.

SAMUELSSON, B., GRANSTROM, E., GREEN, C., HAMBERG, M. and HAMMARSTROM, S. 1975. Prostaglandins. Ann. Rev. Biochem. 44, 669–695.

SANDERSON, G. W. 1975. Black tea aroma formation. In Geruch-und Geschmackstoffe, F. Drawert (Editor). Verlag Hans Carl, Nurenberg.

SANDERSON, G. W., BERKOWITZ, J. E., CO, H. and GRAHAM, H. N. 1972. Biochemistry of tea fermentation: products of the oxidation of tea flavanols in a model tea fermentation system. J. Food Sci. *37*, 399–404.

SCHLUETER, M. and GMELIN, R. 1972. Abnormal enzymatic splitting of 4-methylthioglucosinolate in fresh plants from *Eruca sativa*. Phytochem. *11*, 3427–3431. (Ger.)

SCHWIMMER, S. 1960. Myrosin-catalyzed formation of turbidity and hydrogen sulfide from sinigrin. Acta Chem. Scand. *14*, 1439–1441.

SCHWIMMER, S. 1961. Spectral changes during the action of myrosinase on sinigrin. Acta Chem. Scand. *15*, 535–544.

SCHWIMMER, S. 1962. Theory of double pH optima of enzymes. J. Theor. Biol. *3*, 102–110.

SCHWIMMER, S. 1969. Characterization of S-propenyl L-cysteine sulfoxide as the principal endogenous substrate of L-cysteine sulfoxide lyases of onion. Arch. Biochem. Biophys. *130*, 312–320.

SCHWIMMER, S. 1972. Biochemical control systems: cell disruption and its consequences in food processing. J. Food Sci. *37*, 530–535.

SCHWIMMER, S. 1973. Flavor enhancement of Allium products. U.S. pat. 3,725,085. Apr. 3.

SCHWIMMER, S. 1975. Effect of enzymes on nutritional quality and availability of proteins. *In* Nutritional Quality of Foods and Feeds. Part 2. Quality Factors—Plant Breeding Processing and Antinutrients, M. Friedman (Editor). Marcel Dekker, Inc., New York.

SCHWIMMER, S. 1979. Food Enzymology. Avi Publishing Co., Westport, Conn. In preparation.

SCHWIMMER, S. and AUSTIN, S. J. 1971. Gamma glutamyl transpeptidase of sprouted onion. J. Food Sci. *36*, 807–811.

SCHWIMMER, S. and FRIEDMAN, M. 1972. Genesis of volatile sulphur-containing food flavors. Flavour Ind. *3*, 137–145.

SHIMAZONO, H. 1964. Distribution of 5'-nucleotides in foods. Food Technol. *18*, 294, 299–303.

SHINE, W. E. and STUMPF, P. K. 1974. Fat metabolism in higher plants. Recent studies on plant -oxidation systems. Arch. Biochem. Biophys. *162*, 147–157.

SINGH, R. and SHEORAN, I. S. 1972. Enzymatic browning of whole wheat flour. J. Sci. Food Agric. *23*, 121–125.

SINGLETON, J. A., PATTEE, H. E. and SANDERS, T. H. 1975. Some parameters affecting volatile production in peanut homogenates. J. Food Sci. *40*, 386–389.

SKINNER, K. J. 1975. Radical probe in prostaglandin origins probed. Chem. Eng. News. *53*, No. 48, *22*–23.

STONE, E. J., HALL, R. M. and KAZENIAC, S. J. 1975. Formation of aldehydes and alcohols in tomato fruits from $U^{14}C$ labeled linolenic and linoleic acids. J. Food Sci. *40*, 1138–1141.

TOOKEY, H. L., VAN ETTEN, C. H. and DAXENBICHLER, M. E. 1977. *In* Toxic Constituents of Plant Foodstuffs, 2nd ed., I. F. Liener (Editor). Academic Press, New York.

VAN ETTEN, C. H., DAXENBICHLER, M. E. and WOLFF, I. A. 1969. Natural glucosinolates (thioglucosides) in foods and feeds. J. Agric. Food Chem. *17*, 483–491.

WALLACE, J. M. and WHEELER, E. L. 1972. Lipoxygenase inactivation in wheat protein concentrate by heat-moisture treatments. Cereal Chem. *49*, 92–98.

WALSH, D. E., YOUNGS, V. L. and GILLES, K. A. 1970. Inhibition of durum wheat lipoxidase with ascorbic acid. Cereal Chem. *47*, 119–125.

WARDALE, D. A. and GALLIARD, T. 1975. Subcellular localization of lipoxygenase and lipolytic acyl hydrolase enzymes in plants. Phytochem. *14*, 2323–2329.

WHITAKER, J. R. 1976. Development of flavor, odor, and pungency in onion and garlic. Adv. Food Res. *22*, 73–133.

WICKREMASINGHE, R. L. 1974. The mechanism of operation of climatic factors in the biogenesis of tea flavor. Phytochem. *13*, 2057–2063.

WINTER, F. H. 1976. Critical points in food processing where enzyme action can occur—practical examples. Proceedings of the Symposium on Enzymes in the Food Processing Industry. University of California, Davis. Jan. 14, pp. 26–31.

Quality Attributes of Edible Plant Tissues

CHAPTER 13

TOXINS IN LEGUMES: THEIR POSSIBLE NUTRITIONAL SIGNIFICANCE IN THE DIET OF MAN

IRVIN E. LIENER

Department of Biochemistry
College of Biological Sciences
University of Minnesota
St. Paul, MN 55108

ABSTRACT

Two categories of legume toxins will be considered: those whose effects have been extensively studied in experimental animals but whose significance in man must remain open to conjecture; and those which are known to produce toxic effects in man but in which the causative factors have been difficult to identify because similar effects are not readily reproduced in animal models. In the first category, evidence pertaining to the role which protease inhibitors and hemagglutinins play in determining the nutritive value of leguminous proteins in animal studies will be reviewed. In attempting to relate such information to man, differences between man's probable response to these factors as compared to that of animals will be pointed out. Lathyrism and favism are diseases in man which are associated with the consumption of Lathyrus sativus *and* Vicia faba *respectively. Evidence leading to the probable identification of the causative factors of these diseases and the steps necessary for their elimination will be discussed.*

It is well recognized that, for whatever reason, nature has seen fit to endow plants with the capacity to synthesize substances which may have an adverse physiological effect on animals (Liener 1969). The very fact that we are having this symposium bears testimony to the importance that is being attached to use of plant materials as food component of the human diet, so it is understandable that we should give some consideration to the question as to whether these so-called toxic constituents of plants are of any significance with respect to man. In general one may say that there are two categories of plant toxins: those whose effects have been tested extensively in a wide variety of animals, but whose effects in humans have not been tested for obvious reasons and must therefore remain open to conjecture; and those foods of plant origin which are known to produce toxic reactions in humans, but in which the causative factor has been difficult to identify because the same disease is difficult and sometimes impossible to reproduce in animals.

Since time does not permit one to cover all of the many constituents which are present in plant foodstuffs (Liener 1969), I would like to confine myself to a consideration of some selected examples of each of the two categories of food toxicants referred to above. In the first category, I shall undertake the foolhardy task of trying to evaluate the significance of trypsin inhibitors and hemagglutinins in the diet of man, since these substances are the principle antinutritional factors associated with the legumes most commonly consumed by man. In the second category, I shall briefly summarize our current knowledge concerning two diseases in man associated with the consumption of specific legumes, namely lathyrism and favism.

TRYPSIN INHIBITORS

The trypsin inhibitors are probably the best known, and certainly the most studied, of all of the antinutritional factors since they are so widely distributed among the legumes, particularly the soybeans. It is well known of course that the trypsin inhibitors are inactivated by heat treatment and that their destruction is accompanied by a marked enhancement in the nutritive quality of the protein as measured in experimental animals such as the rat or the chick (Fig. 13.1). It is sometimes concluded from data such as this that the trypsin inhibitor is the main cause of growth inhibition, but several lines of evidence indicate that the trypsin inhibitors are only partially responsible for the poor nutritive value of raw soybeans. For example if rats are fed heated soybeans to which has been added the isolated trypsin inhibitors so as to provide the same level of antitryptic activity which is present in the unheated beans, the reduction in growth observed falls short of that of the raw bean (Table 13.1). An investigation of a larger number of different varieties of soybeans with respect to the growth-promoting quality of their protein in rats (PER) and their trypsin-inhibitory activity revealed no correlation between these two parameters (Fig. 13.2). Finally, if the trypsin inhibitor activity of a crude extract of soybeans is eliminated by affinity chromatography with sepharose-bound trypsin (Fig. 13.3) the resulting extract is still capable of causing growth inhibition and pancreatic hypertrophy (Table 13.2). It may be estimated from these data that the trypsin inhibitor accounts for about 40% of the growth inhibition observed with raw soybeans. It may be significant to note that only about 40% of the enlargement of the pancreas caused by the ingestion of raw soybeans is also accounted for by the trypsin inhibitors.

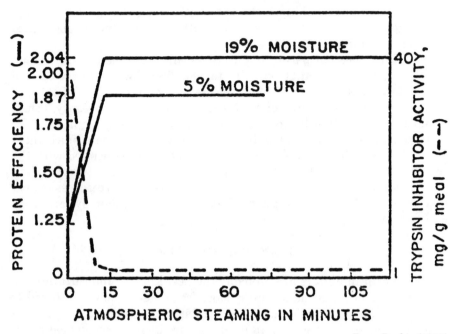

From Rackis (1972)

FIG. 13.1.   EFFECT OF HEAT TREATMENT ON TRYPSIN INHIBITOR AC-
TIVITY AND PROTEIN EFFICIENCY RATIO OF SOYBEAN PROTEIN

TABLE 13.1
EFFECT OF ADDING PARTIALLY PURIFIED SOYBEAN
TRYPSIN INHIBITOR (STI) TO DIETS CONTAINING HEATED
SOYBEAN MEAL IN THE PRESENCE AND ABSENCE OF METHIONINE[1]

| | PER | |
Diet	−met	+met[2]
Raw soybeans	1.40	2.42
Heated soybeans[3]	2.63	2.99
Heated soybeans + 1.8% STI	1.95	2.63

[1]Source: Liener *et al.* (1949).
[2]Diets were supplemented with 0.6% methionine.
[3]Autoclaved at 15 lb pressure (115° C) for 20 min.

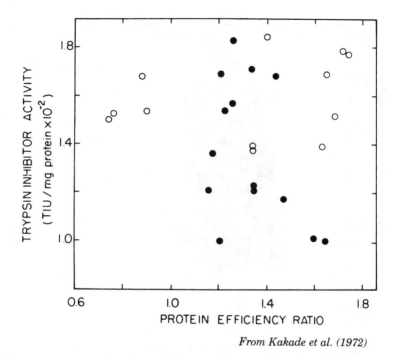

From Kakade et al. (1972)

FIG. 13.2.   RELATIONSHIP OF TRYPSIN INHIBITORY ACTIVITY
OF SOYBEAN VARIETIES TO PER

These findings of course raise the question as to what is responsible for the remaining 60% of the growth-retarding effect of raw soybeans. A possible clue comes from experiments in which the crude soybean protein extract from which the inhibitor had been removed was subjected to digestion with trypsin *in vitro* (Fig. 13.4). Heat treatment of such a preparation of soybean protein produced an increase in the digestibility of the protein over and above the digestibility of a similar sample from which the inhibitor had not been removed. This observation suggests that native, undenatured soybean protein is in itself refractory to enzymatic attack unless denatured by heat, and this may very well account for the growth inhibition seen with inhibitor-less soybean protein. Since the level of active trypsin in the intestine is believed to control the size of the pancreas (Green and Lyman 1972; Niess *et al.* 1972), the undenatured protein may in fact act as a competitive inhibitor of trypsin (Thompson and Liener 1977) and thus serve to reduce the level of active trypsin in the intestines causing an enlargement of the pancreas.

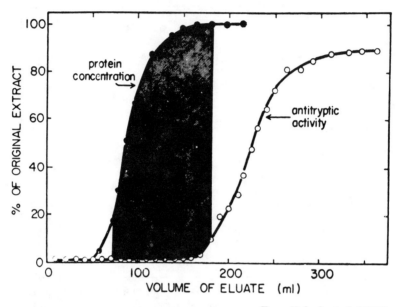

From Kakade et al. (1972)

FIG. 13.3.   REMOVAL OF TRYPSIN INHIBITOR FROM SOY-
BEAN EXTRACT BY AFFINITY CHROMATOGRAPHY ON
SEPHAROSE/TRYPSIN

TABLE 13.2

CONTRIBUTION OF TRYPSIN INHIBITORS TO THE GROWTH INHIBITION
AND PANCREATIC HYPERTROPHY INDUCED IN RATS BY DIETS
CONTAINING UNHEATED SOYBEAN PROTEIN[1]

Source of Protein	PER	Wt of Pancreas g/100g Body Wt
Soy flour extract, unheated	1.4	0.71
Soy flour extract, heated	2.7	0.57
Soy flour extract minus inhibitor[2]	1.9	0.65
Percent change due to removal of inhibitor	+38	−41

[1]Source: Kakade et al. (1973).
[2]Trypsin inhibitors removed by passage of unheated soy flour extract through a column of Sepharose-trypsin.

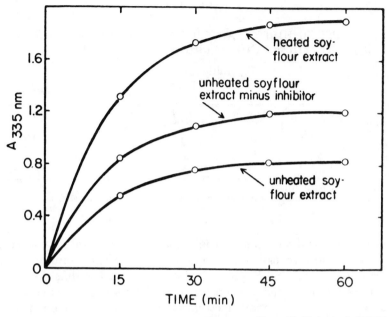

From Kakade et al. (1972)

FIG. 13.4.   *IN VITRO* DIGESTIBILITY BY TRYPSIN OF SOYBEAN
EXTRACT WITH AND WITHOUT THE TRYPSIN INHIBITOR RE-
MOVED COMPARED TO THE HEATED EXTRACT

It should be appreciated of course that all of the experiments I have
described thus far were conducted with rats or chicks as the experimental
model. We thus come back to our original question as to the significance of
these observations with respect to humans. As a basis for speculation on
this point, and it must be regarded only as speculation, allow me to cite the
following lines of evidence which suggest that the trypsin inhibitors are
most likely of little consequence in human nutrition and hence of minor
concern to the food industry.

Many of the soybean products on the market today have been made from
protein isolates which may contain as much as 30% of the trypsin inhibitor
activity of raw soybean meal. An examination of trypsin inhibitor activity
of several textured meat analogs during the various stages of their manufacture reveals that, although the protein isolate may be rich in antitryptic activity, the latter is reduced to very low levels in the final product
(Table 13.3). Household cooking of such products would be expected to
reduce these levels even further. Kotter *et al.* (1970) have reported that

TABLE 13.3

TRYPSIN INHIBITOR ACTIVITIES OF SOYBEAN FLOUR,
ISOLATE, FIBER, AND FINISHED TEXTURED PRODUCTS[1]

	Antitrypsin Activity TIU[2]/g Dry Solids $\times 10^{-3}$	% of Soy Flour
Soy flour (unheated)	86.4	100
Soybean isolate	25.5	30
Soybean fiber	12.3	14
Chicken analog	6.9	8
Ham analog	10.2	12
Beef analog	6.5	7

[1]Source: Liener, unpublished data.
[2]TIU = trypsin inhibitor units as determined by the method of Kakade *et al.* (1969).

canned frankfurter-type sausage containing 1.5% soy isolate was essentially devoid of any trypsin inhibitor activity after the canning process. Furthermore, Nordal and Fossum (1974) have reported that the trypsin inhibitor activity provided by soybean protein in meat products was actually made more labile to heat inactivation due to some component in the meat ingredients. They postulated that this factor increased the sensitivity of the trypsin inhibitors to heat inactivation by causing a rupture of disulfide bonds in the inhibitor molecule, particularly the Bowman-Birk inhibitor which is rich in disulfide bonds. Of particular concern to the drug industry is the possibility that infants fed soy milk manufactured from soy isolates might be more sensitive to the physiological effects of the trypsin inhibitors. Churella *et al.* (1976), however, have recently demonstrated that the heat treatment involved in the processing and sterilization of soy formulas based on inhibitor-rich protein isolates prior to marketing reduces the trypsin inhibitor to less than 10% of the original activity. This residual level of activity did not produce any weight reduction or pancreatic hypertrophy in rats. These observations are consistent with the findings of Rackis *et al.* (1975) who found that only 70–80% of the trypsin inhibitory activity need be destroyed in order to achieve maximum weight gains and PER with rats, and only 40–70% destruction is needed to eliminate pancreatic hypertrophy (Table 13.4).

Assuming for the moment that processing conditions may have been inadequate to reduce the level of trypsin inhibitory activity below that of the threshold level established for rats, would the residual trypsin inhibitor activity still pose a risk to human health? Let us first address ourselves to the more basic question of whether the soybean inhibitors do in fact inhibit human trypsin. Trypsin inhibitor activity is invariably measured *in vitro* on the basis of the ability of soybean preparations to

TABLE 13.4

EFFECT OF SOY FLOUR CONTAINING VARIOUS LEVELS OF
TRYPSIN INHIBITOR ON GROWTH AND SIZE OF PANCREAS OF RATS[1]

Trypsin Inhibitor Content		Body Wt		Pancreas Wt
mg/100g diet	% Destruction	g	PER	g/100g Body Wt
887	0	79	1.59	0.70
532	40	111	2.37	0.56
282	68	121	2.78	0.50
157	82	134	2.97	0.49
119	87	148	3.08	0.47
71	92	142	3.03	0.45
Casein	—	145	3.35	0.55

[1]Source: Rackis *et al.* (1975).

inhibit *bovine* or *porcine* trypsin since the latter are readily available
commercially in "pure" crystalline form. Human trypsin is known to exist
in two forms, a cationic species which constitutes the major component and
an anionic species which accounts for only about 10–20% of the total
trypsin activity (Mallory and Travis 1973; Figarella *et al.* 1975; Robinson
et al. 1972). While the latter is fully inactivated by the soybean inhibitor
(Mallory and Travis 1973, 1975) the cationic form of trypsin which comprises about 80–90% of the potential tryptic activity of human pancreatic
juice is only weakly inhibited (Feeney *et al.* 1969; Figarella *et al.* 1975;
Travis and Roberts 1969).

In further support of the probability that the soybean inhibitors are
ineffective against human trypsin is the rather interesting relationship
that seems to exist between the size of the pancreas of various species of
animals and their hypertrophic response to raw soybeans or the inhibitor
purified therefrom (Table 13.5). There appears to be a direct relationship
between the size of the pancreas and the sensitivity of response to raw
soybeans or the trypsin inhibitor. The pancreas of those species of animals
whose weights exceed 0.3% become hypertrophic when fed raw soybeans,
whereas those whose weights are below this value do not respond to the
hypertrophic effects of the trypsin inhibitor; the guinea pig would appear
to be on the border line. One would predict from this relationship that the
human pancreas would not be sensitive to the trypsin inhibitor, although
there is no direct experimental evidence bearing on this point.

HEMAGGLUTININS

It is well recognized that, in addition to trypsin inhibitors, most legumes
and cereals contain substances, the so-called phytohemagglutinins or lec-

TABLE 13.5

RELATIONSHIP BETWEEN SIZE OF PANCREAS OF VARIOUS SPECIES OF
ANIMALS AND THE RESPONSE OF THE PANCREAS TO RAW SOYBEANS
OR TRYPSIN INHIBITOR

Species	Size of pancreas (% of body wt)	Pancreatic hypertrophy	References
mouse	0.6–0.8	+	Schingoethe et al. (1970)
rat	0.5–0.6	+	Liener and Kakade (1969)
chick	0.4–0.6	+	Liener and Kakade (1969)
guinea pig	0.29	±[1]	Patten et al. (1973)
dog	0.21–0.24	−	Patten et al. (1971)
pig	0.10–0.12	−	Yen et al. (1977)
human	0.09–0.12[2]	(−)[3]	
calf	0.06–0.08	−	Kakade et al. (1976)

[1]Observed in young guinea pigs but not in adults.
[2]Taken from Long (1961).
[3]*Predicted* response.

tins, which have the unique property of binding carbohydrate-containing
biological components (Jaffé 1969; Liener 1976). With red blood cells the
interaction of lectins with glycoprotein receptor sites of the cell membrane
is manifested *in vitro* by an agglutination of the cells. Ever since the time of
Ehrlich, it has been known that some of these lectins, such as ricin from the
castor bean, are extremely toxic to animals. Little is known, even now,
concerning the extent to which these substances might contribute to the
nutritive value of the more common legumes which are consumed by man.
The lectins, like the trypsin inhibitors, are readily destroyed by heat, and
their destruction is accompanied by a dramatic improvement in the biological value of the protein (Fig. 13.5). When the isolated soybean lectin was
fed to rats, the results obtained were somewhat ambiguous (Fig. 13.6). As
long as the animals were allowed free access to their food, there was an
apparent depression in growth. However, since this growth depression was
accompanied by a concomitant decrease in food consumption, it was not
clear whether the failure of the animals to grow was a consequence of
lowered food intake or whether the lower food consumption was the result
of depressed growth. When the food intake was equalized, however, the
soybean lectin had little effect on growth. This conclusion was further
supported when soybean extracts from which the soybean lectin had been
removed by affinity binding to a column of Sepharose-concanavalin A (Fig.
13.7) was fed to rats. As shown in Table 13.6 rats fed protein from which
over 90% of the soybean lectin had been removed grew just as poorly as
those fed the original crude protein extract. Although these results

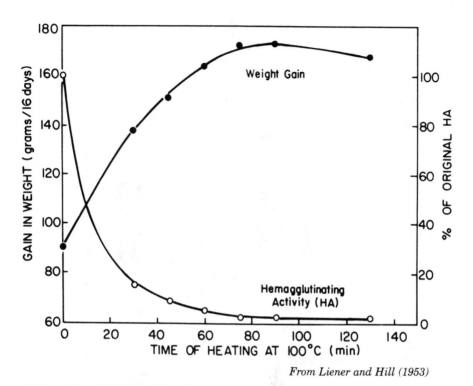

From Liener and Hill (1953)

FIG. 13.5. EFFECT OF HEAT TREATMENT OF SOYBEANS ON HEMAGGLUTINATING ACTIVITY AND GROWTH RESPONSE OF CHICKS

suggest that the soybean lectin is innocuous, one cannot rigorously exclude the possibility that any beneficial effect resulting from removal of the lectin may have been masked by the deleterious effects of harmful components still remaining in the unheated crude soybean extract.

Since these lectins are also present in many other commonly consumed legumes and since lectins are known to be quite diverse in their physicochemical and biological properties, we did not feel we could necessarily conclude that the lectins of other legumes were as innocuous as the soybean lectin. Table 13.7 shows the effect of heat on the growth-promoting property of a number of legumes which enjoy popular consumption in various parts of the world. It is evident that the two beans which are botanically classified as *Phaseolus vulgaris* are quite toxic to rats unless subjected to heat treatment. These two beans likewise display extremely high levels of hemagglutinating activity compared to those which do not respond to heat treatment. When purified preparations of the lectins from

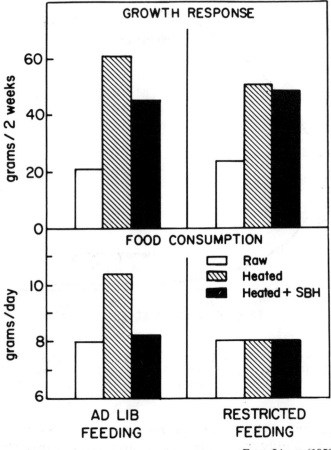

From Liener (1953)

FIG. 13.6.   EFFECT OF SOYBEAN HEMAGGLUTININ (SBH)
ON GROWTH AND FOOD CONSUMPTION OF RATS

the black bean and kidney bean were fed to rats, growth depression was noted which became more marked as the level of lectin in the diet increased (Fig. 13.8). In fact, at the higher levels of lectin, greater than 1% of the diet, a high incidence of mortality was observed. Autoclaving of the lectins for 20 min, however, destroyed the toxicity of these lectins.

It is difficult, of course, to assess the significance of these results with respect to the consumption of these beans in the human diet. As long as sufficient heat treatment has been applied to insure destruction of the lectins, there would appear to be little cause for concern. Nevertheless, it

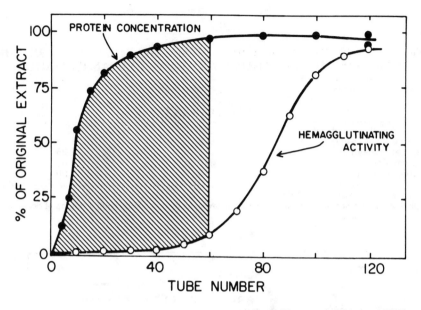

From Turner and Liener (1975)

FIG. 13.7. REMOVAL OF HEMAGGLUTININ FROM CRUDE SOY-
BEAN EXTRACT BY AFFINITY CHROMATOGRAPHY ON
SEPHAROSE-BOUND CONCANAVALIN A

should be recognized that conditions may prevail wherein complete destruction of the lectins may not always be achieved. For example, an
outbreak of massive poisoning occurred in Berlin in 1948 after the consumption of partially cooked bean flakes (Griebel 1950). Mixtures of
ground beans and cereals have been recommended in child-feeding programs in more primitive countries (King *et al.* 1966). Such mixtures can be
prepared locally from easily available materials and can be formulated in
proportions resulting in optimal nutritive value. Cooking such mixtures
requires a short heating time to become palatable so that the lectin may
not be completely destroyed. Primitive cooking is often done in earthen
pots on a wood fire. With a tough viscous mass like cooked beans, heat
transfer is necessarily imperfect, and, in the absence of constant and
vigorous stirring, the temperature reached in parts of the preparation may
well be inadequate for destruction of the lectins. Korte (1972) has in fact
observed that with mixtures of ground beans and ground cereals prepared
under conditions prevailing in Africa, the hemagglutinins were not always destroyed. With a reduction in the boiling point of water such as

TABLE 13.6

EFFECT OF REMOVING SOYBEAN HEMAGGLUTININ (SBH) ON THE
GROWTH-PROMOTING ACTIVITY OF RAW SOYBEAN EXTRACTS[1]

Protein Component of Diet	Hemagglutinating Activity Units/g protein $\times 10^{-3}$	PER
Original soybean extract	324	0.91
Original soybean extract – SBH[2]	29	1.13
Original soybean extract, heated	6	2.25
Raw soy flour	330	1.01
Heated soy flour	13	2.30

[1]Source: Turner and Liener (1975).
[2]SBH was removed from an aqueous extract of soybeans by passage through a column of Sepharose-bound concanavalin A.

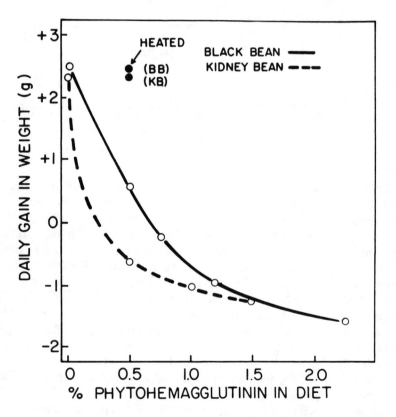

From Honavar et al. (1962)

FIG. 13.8. EFFECT OF BLACK BEAN AND KIDNEY BEAN
HEMAGGLUTININS ON GROWTH OF RATS

TABLE 13.7

EFFECT OF HEAT ON NUTRITIVE VALUE AND HEMAGGLUTINATING ACTIVITY OF SOME LEGUMES[1]

Legume	Gain in Wt. (g/day)		Hemagglutinating Activity (units/g)	
	Raw	Heated	Raw	Heated
Phaseolus vulgaris				
black bean	−1.94(4–5)[1]	+1.61	2450	0
kidney bean	−1.04(11–13)[1]	+1.48	3560	0
Cicer aritinum				
chick pea	+1.25	+1.16	0	0
Cajanus cajan				
Red grain	+1.33	+1.74	0	0
Phaseolus aureus				
mung bean	+1.05	+1.07	0	0

[1]Source: Honavar *et al.* (1962).
[2]100% mortality observed during period (in days) shown in parenthesis.

would be encountered in mountainous regions might also result in incomplete elimination of the hemagglutinins. Whether this residual amount of lectin might be toxic to humans must remain a matter of speculation, but until this question is resolved caution should be exercised in recommending the use of beans under conditions where proper heat treatment may not be insured.

LATHYRISM

Lathyrism, as it is known to occur in humans, is a disease associated with the consumption of a species of legume or peas known as *Lathyrus sativus* (also known as the chickling vetch or *kesari dal*). This disease is particularly prevalent in India, especially during periods of famine resulting from droughts when the crop fields become blighted, and, as an alternate crop this particular legume is cultivated. We are not dealing here with an occasional case of poisoning but a disease which can be almost of epidemic proportions. For as recently as 1975 over 100,000 cases of lathyrism in men between the ages of 15 to 45 years was reported in India (Natarajan 1976). Strangely enough, lathyrism seems to affect only males, particularly young adults. The disease symptoms develop in stages. In the beginning the victim walks in short and jerky steps. Later his knees begin bending and the heels raised, forcing the victim to walk on his toes. In the final stages, paralysis sets in below the hips, and death may result in extreme cases. All of these effects appear to stem from some kind of

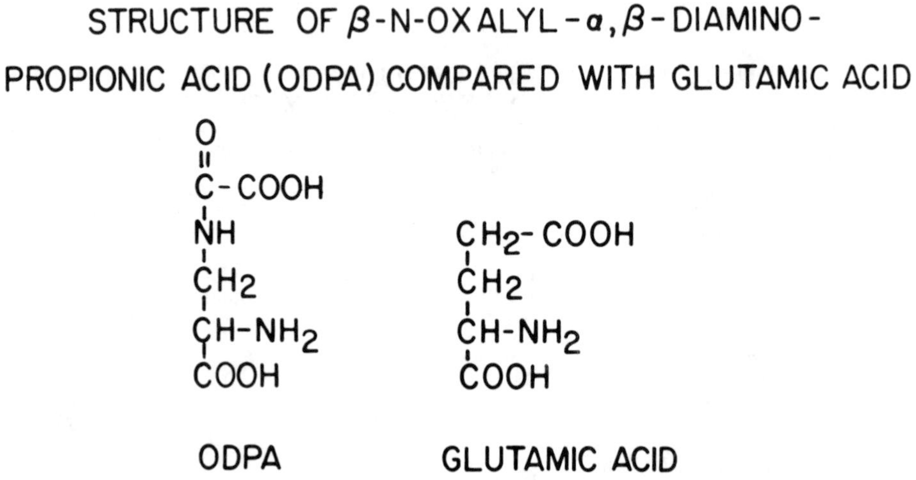

STRUCTURE OF β-N-OXALYL-α,β-DIAMINO-
PROPIONIC ACID (ODPA) COMPARED WITH GLUTAMIC ACID

```
        O
        ‖
        C-COOH
        |
        NH                    CH2-COOH
        |                     |
        CH2                   CH2
        |                     |
        CH-NH2                CH-NH2
        COOH                  COOH


        ODPA              GLUTAMIC ACID
```

FIG. 13.9. STRUCTURE OF NEUROLATHYROGEN
FROM *LATHYRUS SATIVUS* (β-N-OXALYL-α,β-
DIAMINOPROPIONIC ACID) AND ITS RELATIONSHIP
TO GLUTAMIC ACID

neuropathological lesion of the central nervous system. Hence the term "neurolathyrism" is frequently used to differentiate this disease from a form of lathyrism, osteolathyrism, which in animals is associated with the consumption of the sweet pea *(Lathyrus odoratus)* and is characterized by deformities of the bone (Liener 1975).

Progress towards the identification of the causative factor of human lathyrism has been hampered by the inability to reproduce a similar disease in animals. However, a compound, β-N-oxalyl-α,β-diamino-propionic acid (ODAP, Fig. 13.9), has been isolated from *L. sativus* which when injected into young rats, mice, chicks, and monkeys does produce neurotic symptoms closely resembling the symptoms of human lathyrism (Sarma and Padmanaban 1969). Roy (1973) has recently shown that the oral administration of ODAP to baby chicks can evoke neurological symptoms but at a much higher dose than that required by the intraperitoneal route. Because of its structural similarity to glutamic acid, it is not surprising that ODAP has been found to interfere with the role of glutamic acid as an excitory neurotransmitter in brain tissue (Jacob *et al.* 1967; Duque-Magalhaus and Packer 1972; Lakshmanan and Padmanaban 1974).

Despite all of the experimental evidence which would appear to implicate ODAP as the causative factor of neurolathyrism, it remains to be demonstrated that the neurological symptoms of favism can be produced by feeding L. sativus seeds to animals. Assuming for the moment that ODAP is in fact the causative principle of human lathyrism, it may come as a surprise to learn that all of the misery in the past associated with the consumption of L. sativus could have been avoided by relatively simple detoxification procedures (Mohan et al. 1966). Most of the toxin (90% or more) can be effectively removed by either cooking the seeds in excess water which is then discarded or by soaking the bean overnight followed by steaming, roasting, or sun drying. The dried bean can then be ground into flour for making chapatis, an unleavened Indian bread. Actually L. sativus contains 24–28% protein which is rich in lysine (Sarma 1968). Thus, despite the tarnished reputation which this legume has had over the years, it may yet have potential as a protein supplement in a country which sorely needs it.

FAVISM

There has recently been a resurgence of interest in the expanded use of the field bean (*Vicia faba,* also known as the broad bean or horse bean) as an alternate source of protein for livestock and poultry. The field bean is relatively rich in protein (about 28–30%) and, although somewhat deficient in methionine, the protein of the heat-treated bean has a biological value not too different from soybeans (Nitsan 1971; Bell 1973). Like soybeans and other legumes, the field bean contains trypsin inhibitor and hemagglutinin, but these do not appear to be responsible for the poor nutritive value of the raw beans (Abbey et al. 1976; Marquardt et al. 1976). Although the field bean would appear to offer a promising source of protein for human consumption, and has in fact been found to be suitable as a protein supplement in breadmaking (Patel and Johnson 1975), recommendation for its use for human consumption has been tempered by the fear that its consumption could lead to a disease in some susceptible individuals known as favism. This would be particularly true in those countries (i.e. Middle East) where the field bean is a major food staple and where the genetic defect associated with favism (see below) is most prevalent.

Favism is a disease characterized by hemolytic anemia which affects certain individuals following the ingestion of field bean (see reviews by Mager et al. 1969; Belsey 1973). The symptoms which accompany this disease include weakness or fatigue, pallor, jaundice and hemoglobinuria. Favism is confined largely to the inhabitants of the Mediterranean basin, although individuals of the same ethnic background residing in other countries frequently suffer from favism. Although the majority of the cases of favism (about two-thirds) is associated with the consumption of the fresh or dried beans, the remainder of the cases is caused by cooked beans.

One of the main difficulties in trying to elucidate the pathogenesis of favism has been the inability of being able to reproduce this disease in an animal model. Although heating definitely improves the nutritive value of the field bean for experimental animals, no symptoms resembling human favism have been observed with the raw bean.

Extensive clinical studies with favism-prone individuals have revealed that such individuals are genetically deficient in glucose-6-phosphate dehydrogenase (G6PD) and low accompanying levels of reduced glutathione (GSH). The latter is important for maintaining the integrity of the cell membrane, and the role of G6PD is to generate reduced triphosphopyridine nucleotide (NADPH) via the pentose phosphate shunt. NADPH is necessary for the action of glutathione reductase which causes the reduction of oxidized glutathione (GSSG) to GSH.

It follows from the above that substances which are capable of oxidizing GSH, particularly in the absence of G6PD, might be expected to cause hemolysis of red blood cells. The field bean is known to contain such substances in the form of the pyrimidine-o-glycosides, vicine and convicine (Fig. 13.10). *In vitro* experiments have demonstrated that the aglycones of

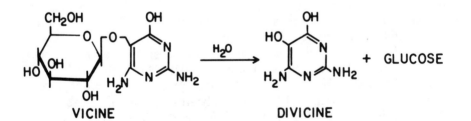

VICINE   DIVICINE

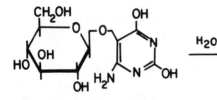
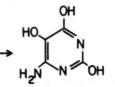

CONVICINE   ISOURAMIL

FIG. 13.10. STRUCTURE OF PROBABLE CAUSATIVE FACTORS OF FAVISM

these glycosides, divicine and isouramil, cause a rapid oxidation of GSH in G6PD-deficient erythrocytes but not in normal cells (Fig. 13.11). It is evident that in order for divicine and isouramil to function as oxidizing agents they must be released from their parent glycosides. How this is accomplished is not known; this could take place through enzymatic hydrolysis in the intestines or in the blood itself (Flohé et al. 1971).

Assuming that the pyrimidine glucosides are in fact the causative agents of favism, it would appear that the only way of diminishing the risk of this disease is to effect their removal by strain selection or by some suitable processing techniques. With the availability of tests for the quantitative estimation of vicine and convicine (Brown and Roberts 1972; Higazi and Read 1974; Collier 1976; Jamalian et al. 1976), it should be possible for the plant geneticist to screen cultivars of the field bean in a search for nontoxic levels of these substances. The search for processing methods which would effect the removal of favism-producing components should be encouraged. Although protein concentrates (Bell 1973) and

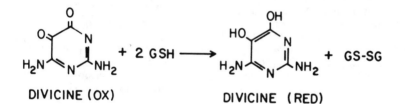

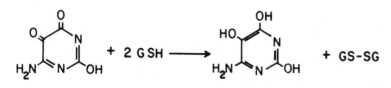

GSH = REDUCED GLUTATHIONE

GS-SG = OXIDIZED   "

FIG. 13.11.   EQUATIONS SHOWING THE MANNER IN WHICH DIVICINE AND ISOURAMIL SERVE TO REDUCE LEVELS OF REDUCE GLUTATHIONINE IN RED BLOOD CELLS

isolates (Duthrie *et al.* 1972) have been prepared from the field bean and have been found to be of good nutritional quality for animals, the extent to which they might still be contaminated with vicine or convicine does not appear to have received the attention this problem deserves if such products are to have any application to their use in human diets.

BIBLIOGRAPHY

ABBEY, B. W., NEALE, R. J. and NORTON, G. 1976. Nutritional effects of field bean *(Vicia faba* L.) protease inhibitors and field beans fed to rats. Proc. Nutr. Soc. *35*, 84A.

BELL, J. M. 1973. Nutritional evaluation of protein concentrates from field peas *(Pisum sativum)* and faba beans *(Vicia faba). In* Nutritional Aspects of Common Beans and Other Legume Seeds for Animal and Human Foods, W. G. Jaffé (Editor). Arch. Latinoamer. Nutr., Caracas, Venezuela, pp. 165–178.

BELSEY, M. A. 1973. The epidemiology of favism. Bull. World Health Org. *48*, 1–13.

BROWN, E. G. and ROBERTS, F. M. 1972. Formation of vicine and convicine by *Vicia faba.* Phytochemistry *11*, 3203–3206.

CHURELLA, H. R., YAO, B. C. and THOMSON, W. A. B. 1976. Soybean trypsin inhibitor activity of soy infant formulas and its nutritional significance for the rat. J. Agr. Food Chem. *24*, 393–397.

COLLIER, H. B. 1976. The estimation of vicine in faba beans by an ultra-violet spectrophotometric method. Can. Inst. Food Sci. Technol. J. *9*, 155–159.

DUQUE-MAGALHAUS, M. C. and PACKER, L. 1972. Action of neurotoxin, β-N-oxalyl-Lα,β-diaminopropionic acid, in glutamate metabolism of brain mitochondria, FEBS Letters *23*, 188–190.

DUTHRIE, I. F., PORTER, P. D. and GADALY, B. 1972. Nutritional value of a protein isolate from the Throws M.S. variety field beans *(Vicia faba* L.) grown in the U.K. Proc. Nutr. Soc. *31*, 80A.

FEENEY, R. E., MEANS, G. E. and BIGLER, J. C. 1969. Inhibition of human trypsin, plasmin and thrombin by naturally occurring inhibitors of proteolytic enzymes. J. Biol. Chem. *244*, 1957–1960.

FIGARELLA, C., NEGRI, G. A. and GUY, O. 1975. The two human trypsinogens. Inhibition spectra of the two human trypsins derived from their purified zymogens. Eur. J. Biochem. *53*, 457–463.

FLOHÉ, L., NIEBCH, G. and REIBER, H. 1971. The effect of divicine on human erythrocytes. Z. Klin. Chem. u. Klin. Biochem. *9*, 431–437.

GREEN, G. M. and LYMAN, R. L. 1972. Feedback regulation of pancreatic enzyme secretion as a mechanism for trypsin inhibitor-induced hypersecretion in rats. Proc. Soc. Exptl. Biol. Med. *140*, 6–12.

GRIEBEL, C. 1950. Erkrankungen durch bohnen flocken *Phaseolus vulgaris* L. and plattererbsen *Lathyrus tingitanus.* Z. Lebensm. Untersuch.-Forsch. *90*, 191–195.

HIGAZI, M. I. and REED, W. W. C. 1974. Method for determination of vicine in plant material and in blood. J. Agr. Food Chem. *22*, 570–571.

HONAVAR, P. M., SHIH, C.-V. and LIENER, I. E. 1962. Inhibition of growth of rats by purified hemagglutinin fractions isolated from *Phaseolus vulgaris*. J. Nutr. *77,* 109–114.

JACOB, E., PATEL, A. S. and RAMAKRISHNAN, C. V. 1967. Effect of neurotoxin from the seeds of *Lathyrus sativus* on glutamate metabolism in chick brain. J. Neurochem. *14,* 1091–1094.

JAFFÉ, W. G. 1969. Hemagglutinins. *In* Toxic Constituents of Plant Foodstuffs, I. E. Liener (Editor). Academic Press, N.Y.

JAMALIAN, J., AYLWARD, F. and HUDSON, B. J. F. 1976. Favism-inducing toxins in broad beans *(Vicia faba):* examination of bean extracts for pyrimidine glucosides. Qual. Plant. Pl. Fds. Hum. Nutr. *26,* 331–339.

KAKADE, M. L., HOFFA, D. and LIENER, I. E. 1973. Contribution of trypsin inhibitors to the deleterious effects of unheated soybeans fed to rats. J. Nutr. *103,* 1772–1778.

KAKADE, M. L., SIMONS, N. and LIENER, I. E. 1969. An evaluation of natural vs. synthetic substrates for measuring the antitryptic activity of soybean samples. Cereal Chem. *46,* 518–526.

KAKADE, M. L., SIMONS, N., LIENER, I. E. and LAMBERT, J. W. 1972. Biochemical and nutritional assessment of different varieties of soybeans. J. Agr. Food Chem. *20,* 87–90.

KAKADE, M. L., THOMPSON, R. D., ENGELSTAD, W. E., BEHRENS, G. C., YODER, R. D. and CRANE, F. M. 1976. Failure of soybean trypsin inhibitor to exert deleterious effects in calves. J. Dairy Sci. *59,* 1484–1489.

KING, K., FOURGERE, W., FOUCALD, J., DOMINIQUE, G. and BEGKIN, I. D. 1966. Response of pre-school children to high intake of Haitian cereal-bean mixture. Arch. Latinamer. Nutr. *16,* 53–64.

KORTE, R. 1972. Heat resistance of phytohemagglutinins in weaning food mixtures containing beans *(Phaseolus vulgaris).* Ecol. Fd. Nutr. *1,* 303–307.

KOTTER, L., PALITZSCH, A., BELITZ, H.-D. and FISCHER, K.-H. 1970. The presence and significance of trypsin inhibitors in isolated soya protein intended for use in the manufacture of meat products which are heated to high temperatures. Die Fleischwirtschaft *8,* 1063–1064.

LAKSHMANAN, J. and PADMANABAN, G. 1974. Effect of β-N-oxalyl-α, β-diaminopropionic acid on glutamate uptake by synaptosomes. Nature *249,* 469–470.

LIENER, I. E. 1953. Soyin, a toxic protein from the soybean. 1. Inhibition of rat growth. J. Nutr. *49,* 527–539.

LIENER, I. E. (Editor) 1969. Toxic Constituents of Plant Foodstuffs. Academic Press, New York, N.Y.

LIENER, I. E. 1975. Antitryptic and other factors. *In* Nutritional Improvement of Food Legumes by Breeding, M. Milner (Editor), United Nations, N.Y.

LIENER, I. E. 1976. Phytohemagglutinins (Phytolectins). Ann. Rev. Plant Physiol. *27,* 291–319.

LIENER, I. E., DEUEL, H. J. JR. and FEVOLD, H. L. 1949. The effect of supplemental methionine on the nutritive value of diets containing concentrates of the soybean trypsin inhibitor. J. Nutr. *39,* 325–339.

LIENER, I. E. and HILL, E. G. 1953. The effect of heat treatment on the nutritive value and hemagglutinating activity of soybean oil meal. J. Nutr. *49,* 609–620.

LIENER, I. E. and KAKADE, M. L. 1969. Protease inhibitors. *In* Toxic Constituents of Plant Foodstuffs, I. E. Liener (Editor), Academic Press, Inc., N.Y.

LONG, C. 1961. Biochemists' Handbook. D. Van Nostrand Co., Inc., Princeton, N.J.

MAGER, J., RAZIN, A. and HERSHKO, A. 1969. Favism. *In* Toxic Constituents of Plant Foodstuffs, I. E. Liener (Editor), Academic Press, Inc., N.Y.

MALLORY, P. A. and TRAVIS, J. 1973. Human pancreatic enzymes. Characterization of anionic trypsin. Biochemistry *12,* 2847–2851.

MALLORY, P. A. and TRAVIS, J. 1975. Inhibition spectra of the human pancreatic endopeptidases. Amer. J. Clin. Nutr. *28,* 823–830.

MARQUARDT, R. R., CAMPBELL, L. D. and WARD, T. 1976. Studies with chicks on the growth depressing factor(s) in faba beans (*Vicia faba* L. var. minor). J. Nutr. *106,* 275–284.

MOHAN, V. S., NAGARAJAN, V. and GOPALAN, C. 1966. Simple practical procedures for the removal of toxic factors in *Lathyrus sativus* seeds (*Khesari dahl*). Indian J. Med. Res. *54,* 410–414.

NATARAJAN, K. R. 1976. India's poison peas. Chemistry *49,* 12–13.

NIESS, E., IVY, C. A. and NESHEIM, M. C. 1972. Stimulation of gall bladder emptying and pancreatic secretion in chicks by soybean whey protein Proc . Soc. Biol. Expt. Med. *140,* 291–295.

NITSAN, Z. 1971. *Vicia faba* beans vs. soybean meal as a source of protein. J. Sci. Fd. Agr. *22,* 252–255.

NORDAL, J. and FOSSUM, K. 1974. The heat stability of some trypsin inhibitors in meat products with special reference to added soybean protein. Z. Lebensm. Unters.-Forsch. *154,* 144–150.

PATEL, K. M. and JOHNSON, J. A. 1975. Horsebean protein supplements in breadmaking. 2. Effect on physical dough properties, baking quality, and amino acid composition. Cereal Chem. *52,* 791–795.

PATTEN, J. R., PATTEN, J. A. and POPE, H., II. 1973. Sensitivity of the guinea-pig to raw soya beans in the diet. Fd. Cosmet. Toxicol. *11,* 577–583.

PATTEN, J. R., RICHARDS, E. A. and WHEELER, J. 1971. The effect of dietary soybean trypsin inhibitor in the histology of dog pancreas. Life Sci. *10,* 145–150.

RACKIS, J. J. 1972. Biologically active components. *In* Soybeans: Chemistry and Technology, A. K. Smith and S. J. Circle (Editors), Avi Publishing Co., Westport, Conn.

RACKIS, J. J., McGHEE, J. E. and BOOTH, A. N. 1975. Biological threshold levels of soybean trypsin inhibitors by rat bioassay. Cereal Chem. *52,* 85–92.

ROBINSON, L. A., KIM, W. J., WHITE, T. T. and HADORN, B. 1972. Trypsins in human pancreatic juice—their distribution as found in 34 specimens. Two human pancreatic trypsinogens. Scand. J. Gastroenterol. *7,* 43–45.

ROY, D. N. 1973. Effect of oral administration of β-N-oxalylamino-L-alanine (BOAA) with or without *Lathyrus sativus* trypsin inhibitor (L-TI) in chicks. Environ. Physiol. Chem. *3,* 192–195.

SARMA, P. S. 1968. Nutritional problems of lathyrism in India. J. Vitaminol. *14,* 53–57.

SARMA, P. S. and PADMANABAN, G. 1969. Lathyrogens. *In* Toxic Constituents in Plant Foodstuffs, I.E. Liener (Editor), Academic Press, Inc., N.Y., pp. 267–291.

SCHINOGOETHE, D. J., AUST, S. D. and THOMAS, J. W. 1970. Separation of a mouse growth inhibitor in soybeans from trypsin inhibitors. J. Nutr. *100,* 739–748.

THOMPSON, R. D. and LIENER, I. E. 1977. Unpublished observations.

TRAVIS, J. and ROBERTS, R. C. 1969. Human trypsin isolation and physiochemical characterization. Biochemistry *8,* 2884–2889.

TURNER, R. H. and LIENER, I. E. 1975. The effect of the selective removal of hemagglutinins on the nutritive value of soybeans. J. Agr. Food Chem. *23*, 484–487.

YEN, J. T., JENSEN, A. H. and SIMON, J. 1977. Effect of dietary raw soybean trypsin inhibitor on trypsin and chymotrypsin activities in the pancreas and in small intestinal juice of growing swine. J. Nutr. *107*, 156–165.

ENZYMES AS QUALITY INDICATORS IN EDIBLE PLANT TISSUES

JOHN P. CHERRY

Southern Regional Research Center
Agricultural Research Service
U.S. Department of Agriculture
New Orleans, LA 70179

ABSTRACT

Enzymes have potential of being utilized as biological indicators to assist investigators in determining and maintaining optimum quality composition of edible substances during various food handling stages. These stages include postharvest, harvest, storage, and processing conditions; the latter two factors include postharvest biology and biotechnology. Enzymes change in activity depending on the physiology of plant tissues, and/or are altered structurally when abused during handling and processing. Gel electrophoretic techniques serve as tools for qualitatively detecting these variations in enzyme activity. Researchers and commercial processors can use these techniques to evaluate plant material for (a) optimum maturity, (b) compositional variations due to genetic and environmental factors, (c) fungal contamination and deterioration during storage intervals, and (d) effect of various conditions normally included in the processing of raw material to finished food products. With respect to processing, for example, enzymes can serve as quality indicators in the maintenance of (a) optimum processing of seed to flour, concentrates, and/or isolates, (b) heat treatment conditions and effects during product formation, and (c) properly identifying constituents during and after formulation of composite foods.

INTRODUCTION

Commercially, enzymes are utilized in such areas of food processing as appearance modification, flavor enhancement, the control of texture and viscosity, and waste management and pollution control. To be specific, carbohydrases and amylases convert starches to sugars, pectinases clarify fruit juices and wines, proteases hydrolyze proteins, and lipases enhance flavor (Grodner 1976; Reed 1976; Weetall 1976; Barfoed 1976). In still another area of application, clinical laboratories utilize enzymes in tests on patients to indicate the existence of specific physiological disorders (Wilkinson 1976). The dramatic increase in the use of enzymes can be attributed to the following: (a) the value of enzymes is recognized in both basic and applied fields; (b) research is continually introducing new enzymes to technology; (c) procedures are being developed to specifically utilize enzymes in unique situations; (d) methods to identify, prepare, and utilize enzymes are undergoing rapid commercialization; and (e) in the food field, enzymes occur naturally in edible products and thus are normally considered to be a part of the food chain.

Gel electrophoretic techniques are excellent examples of modern technology developed to qualitatively detect proteins and, more specifically, enzyme activity. The background and theory of electrophoretic techniques were discussed by Ornstein (1964), and applications of these procedures to analyze and compare enzymes, as well as proteins in general, were presented by a number of investigators (Smithies 1962; Davis 1964; Brewbaker et al. 1968; Scandalios 1969; Gottlieb 1971). Basically, the methods separate partially to highly purified fractions of protein by an electric charge in aqueous extracts in a gel matrix such as polyacrylamide or starch. Protein mobility through electrophoretic media depends upon a combination of protein-related factors, including net charge, molecular size, and conformation. With enzymes, histochemical staining procedures have been developed to detect location of protein in gels according to catalytic activity.

Enzymes detectable on electrophoretic gels can be used as biological indicators for identifying compositional variations of food substances at preharvest, harvest, and postharvest stages in food processing (Cherry 1977A). The quality characteristics of seeds can be determined— immaturity, maturity, and overmaturity— as well as genetic determinants and environmental or agronomic factors. During storage, enzymes in food substances are affected by handling practices and aging and thus can be used to predict molecular changes that may affect product performance. The presence of various fungi in stored foods and the deteriorating changes the fungi induce during development can be detected by enzymes. Quality control of processing steps, e.g., converting seed to flour, heating conditions, and identifying flours, concentrates or isolates in composites, can be maximized with enzymes as indicators. The use of enzymes as biological indicators, therefore, can assist in insuring efficient use of raw material, proper processing conditions, and subsequent production of high quality food products.

HARVEST STAGE

Peroxidases of Developing Seeds

Peroxidase activity has been related to fungal development, cellular injury, hormonal activity, off-flavors, and off-colors of plant tissues (Burnette 1977). In peanuts, most peroxidase activity has been located in the albumin fraction of seed storage proteins (Cherry *et al.* 1973). Zymograms that contain a band at the origin, one in region 0.7 cm, and three in region 3.0–4.0 cm are typical of most patterns of mature peanuts (Fig. 14.1). Examination of peanuts from two cultivars (A, B) produced identical peroxidase patterns for immature, mature, and germinating seeds. A

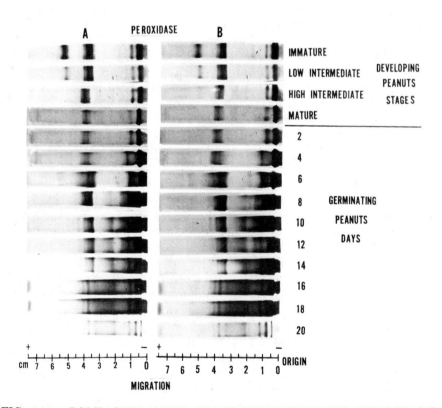

FIG. 14.1. POLYACRYLAMIDE GEL ELECTROPHORETIC ZYMOGRAMS OF PEROXIDASES IN PEANUT SEEDS FROM TWO CULTIVARS (A IS FLORUNNER, B IS SPANCROSS) DURING DEVELOPING, MATURE, AND GERMINATING STAGES

peroxidase band in region 5.0 cm distinguished immature and low intermediate seeds from high intermediate and mature peanuts in both cultivars. Germinating peanuts showed increased peroxidase activity in electrophoretic gels after 2–4 days, that became prevalent in region 0–3.0 cm between 8 and 18 days. An increase in peroxidase activity was noted in region 3.0–4.0 cm at days 6 and 8, decreasing slowly thereafter as germination approached 20 days.

This study showed that immature, mature, and germinating stages in peanut development can be distinguished by separation of peroxidase activity on electrophoretic gels. In addition, plant breeders have developed uniformity in seed development for these two cultivars. Ideally, methods for detecting seed maturity, and thus harvesting and processing conditions, should be readily interchangeable for both cultivars.

This similarity of peroxidase zymograms for two cultivars supports the contention of many geneticists that the genetic base, or gene pool, of cultivated plants does not have the reserve germ plasm needed to resist many of the new agricultural problems brought on by pollution, dwindling water supplies, and the necessity for biological control methods against insects and plant pathogens (Cherry 1977B).

Leucine Aminopeptidases and Proteins in Developing Seeds

The leucine aminopeptidases are useful enzymes in electrophoretic studies of protein polymorphism, seed maturity, plant hybridization, and polypeptide hydrolysis (Manwell and Baker 1970). Zymograms of leucine aminopeptidase activity distinguished peanuts of a particular cultivar at the high intermediate and mature stages (Fig. 14.2). On electrophoretic gels, the activity of this enzyme in overmature peanuts was distinguishable from that of mature seeds.

Changes in leucine aminopeptidase activity relative to protein content were examined in cotyledonary extracts of a peanut cultivar on electrophoretic gels during extended seed germination (Fig. 14.3). In these preparations, the peptidase activity with mobility in the region 3.5–4.5 cm remained consistently high during early stages of germination to day 6, after which it decreased. By day 12, most of this activity could not be detected. This observation suggested that peptidase activity may be partially responsible for the breakdown of reserve protein during early stages of germination. Moreover, these enzymes from germinating peanuts could be used in food studies to hydrolyze proteins in flours and isolates of mature seeds, thereby altering the functional properties of these products.

Protein extracts of ungerminated peanuts (0 day) consist mainly of the major storage globulin, arachin, in region 0–2.0 cm, and the nonarachin or conarachin proteins in region 2.0–4.0 cm (Fig. 14.3; Cherry et al. 1973;

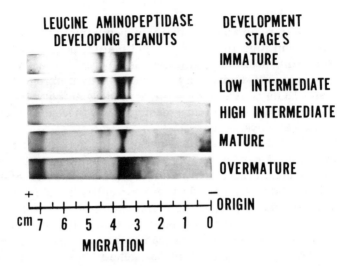

FIG. 14.2. POLYACRYLAMIDE GEL ELECTROPHORETIC ZYMOGRAMS OF LEUCINE AMINOPEPTIDASES IN PEANUTS FROM ONE CULTIVAR DURING DEVELOPING, MATURE AND OVERMATURE STAGES

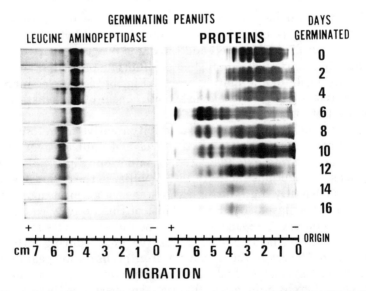

FIG. 14.3. EVALUATION OF POLYACRYLAMIDE GEL ELECTROPHORETIC PATTERNS OF LEUCINE AMINOPEPTIDASES AND AQUEOUS SOLUBLE PROTEINS IN GERMINATING PEANUTS

Basha and Cherry 1976). Major changes in protein composition of peanut extracts were shown at and after day 4. The arachin components and the bands in region 2.0–3.5 cm decreased quantitatively in the gel patterns, and simultaneously numerous proteins appeared in region 2.5–7.0 cm. The appearance of small protein components in the gel patterns indicates hydrolysis of the major storage proteins of peanuts by proteases and peptidases (located in region 3.5–4.5 cm) to polypeptides of various sizes and to free amino acids. These data further support the hypothesis that these enzymes could be used to hydrolyze proteins in flours and isolates of mature seeds to alter their food functional properties.

During the interval of days 6–8, there was a major change in type of leucine aminopeptidase activity (region 3.5–4.5 cm activity decreased, whereas the band in region 5.0–6.0 cm increased). After this interval most of the remaining activity was associated with these latter bands, suggesting that they may have been activated or synthesized *de novo* during this period to the various sizes and/or charged groups in region 2.5–7.0 cm. During days 14–16, only minor protein constituents could be detected on electrophoretic gels. This finding coincided with the consistently low quantity of protein and peptidase activity that remained in cotyledonary extracts between days 12 and 16 after germination began.

STORAGE STAGE

Fungal Contamination of Seeds

Studies have shown that standard enzyme electrophoretic patterns of peanuts are distinctly modified by infection of the seed with various saprophytic fungal species (Cherry 1977B; Cherry et al. 1974). The sequence of biochemical changes which is distinguishable from standard profiles of uninoculated seeds include: (a) decomposition of proteins to low molecular weight components; (b) quantitative depletion of low molecular weight components; (c) depletion of some enzymes, intensification of others, and/or production of new multiple forms of enzymes; and (d) production of aflatoxins by microorganisms. These changes suggest that the biochemical mechanisms operative in the saprophyte-seed interrelationship function very efficiently and systematically for the growth of the fungus at the expense of the seed. Gel electrophoretically detected transformations, such as catabolism of peanut proteins, coincided with enzyme changes in extracts of fungal tissue from the seed surface. Moreover, many of the multiple forms of enzymes of peanuts remained active during the infection period. Examples of enzyme and protein changes in peanuts

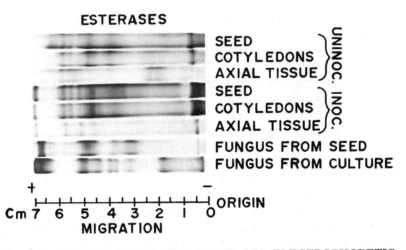

FIG. 14.4. TYPICAL POLYACRYLAMIDE GEL ELECTROPHORETIC PATTERNS OF ESTERASES EXTRACTED FROM PEANUT SEED, COTYLEDON, AND AXIAL TISSUES AFTER INCUBATION FOR 5 DAYS IN THE ABSENCE (UNINOC) AND PRESENCE (INOC) OF *A. parasiticus;* AND MYCELIAL TISSUE COLLECTED FROM THE EXTERIOR SURFACE OF SEEDS AND FROM CZAPEK'S SOLUTION

caused by infection by various fungi that are detectable by polyacrylamide disc and starch slab gel electrophoresis are discussed below.

Esterases

Peanuts contaminated for 5 days with *Aspergillus parasiticus* had esterase changes detectable in electrophoretic gels (region 3.0–7.0 cm) that were not present in those of uninoculated seeds (Fig. 14.4). Esterase activity increased both quantitatively and qualitatively in regions 4.0–5.0 cm and 6.5–7.0 cm of polyacrylamide gels of infected seeds. These esterase bands from infected peanuts were similar to those of fungal tissue collected from the exterior surface of peanuts. However, bands in region 1.5 and 2.5–4.5 cm distinguish fungal tissue grown on synthetic medium (Czapek's solution) from that collected from the surface of peanuts.

The esterase changes in peanuts caused by five *A. flavus* strains were distinguished by polyacrylamide gel electrophoresis (Fig. 14.5). The total esterase gel patterns of *A. flavus*-infected peanuts were different from those of *A. parasiticus*. A major (dark staining) esterase band in region 4.5–5.0 cm was present in extracts of both contaminated sources (compare Fig. 14.4 and 14.5). Most *A. flavus*-infected peanut extracts contained a major esterase band in region 6.5–7.0 cm that was not clearly discernible in material contaminated with *A. parasiticus*.

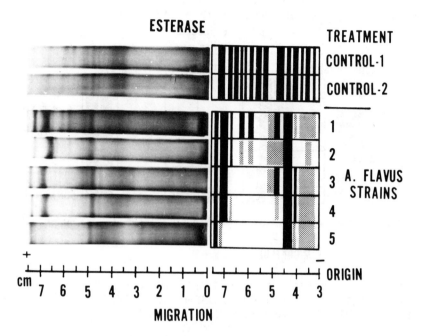

FIG. 14.5. POLYACRYLAMIDE GEL ELECTROPHORETIC PAT-
TERNS OF ESTERASES IN NONINFECTED (CONTROLS 1, 2) AND
A. flavus-INFECTED PEANUTS AFTER 4 DAYS
Controls 1 and 2 represent peanuts at day 0 and after incubation for 4
days, respectively, at 29° C in a high humidity chamber. A. flavus strains
from the Northern Regional Research Center, USDA-ARS, Peoria, IL,
were labeled as follows: 1. NRRL 5220, 2. NRRL 5518, 3. A-14152, 4.
NRRL 3517, and 5. A-62462.

**Catalases, Phosphogluconate Dehydrogenases, Alcohol Dehydrogenases,
Glucose-6-phosphate Dehydrogenases, Acid Phosphatases and Alkaline
Phosphatases**

Examples of other enzyme changes in peanuts infected for 4 days with
various A. flavus strains are shown on starch slab electrophoretic gels in
Fig. 14.6 and 14.7. No differences were noted between starch gel patterns
of catalase from noninfected and A. flavus-infected peanuts (Fig. 14.6); two
bands were consistently observed in all zymograms. In general, the activities of dehydrogenases such as phosphogluconate, alcohol, and

glucose-6-phosphate dehydrogenases, which were distinguished on starch electrophoretic gels of noninfected peanuts, became difficult to discern in extracts of *A. flavus*-infected seeds (Fig. 14.6, 14.7). Small amounts of

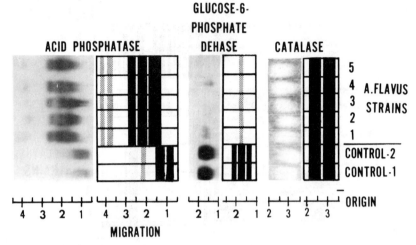

FIG. 14.6. STARCH GEL ELECTROPHORETIC PATTERNS OF ACID PHOSPHATASE, GLUCOSE-6-PHOSPHATE DEHYDROGENASE (DEHASE), AND CATALASE OF NONINFECTED AND *A. flavus*-INFECTED PEANUTS

Gel descriptions are given in Fig. 14.5.

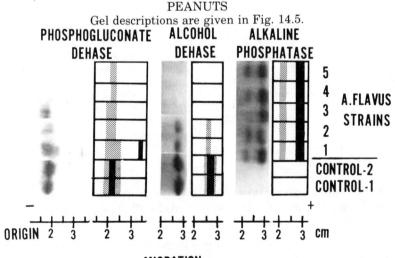

FIG. 14.7. STARCH GEL ELECTROPHORETIC PATTERNS OF PHOSPHOGLUCONATE AND ALCOHOL DEHYDROGENASE (DEHASE), AND ALKALINE PHOSPHATASE OF NONINFECTED AND *A. flavus*-INFECTED PEANUTS

activities were observed in certain *A. flavus*-infected seeds (alcohol dehydrogenase, strains 1 and 2; phosphogluconate dehydrogenase, strains 1, 2 and 3; glucose-6-phosphate dehydrogenase, strain 1). A phosphogluconate dehydrogenase band in region 3.5 cm specifically distinguished peanuts infected with *A. flavus,* strain 1, from seed infected with all other contaminants.

In contrast, the acid phosphatase starch gel zymograms distinguished extracts of *A. flavus*-infected peanuts from those of noninfected seeds (Fig. 14.6). Alkaline phosphatase was detected only in *A. flavus*-infected peanuts (Fig. 14.7); two bands of similar level activity were noted in all contaminated peanuts.

Since alkaline phosphatase is not normally present in peanuts, this enzyme could be used as an indicator of *A. flavus* contamination in peanut protein preparations. Distinct changes in acid phosphatase, as shown after infection of peanuts with *A. flavus,* could also be an indicator of fungal contamination. This would be especially suggestive of compositional changes in proteins or other storage constituents and products that could affect peanut functional and nutritional properties and use in food products. Preliminary tests for phosphatase activity (acid, alkaline) in peanut products prior to their utilization as food ingredients could be included as part of quality control measures normally used by food processors. The absence of activity of other enzymes (e.g., dehydrogenase activity) could add supportive information to such test procedures.

Proteins

Gel electrophoretic patterns of proteins in aqueous extracts of peanuts showed that fungi such as *A. flavus, A. parasiticus, A. oryzae, Rhizopus oligosporus,* and *Neurospora sitophila* altered extractability of storage constituents compared to that of control seeds (Fig. 14.8, 14.9). These changes coincided with alterations in protein solubility properties during the test with select fungi (Fig. 14.10).

In general, protein patterns of seeds infected by the different fungi, when compared to those of the noninfected control, showed new polypeptide components in region 0–1.0 cm and increased mobility and poor resolution of the bands in region 1.0–2.0 cm as the infection progressed. At the same time, bands normally located in region 2.0–3.5 cm disappeared, and a new group of polypeptides appeared in region 3.5–7.0 cm. All five *A. flavus* strains produced similar effects on water-soluble proteins of peanuts infected for 4 days. By day 7, after the seeds were inoculated with either *A. parasiticus* or *A. oryzae,* many of the proteins in the lower half of the gel patterns were difficult to distinguish.

Mean percentages of crude protein in aqueous soluble and insoluble

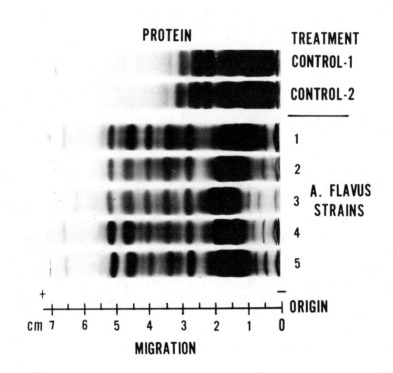

FIG. 14.8. POLYACRYLAMIDE GEL ELECTROPHORETIC
PATTERNS OF WATER-SOLUBLE PROTEINS OF NONIN-
FECTED AND *A. flavus*-INFECTED PEANUTS
Gel descriptions are given in Fig. 14.5.

fractions of noninfected peanuts during a 7-day test period remained
relatively unchanged at approximately 60.5% and 34.2%, respectively
(Fig. 14.10). Soluble and insoluble fractions of seeds infected with *A.
parasiticus, R. oligosporus,* and *N. sitophila* contained significantly lower
and higher protein percentages, respectively, than those of noninfected
preparations during days 4–7; at day 2, all values were between 40% and
55%. Percentage of soluble protein in *A. oryzae*-infected seeds decreased at
day 2, then increased continuously until day 7 to a value (60.5%) similar to
that of the control.

    Previous to these studies with raw peanut seeds infected with various
fungi, most research on this subject was on the proximate composition of
finished fermented products compared to nonfermented substrates (Gray

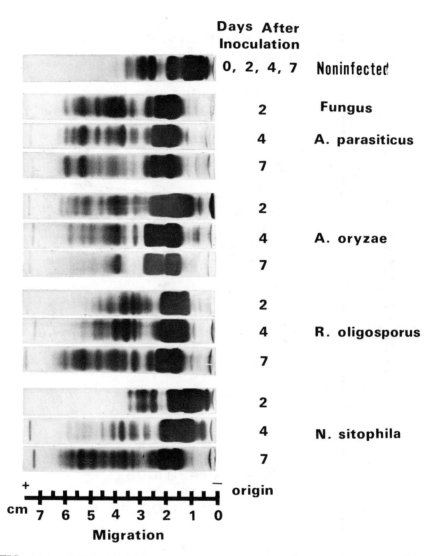

Days After
Inoculation

0, 2, 4, 7   Noninfected

2   Fungus

4   A. parasiticus

7

2

4   A. oryzae

7

2

4   R. oligosporus

7

2

4   N. sitophila

7

+ _____ – origin

cm 7 6 5 4 3 2 1 0

Migration

FIG. 14.9. POLYACRYLAMIDE GEL ELECTROPHORETIC PATTERNS
OF WATER-SOLUBLE PROTEINS OF NONINFECTED AND SELECT
FUNGI-INFECTED PEANUTS FROM A 0- TO 7-DAY TEST PERIOD

1970). The present work shows that techniques normally used to prepare
protein extracts (concentrates, isolates, etc.) from high quality peanuts
will not necessarily produce fractions similar to those of seeds infected

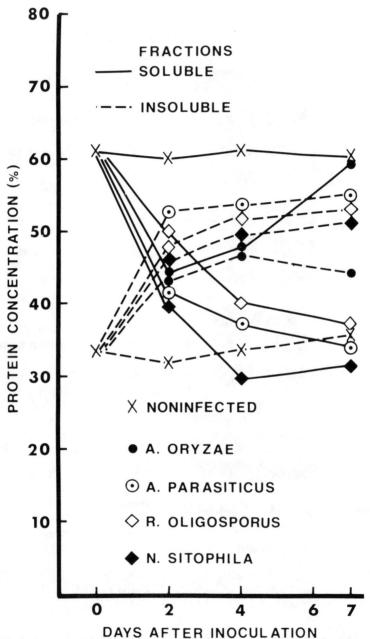

FIG. 14.10.   PERCENTAGE PROTEIN CHANGES IN WATER-
SOLUBLE AND INSOLUBLE NONINFECTED AND SELECT
FUNGI-INFECTED PEANUTS FROM A 0- TO 7-DAY TEST
PERIOD

with various fungi. In fact, the resulting extracts will depend on the species of fungus growing on the seeds and the length of the infection period. Since these conditions affect the type and quantity of proteins in various extracts (Fig. 14.8, 14.9, 14.10) of aqueous peanut preparations, they should alter the nutritional and functional (Fig. 14.10) properties of protein derivatives to different forms from those of high quality seeds. These factors need to be considered in research to expand utilization, in foods, of protein isolates or concentrates from various fungi-infected seeds and ferments.

PROCESSING

Protein Fractionation

Gel electrophoretic techniques can be used to follow the steps in the purification of various protein fractions from seeds. For example, maximum recovery of both arachin and nonarachin proteins in peanuts was accomplished with 1M NaCl-20 mM sodium phosphate buffer (pH 7.0) (Basha and Cherry 1976). Expanding this methodology resulted in the development of a procedure to prepare relatively pure isolates of arachin and nonarachin proteins from the total aqueous extract with a series of simple steps involving differential solubility, cryoprecipitation, and dialysis methods (Fig. 14.11). The degree of purity of each isolate was shown by polyacrylamide gel electrophoresis of the proteins.

Enzymes were used as indicators of purity of the protein preparation; i.e., certain enzymes may not be associated with a protein fraction, and, therefore, the degree of purity is a function of the absence of their activity. For example, initial studies of esterase activity in the total peanut extract produced unclear and variable enzyme patterns (Fig. 14.12; total protein fraction). Ammonium sulfate (40% saturation) precipitation of the large molecular weight storage protein, arachin, from the total extract resulted in clear and repeatable enzyme patterns on polyacrylamide electrophoretic gels (arachin precipitated fraction). On further examination of the nonarachin fraction, a good esterase pattern was produced which showed the effect of arachin removal during extraction and purification steps (nonarachin fraction). The arachin fraction from this purification exhibited some esterase activity on electrophoretic gels (Fig. 14.12), but further purification of this fraction removed most of the activity.

Thiol-reducing Reagents

Dithiothreitol and 2-mercaptoethanol have been used extensively in extracts of proteins and enzymes to prevent oxidation of sulfhydryl groups,

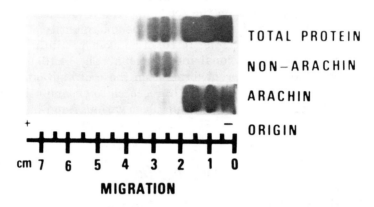

FIG. 14.11. POLYACRYLAMIDE GEL ELECTROPHORETIC
PATTERNS OF THE TOTAL PROTEINS EXTRACTED WITH 1
M NaCl-20 mM SODIUM PHOSPHATE BUFFER (pH 7.0) AT A
MEAL-TO-SALT BUFFER RATIO OF 1:18, (w/v) AND THE
FINAL PURIFIED FRACTIONS OF NONARACHIN AND
ARACHIN PROTEINS MADE FROM THIS PREPARATION

decrease the amount of protein-protein interaction, and improve food
stability (Cherry and Ory 1973; Jones and Carnegie 1971; Anderson 1968).
However, disulfide bridges, which maintain conformational and structural integrity of proteins, can sometimes be reduced by these compounds.

Fig. 14.12 contains zymograms of esterases which were altered by varying levels of thiol-reducing reagent. Increasing the concentration of thiol
reagent in extracts altered enzyme patterns both quantitatively and qualitatively. These data show that thiol-reducing reagents are a source of
variability in studies on enzymes; they also affect gel electrophoretic data
of proteins (Cherry and Ory 1973). Thus, results obtained with these
compounds as "SH-protective reagents" should be interpreted only after
comparisons with different concentrations and appropriate controls.

Processing Flour

The liquid cyclone process was engineered to remove gossypol-
containing pigment glands from cottonseed meal and produce a high
protein, edible-grade, flour (Gardner *et al.* 1976). In general, the process
uses differential settling in hexane to separate the particles of ground
cottonseed meal into edible-grade, low gossypol overflow and high gossypol underflow fractions. Samples were collected from each step in the
process and their proteins, esterases, and leucine aminopeptidases were

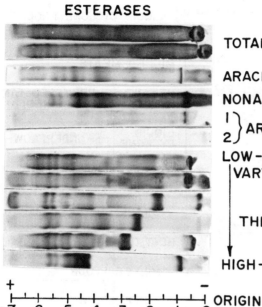

FIG. 14.12.   POLYACRYLAMIDE GEL ELECTROPHORETIC ZYMOGRAMS
OF ESTERASES IN THE TOTAL PROTEIN EXTRACT AND NONARACHIN
AND ARACHIN FRACTIONS
The effects of varying levels of thiol-reducing reagent on esterase activity in total
extracts are presented.

evaluated on polyacrylamide electrophoretic gels (Fig. 14.13). The gel
patterns show that these constituents are not denatured during liquid
cyclone processing of glanded cottonseed into edible flour. Only the protein
patterns distinguished the underflow fraction from the overflow portion,
suggesting that the process does not completely fractionate all protein
components.

The step to desolventize (remove residual hexane) liquid cyclone processed flour involves heat treatment at approximately 100° C. No detectable changes in proteins, esterases, and leucine aminopeptidases were
observed on electrophoretic gels during this step in the process (Fig. 14.13).
These data suggest that, at the level of detection of proteins and enzymes
by the polyacrylamide gel electrophoretic technique used in this study,
these seed storage constituents are not affected by the desolventizing
process. Further studies involving more extensive dry heating (150° C for

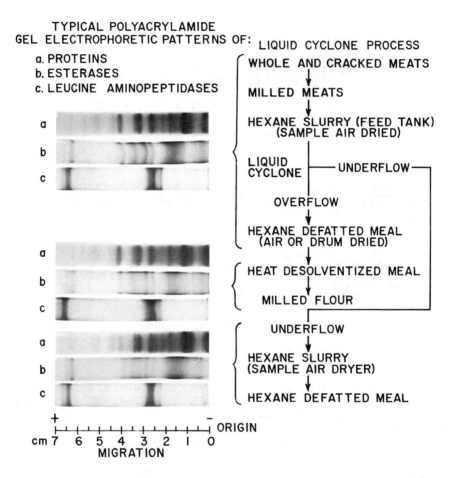

FIG. 14.13. POLYACRYLAMIDE GEL ELECTROPHORETIC PATTERNS OF
PROTEINS, ESTERASES, AND LEUCINE AMINOPEPTIDASES OF SAMPLES
COLLECTED FROM VARIOUS STEPS IN THE LIQUID CYCLONE PROCESS
FOR MAKING COTTONSEED FLOUR FROM GLANDED COTTON

5–120 min) of liquid cyclone processed flour showed changes in the protein
and enzyme patterns (Fig. 14.14). Gel patterns of flour heated between 5
and 120 min showed both qualitative and quantitative changes in certain
protein, esterase, and leucine amino-peptidase bands present in region
0–4.5 cm. A band in region 6.0–6.5 cm intensified in gels of all three
components analyzed during this heating period. Interestingly, esterase
and leucine aminopeptidase activity increased in region 2.5–4.5 and 2.0–
3.0 cm, respectively, at 15 min coinciding with a slight increase in protein
extractability at this time.

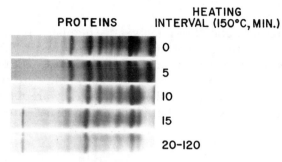

PROTEINS

HEATING
INTERVAL (150°C, MIN.)

0

5

10

15

20-120

ESTERASES

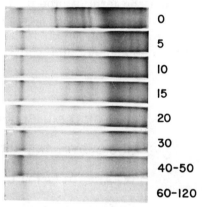

0

5

10

15

20

30

40-50

60-120

LEUCINE AMINOPEPTIDASES

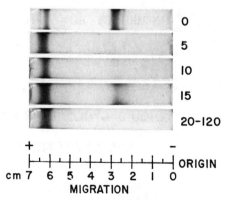

0

5

10

15

20-120

+   −

cm 7 6 5 4 3 2 1 0   ORIGIN

MIGRATION

FIG. 14.14. POLYACRYLAMIDE GEL ELECTRO-
PHORETIC PATTERNS OF PROTEINS, ESTERASES,
AND LEUCINE AMINOPEPTIDASES IN LIQUID
CYCLONE PROCESSED GLANDED COTTONSEED
FLOUR, DRY HEATED AT 150° C FOR VARIOUS TIME
INTERVALS

A hexane-defatted glandless flour was distinguished from liquid cyclone processed flour by the gel patterns of proteins (region 0.5–2.0 cm), esterases (2.0–4.0 cm) and leucine aminopeptidases (2.0–3.0 cm) (Fig. 14.15). These differences were present in glanded and glandless cottonseed before they were processed into flour, showing that they are related to genetic variations in the two cotton cultivars and are not due to processing procedures.

Heat-related Processing

Kjeldahl analyses showed that the level of proteins in the insoluble material from wet-heated (temperature-controlled steam retort) peanuts changed inversely with that of the aqueous soluble fraction (Fig. 14.16). For example, at 100° C, between 45 and 210 min, percentage protein of

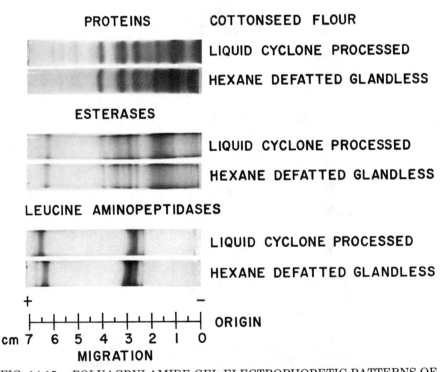

FIG. 14.15. POLYACRYLAMIDE GEL ELECTROPHORETIC PATTERNS OF PROTEINS, ESTERASES, AND LEUCINE AMINOPEPTIDASES IN HEXANE-DEFATTED GLANDLESS AND LIQUID CYCLONE PROCESSED GLANDED COTTONSEED FLOURS

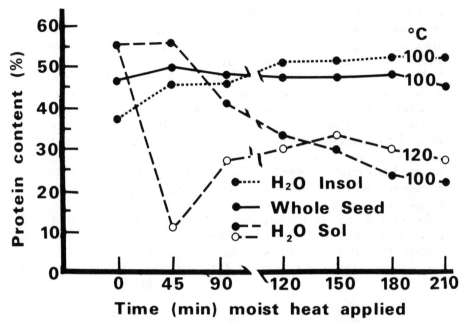

FIG. 14.16.   QUANTITY OF WATER INSOLUBLE AND SOLUBLE PROTEIN
FRACTIONS AND WHOLE SEED PROTEINS IN PEANUTS AFTER MOIST
HEAT TREATMENT AT 100° C AND 120° C FOR INTERVALS OF 0–210 MIN

soluble fractions declined from 55.7% to 21.8%. During this same period,
percentage protein in insoluble portions increased from 45.9% to 52.2%. At
120° C, percentage protein in soluble fractions was very low (10.9%) up to
45 min heating, then increased (27.3% to 32.8%) as heating was extended
through 210 min. The effect of heat on other constituents, such as lipids,
carbohydrates and nonprotein nitrogen containing substances, which vary
quantitatively in their extractability to soluble and insoluble fractions,
could influence the percentage protein values for extracts of heated
peanuts (e.g., at 120° C, 45–120 min, protein values of the soluble fractions
changed from 10.9% to 27.3%, whereas those of the insoluble material
were not altered as much, averaging 54.1% to 55.5%, suggesting that
quantities of other seed storage constituents extractable in the soluble
fraction were becoming insoluble).

Gel electrophoresis of peanuts wet-heated at 100° C for 15 min showed
that most of the activity of enzymes selected for this study could be
detected initially, but decreased rapidly when heating was continued to 45
min. The enzymes included esterase, leucine amino-peptidase, malate

dehydrogenase, phosphogluconate dehydrogenase, glutamate dehydrogenase, alcohol dehydrogenase, glucose-6-phosphate dehydrogenase, alkaline phosphatase, acid phosphatase, catalase, and peroxidase. At 120° C, these enzymes became inactive during the first 15 min of heating.

Changes in protein solubility and inactivation of enzymes of peanuts as a result of moist heat treatments suggested that alterations in protein structure were also occurring. Jacks *et al.* (1975) examined ultraviolet circular dichroic spectra of peanut arachin dry heated at various temperatures. These data showed many changes in conformation modes of heated arachin that were related to increased content of unordered structure from lessened amounts of helical and pleated sheet modes. In the following discussion, gel electrophoretic techniques show that moist heat causes many changes in molecular properties of peanut proteins.

Proteins of peanuts moist-heated at 100° C and 120° C for 30–180 min and 15–210 min, respectively, showed the greatest changes in electrophoretic patterns compared to those heated for short times or at low temperatures (Fig. 14.17 and 14.18). At 100° C, the major globulin

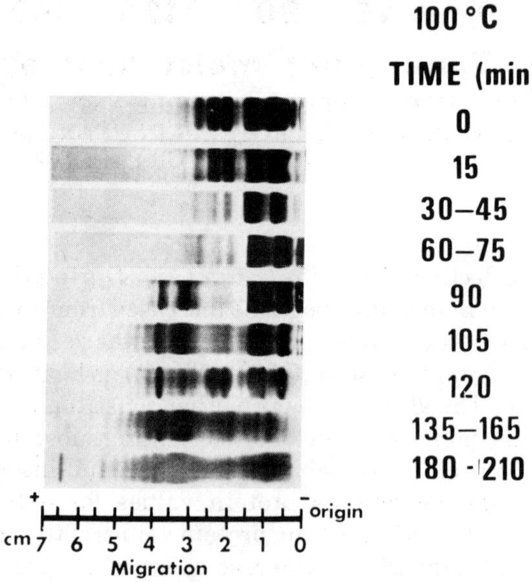

FIG. 14.17. TYPICAL POLYACRYLAMIDE GEL ELECTROPHORETIC PATTERNS OF WATER SOLUBLE PROTEINS IN PEANUTS AFTER MOIST HEAT TREATMENT AT 100° C FOR INTERVALS OF 0–210 MIN

Peanuts are from a 1973 crop. In the text, these patterns are compared to peanuts of a 1974 crop, similarly moist-heated at 100° C for intervals of 0–210 min.

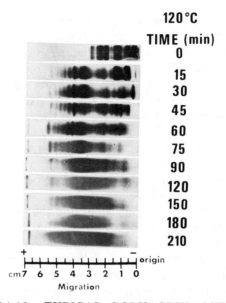

120 °C

TIME (min)

0

15

30

45

60

75

90

120

150

180

210

+   − origin

cm 7   6   5   4   3   2   1   0

Migration

FIG. 14.18. TYPICAL POLYACRYLAMIDE GEL ELECTROPHORETIC PATTERNS OF WATER SOLUBLE PROTEINS IN PEANUTS AFTER MOIST HEAT TREATMENT AT 120° C FOR INTERVALS OF 0–210 MIN

(arachin in region 0.5–1.5 cm) of peanuts heated at 100° C for 15–90 min remained unchanged (Fig. 14.17). The quantity of protein in the bands of nonarachin fractions (region 2.0–3.0 cm) decreased between 30–75 min and during 90–180 min the number of protein bands increased in region 3.0–6.5 cm. The quantity of arachin in region 0.5–1.5 cm declined between 105 and 180 min. Simultaneously, the number of protein bands increased in regions 2.0–2.5 and 3.0–3.5 cm.

Electrophoretically detected changes in the soluble proteins of peanuts heated at 120° C were clear at 15 min (Fig. 14.18). These changes showed a decrease in the quantity of nonarachin protein in region 1.5–2.5 cm and a simultaneous increase in the number of small molecular weight components in region 3.0–6.5 cm; the gel patterns resembled those of peanuts moist-heated at 100° C for 90–105 min. Between 30 and 210 min, arachin (0.5–2.0 cm) and the proteins in region 5.0–6.5 cm became difficult to discern. At the same time, a broad diffuse band in region 6.5–7.0 cm became the major component in the gel patterns.

These data show that, in addition to denaturing proteins to insoluble forms, moist heating of peanuts submerged in water at high temperatures for various intervals sequentially alters these nutritious components to subunit forms or fragments, then to aggregates, and finally to insoluble components (Figs. 14.17 and 14.18). When peanuts are moist-heated at extreme conditions (120° C, 90–210 min), more soluble protein was found at the longer, rather than at the shorter, intervals (Fig. 14.16). This increase in solubility coincided with the formation of a major diffuse component or aggregate that was detectable by gel electrophoresis.

Recent studies (Wolf and Cowan 1971) showed that the extent of denaturation of soybean products determined their physiochemical, functional, and nutritional properties in foods. These imposed characteristics govern to a large extent the ultimate application of soybeans in processed products. The preparation and characterization of seeds with different degrees of heat treatment for various times and the development of proper applications of these by-products, could expand utilization in the food industry of edible material from these sources.

In further studies, gel electrophoresis showed that quantitative changes in proteins from moist-heated peanuts harvested in 1974 differed from those of the 1973 crop shown in Fig. 14.17 (Cherry *et al.* 1975; Cherry and McWatters 1975). Between 120 and 210 min, differences in protein patterns of these two sources of peanuts were noted in regions 0.5–3.0 and 4.0–6.5 cm. The large increase in the number of bands in regions 1.5–2.5 and 4.0–5.5 cm of soluble extracts from heated 1973 peanuts (Fig. 14.17) were not clearly shown in the 1974 sample. The arachin constituents in extracts of 1974 peanuts decreased continuously between 120 and 190 min. In addition, these moist-heated peanuts showed two major protein components that were continuously detected in region 3.0–4.0 cm during the heating interval of 75–210 min.

Proteins in peanuts grown at different locations and/or stored for various time intervals are affected differently by moist heat processing at 100° C. This agrees with evidence that genetic, environmental, agronomic, and handling practices contribute to variations in peanut proteins and other constituents. In the present studies, the quantitative and qualitative differences in soluble proteins caused by these factors were shown to be influenced further by moist heat processing. It is necessary, therefore, to consider the effects of genetic and agronomic variables, as well as processing techniques, in characterizing the functional properties of peanuts and their products (McWatters and Cherry 1975). Such information is basic in efforts to develop more efficient utilization of seeds in food formulations.

Relationship of Seed Quality to Product Quality

Cultivar (genetic composition) and growing location (agronomic factors) affect protein quantity of cottonseed, and, as a result, their processability

(Cherry *et al.* 1977). This has become a major area of research since the development of glandless cotton cultivars (Hess 1976) and the liquid cyclone process (Gardner *et al.* 1976), which have lowered gossypol content in cottonseed and cottonseed flour and paved the way for use of these products to supply protein in food. Possibly breeding and agronomic practices could improve protein content in cottonseed. Higher quality should enhance development of methods to efficiently produce from cottonseed protein products optimum nutritional and functional properties for use in foods.

Gel electrophoretic patterns of proteins in aqueous extracts of cottonseed from glanded and glandless cottons lack qualitative variability (Fig. 14.19). This is further supported by the lack of gel electrophoretic variabil-

PROTEINS
WATER EXTRACTS-TEXAS COTTONSEED

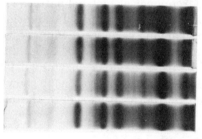

ACALA 1517-70

COKER 310

DELTAPINE 16

LOCKETT 4789A

WATER EXTRACTS-CALIFORNIA COTTONSEED

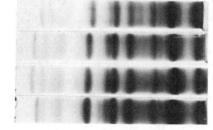

ACALA SJ-2

ACALA SJ-4

ACALA T5690

ACALA G8160

+   − ORIGIN

cm 7 6 5 4 3 2 1 0

MIGRATION

FIG. 14.19.   TYPICAL POLYACRYLAMIDE GEL ELECTROPHORETIC PATTERNS OF NONSTORAGE PROTEINS IN WATER EXTRACTS OF COTTONSEED FROM VARIOUS CULTIVARS GROWN IN TEXAS AND CALIFORNIA

Acala G8160 from California is a glandless cultivar.

ity in esterase and leucine aminopeptidase activity among glanded cultivars grown in various locations (Fig. 14.20). Enzyme patterns do distinguish glanded and glandless cultivars (Fig. 14.15, 14.20); on the other hand, quantity of protein in cottonseed varied significantly relative to genetic and agronomic factors (Cherry *et al.* 1977).

Although protein quantity of cottonseed of different cultivars differs and can be influenced by genetic and agronomic practices, the gel electrophoretic data suggest that this variability is not reflected in types of cottonseed protein. Thus, certain proteins which may be low in specific essential amino acids are unknowingly selected for present day breeding and agronomic programs to improve fiber quality, thereby perpetuating nutritional imbalance in cottonseed of new commercial cultivars.

On the other hand, much esterase variability can be detected among individual seeds of wild species by gel electrophoresis (Fig. 14.21, Cherry 1977B). For example, six different zymograms (A–F) were observed for the cotton species, *G. thurberi*. The frequency of each gel pattern ranged from 1.0–49.0% of the total seed population examined. In addition, the frequency of this variability differed among various growing locations where seeds were collected (Cherry and Katterman 1971). These data showed that gel electrophoresis could be used to detect enzyme variability if it existed within a particular species (genetic), as well as to show that these differences could be influenced by the location where the plants grow (environment and/or agronomic).

Identification of Composite Products

Esterase patterns of meals made separately from cottonseed of two species, *Gossypium arboreum* and *G. thurberi* were compared to those of composites made by mixing these meals in a 1:1 ratio (Fig. 14.22). The zymogram of the blended meal or mixture contained an additive pattern of both sources. Esterases from seeds of a viable hybrid plant (AZ239) made by genetically crossing these two species were examined by gel electrophoresis. The additive zymogram of the synthetic composite compared closely to that of the hybrid AZ239. These tests show that it is possible to use gel electrophoresis to determine the composition of flour and meat mixtures which is shown by the example in Fig. 14.23. Such information could assist in setting up guidelines for proper use of vegetable proteins in blends with various food products (e.g., meat) in the processing industry.

CONCLUSIONS

In this chapter, I have attempted to show how enzymes have the potential for utilization as biological indicators to assist investigators in deter-

ESTERASES
TEXAS COTTONSEED

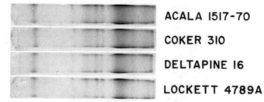

ACALA 1517-70

COKER 310

DELTAPINE 16

LOCKETT 4789A

CALIFORNIA COTTONSEED

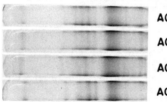

ACALA SJ-2

ACALA SJ-4

ACALA T5690

ACALA G8160

LEUCINE AMINOPEPTIDASES
TEXAS COTTONSEED

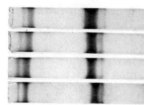

ACALA 1517-70

COKER 310

DELTAPINE 16

LOCKETT 4789A

CALIFORNIA COTTONSEED

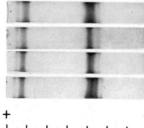

ACALA SJ-2

ACALA SJ-4

ACALA T5690

ACALA G8160

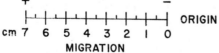

+   −

ORIGIN

cm 7 6 5 4 3 2 1 0

MIGRATION

FIG. 14.20. TYPICAL POLYACRYLAMIDE GEL ELECTROPHORETIC PATTERNS OF ESTERASES AND LEUCINE AMINOPEPTIDASES IN WATER EXTRACTS OF COTTONSEED FROM VARIOUS CULTIVARS GROWN IN TEXAS AND CALIFORNIA
Acala G8160 from California is a glandless cultivar.

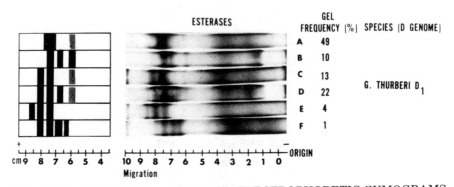

FIG. 14.21. POLYACRYLAMIDE GEL ELECTROPHORETIC ZYMOGRAMS
OF COTTONSEED ESTERASES OF A SPECIES, *G. thurberi*, WITHIN THE D
GENOME OF THE GENUS *Gossypium*
Shown are frequencies of seed having a particular zymogram.

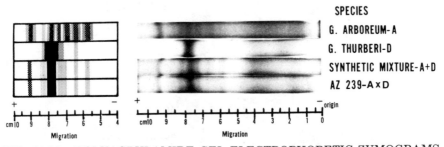

FIG. 14.22. POLYACRYLAMIDE GEL ELECTROPHORETIC ZYMOGRAMS
OF COTTONSEED ESTERASES IN A COMPARISON BETWEEN PARENTALS
(G. arboreum, G. thurberi), THEIR MIXTURE (A + B), AND THEIR HYBRID
OFFSPRING
(AZ239; A x D)

mining and maintaining optimum quality composition of edible substances during various stages of food handling. These stages include
preharvest, harvest, storage, and processing, the latter two being
categorized as postharvest biology and biotechnology. Gel electrophoretic
techniques serve as tools to detect enzyme activity qualitatively, allowing

ESTERASES

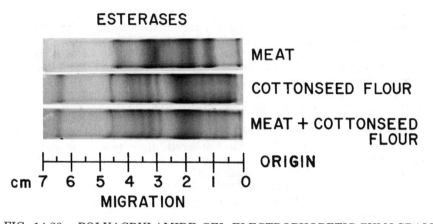

MEAT

COTTONSEED FLOUR

MEAT + COTTONSEED
FLOUR

ORIGIN

cm 7 6   5   4   3   2   1   0

MIGRATION

FIG. 14.23.   POLYACRYLAMIDE GEL ELECTROPHORETIC ZYMOGRAMS
OF ESTERASES IN A COMPARISON OF MEAT, COTTONSEED FLOUR, AND
A MEAT: COTTONSEED FLOUR BLEND (80:20; w:w)

researchers and commercial processors to evaluate seed material for optimum maturity, genetic and environmentally (agronomic) related compositional variations, fungal contamination and deterioration during
storage, and effect of various conditions normally included in the processing of raw material into finished food products. In processing, enzymes
can serve as diagnostic tools to assist in quality control practices for
maintenance of optimum processing conditions.

BIBLIOGRAPHY

ANDERSON, J. W. 1968. Extraction of enzymes and subcellular organelles from
    plant tissues. Phytochem. 7, 1973–1988.
BARFOED, H. C. 1976. Enzymes in starch processing. Cereal Fds. World 11,
    588–593, 604.
BASHA, S. M. M. and CHERRY, J. P. 1976. Composition, solubility, and gel
    electrophoretic properties of proteins isolated from Florunner (Arachis hypogaea
    L.) peanut seeds. J. Agric. Fd. Chem. 24, 359–365.
BREWBAKER, J. L., UPADHYA, M. D., MAKINEN, Y. and MAC DONALD, T.
    1968. Isoenzyme polymorphism in flowering plants. 3. Gel electrophoretic
    methods and applications. Physiol. Plant. 21, 930–948.
BURNETTE, F. S. 1977. Peroxidase and its relationship to food flavor and quality:
    a review. J. Food Sci. 42, 1–6.
CHERRY, J. P. 1977A. Potential sources of peanut seed proteins and oil in the
    genus Arachis. J. Agric. Fd. Chem. 25, 186–193.
CHERRY, J. P. 1977B. Oilseed enzymes as biological indicators for food uses and

applications. *In* Enzymes in Food and Beverage Processing, R. L. Ory and A. J. St. Angelo (Editors). American Chemical Society. Washington, D.C.

CHERRY, J. P., DECHARY, J. M. and ORY, R. L. 1973. Gel electrophoretic analysis of peanut proteins and enzymes. 1. Characterization of DEAE-cellulose separated fractions, J. Agric. Fd. Chem. *21*, 652–655.

CHERRY, J. P. and KATTERMAN, F. R. H. 1971. Nonspecific esterase isozyme polymorphism in natural populations of *Gossypium thurberi*. Phytochem. *10*, 141–145.

CHERRY, J. P., MAYNE, R. Y. and ORY, R. L. 1974. Proteins and enzymes from seeds of *Arachis hypogaea* L. 9. Electrophoretically detected changes in 15 peanut cultivars grown in different areas after inoculation with *Aspergillus parasiticus*. Physiol. Plant Path. *4*, 425–434.

CHERRY, J. P. and MC WATTERS, K. M. 1975. Solubility properties of proteins relative to environmental effects and moist heat treatment of full-fat peanuts. J. Food Sci. *40*, 1257–1259.

CHERRY, J. P., MC WATTERS, K. H. and HOLMES, M. R. 1975. Effect of moist heat on solubility and structural components of peanut proteins. J. Food. Sci. *40*, 1199–1204.

CHERRY, J. P. and ORY, R. L. 1973. Gel electrophoretic analysis of peanut proteins and enzymes. 2. Effects of thiol reagents and frozen storage. J. Agric. Fd. Chem. *21*, 656–660.

CHERRY, J. P., SIMMONS, J. G. and TALLANT, J. D. 1977. Cottonseed protein composition and quality of *Gossypium* species and cultivars. Proc. Beltwide Cotton Conf. *31*, 46–49.

DAVIS, B. J. 1964. Disc electrophoresis. 2. Method and application to human serum proteins. Ann. N.Y. Acad. Sci. *121*, 404–427.

GARDNER, H. K., HRON, R. J. and VIX, H. L. E. 1976. Removal of pigment glands (gossypol) from cottonseed. Cereal Chem. *53*, 549–560.

GOTTLIEB, L. D. 1971. Gel electrophoresis. New approach to the study of evolution. BioScience *21*, 939–944.

GRAY, W. D. 1970. The use of fungi as food and in food processing. Crit. Rev. Food Technol. *1*, 225–329.

GRODNER, R. M. 1976. Enzymes: past, present, and future. Cereal Fds. World *11*, 574–576.

HESS, D. C. 1976. Prospects for glandless cottonseed. Oil Mill Gaz. *81*, 20–26.

JACKS, T. J., NEUCERE, N. J. and MC CALL, E. R. 1975. Thermally induced permutations of antigenicity and conformation of arachin (peanut globular protein). Int. J. Peptide Protein Res. *7*, 153–157.

JONES, I. K. and CARNEGIE, P. R. 1971. Binding of oxidized glutathione to dough proteins and a new explanation, involving thio-disulphide exchange, of the physical properties of dough. J. Sci. Fd. Agric. *22*, 358–364.

MANWELL, C. and BAKER, C. M. A. 1970. Molecular biology and the origin of species: heterosis, protein polymorphism, and animal breeding. University of Washington Press, Seattle, Wash.

MC WATTERS, K. H. and CHERRY, J. P. 1975. Functional properties of peanut paste as affected by moist heat treatment of full-fat peanuts. J. Food Sci. *40*, 1205–1209.

ORNSTEIN, L. 1964. Disc electrophoresis. I. Background and theory. Ann. N.Y. Acad. Sci. *121*, 321–349.

REED, G. 1976. The utility of enzymatic processing. Cereal Fds. World *11*, 578–580, 599.

SCANDALIOS, J. G. 1969. Genetic control of multiple molecular forms of enzymes in plants: A review. Biochem. Genet. *3,* 37–79.

SMITHIES, O. 1962. Molecular size and starch-gel electrophoresis. Arch. Biochem. Biophys. Suppl. *1,* 125–131.

WEETALL, H. H. 1976. Immobilized enzyme technology. Cereal Fds. World *11,* 581–587.

WILKINSON, J. H. 1976. Chemical enzymology: the state of the art. Lab. Manag. 14, 21–24.

WOLF, W. J. and COWAN, J. C. 1971. Soybeans as a food source. CRC Crit. Rev. Fd. Tech. *2,* 81–158.

BIOGENESIS OF FLAVOR COMPONENTS: VOLATILE CARBONYL COMPOUNDS AND MONOTERPENOIDS

RODNEY CROTEAU

Department of Agricultural Chemistry and
Program in Biochemistry and Biophysics
Washington State University
Pullman, WA 99164

ABSTRACT

Recent interest in flavor research has been directed toward elucidating the origins and fates of flavor compounds in fruits and vegetables. In this review, current studies on the biogenesis of two important families of flavor constituents, the volatile carbonyl compounds and the monoterpenes, are described. The major pathways and enzymes involved in the oxidative conversion of fatty acids and amino acids to short-chain acyl-CoA derivatives, and in the lipoxygenase mediated conversion of unsaturated fatty acids to C6 and C9 aldehydes and alcohols, are reviewed. The interrelationships and interconversions of these short-chain compounds, and the formation of acyl and aroyl esters, are discussed. The biosynthesis of monoterpenes is described in a general context, and recent studies on the enzymes and mechanisms involved in the formation of key monocyclic, heterocyclic, bicyclic, and aromatic monoterpenes are reviewed. Some comments relating to the future directions of flavor biogenesis research, and some speculations concerning the possible impact of such studies, are made.

INTRODUCTION AND GENERAL CONSIDERATIONS

One important objective of flavor research is the isolation and identification of those volatile compounds responsible for the characteristic aroma of fruits and vegetables. With the application of powerful analytical tools, such as the combined gas chromatograph-mass spectrometer, this objective is being quite successfully met. Indeed, the number of plant foodstuffs that have been examined in detail, and the number and variety of flavor components that have been identified, are very impressive. More recently, however, interest in flavor research has begun to shift from isolation and identification to the investigation of the origins and fates of volatile aroma compounds in food. Because the types of organic compounds that contribute to the aroma of plant foods is so diverse, ranging from sulfur compounds to simple esters, pyrazines, and terpenoids, this relatively new field, which I shall call "flavor biochemistry," is a very broad one. As it is not possible to cover all phases of research relating to the biogenesis of flavor, this review will emphasize those two major families of flavor substances found in plant foods: the volatile carbonyl compounds and the monoterpenes.

That the formation of volatile carbonyl compounds and monoterpenes in fruits and vegetables is largely an enzymatic process, has, of course, been appreciated for many years. Studies on the biosynthetic origins of these substances, however, have been hampered by the same difficulties faced by the flavor chemist; the analysis of complex mixtures of volatile, often labile, substances that are usually present at trace levels. This latter difficulty, that flavor compounds are often formed at the parts per million level, is particularly important to the biochemist, because it implies that the enzymes involved in flavor biosynthesis may not be very active, or may not be present in very high concentration. Thus, the isolation and characterization of such enzymes can be like searching for a needle in a "metabolic haystack." In spite of such difficulties, considerable progress has been made over the last few years in studying the origin of volatile carbonyl compounds and monoterpenoids in a variety of different commodities. I shall not emphasize any particular commodity in the review, but rather will attempt to provide a more general approach to flavor biosynthesis by concentrating on the basic pathways, enzymes, and mechanisms involved.

ORIGIN OF VOLATILE CARBONYL COMPOUNDS

Any survey of the constituents of the aroma complexes of fruits and vegetables will reveal the presence of a vast array of volatile compounds including acids, esters, aldehydes and ketones (and the corresponding alcohols) (Nursten 1970; Salunkhe and Do 1977). While the variety of structural types of compounds (straight-chain, branched, unsaturated, terpenyl, aromatic, etc.) encountered in plant aroma complexes is great, the structural types encountered in any single species may be rather limited, and no doubt reflect the particular metabolic make-up of the plant. A number of such carbonyl compounds, or closely related families of compounds, have been shown to possess aroma character impact, and some examples are given in Table 15.1. It is obvious from even this brief list that the biosynthetic origins of such compounds must be rather diverse, and it is, therefore, tempting to focus in detail on only one or two commodities or

TABLE 15.1

FRUITS AND VEGETABLES IN WHICH AROMA CHARACTER
IMPACT COMPOUNDS HAVE BEEN IDENTIFIED[1]

Apple	Ethyl-2-methylbutyrate
Banana	3-Methyl-1-butyl acetate
Bilberry	Ethyl-3-methylbutyrate
Blueberry	*trans*-2-Hexenal (and alcohol)
Celery	*cis*-3-Hexen-1-yl pyruvate
Cranberry	Benzoate and benzyl esters
Cucumber	*trans*-2-,*cis*-6-Nonadienal
Grape	Methyl anthranilate
Mushroom	1-Octen-3-one (and alcohol)
Pear	Methyl-*trans*-2,*cis*-4-decadienoate

[1]Compiled from Salunkhe and Do (1976) and references therein.

classes of compounds. For the purpose of this book, however, it seems more appropriate to describe in broad terms those key pathways by which many of the known aroma-bearing carbonyl compounds are probably produced: the conversion of amino acids to their corresponding C_{n-1} acyl derivatives via transamination and oxidative-decarboxylation; the production of short-chain acids, aldehydes, and ketones via α- and β-oxidation of fatty acids; the lipoxygenase-mediated conversion of unsaturated fatty acids into short-chain aldehydes and oxo-acids; and then, finally, to review the subsequent transformations of these compounds in the general context of ester formation.

VOLATILE CARBONYL COMPOUNDS DERIVED FROM LIPIDS

Many volatile carbonyl compounds from a wide range of plant tissues would, on the basis of their structures, appear to be derived by oxidative degradation of endogenous lipids (Salunkhe and Do 1977). Most enzymatic oxidative processes utilize free fatty acids as substrates (Galliard 1975), yet free fatty acids generally occur in very low concentrations in the intact plant. Thus, the first step in the degradation of fatty acids is likely to be the release of the fatty acids from glyceryl lipids by the action of acyl hydrolases. The importance of such acyl hydrolases in initiating flavor development in macerated cucumber, for example, has recently been demonstrated (Hatanaka *et al*. 1975; Galliard *et al*. 1976B). The free fatty acids are then subject to oxidative degradation as outlined in Fig. 15.1.

The enzymatic conversion of unsaturated fatty acids to C_6 and C_9 aldehydes via an oxidative process has been demonstrated in a number of

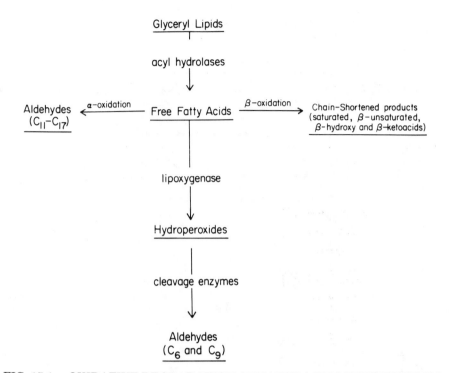

FIG. 15.1.   OXIDATIVE DEGRADATION OF FATTY ACIDS DERIVED FROM
GLYCERYL LIPIDS

fruit and vegetable tissues (see Eriksson 1975; Gardner 1975; Vick and
Zimmerman 1976; Galliard and Matthew 1977; and references therein).
Tressl and Drawert (1973), among others, suggested that fatty acid hydroperoxides, formed by the action of lipoxygenase, were intermediates in
this reaction, and that these hydroperoxides were cleaved to aldehydes by
"hydroperoxide lyases." The direct conversion of the C13 hydroperoxide of
linoleic acid to hexanal by a cell-free lyase preparation from watermelon
seedlings was demonstrated by Vick and Zimmerman (1976), and subsequent studies by Galliard and associates (Galliard and Phillips 1976;
Galliard and Matthew 1977; Galliard et al. 1976A) with cucumber and
tomato preparations, and by Hatanaka and coworkers (1976A, B, C) with
preparations from various leaf tissues have revealed further details about
this enzyme system. The conversion of linoleic acid to hexanal and nonenal
is shown schematically in Fig. 15.2. The hydroperoxides formed from
linoleic acid by plant lipoxygenases exist in two optically active isomers

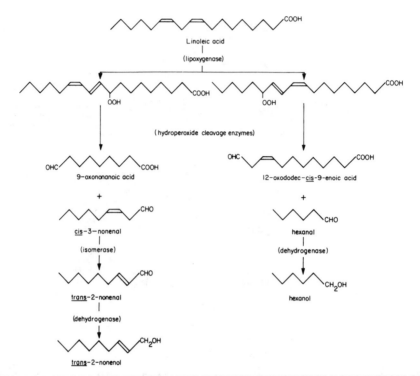

FIG. 15.2. PROPOSED ENZYMATIC SEQUENCE FOR THE FORMATION OF
C₆ AND C₉ ALDEHYDES AND ALCOHOLS FROM LINOLEIC ACID

with the hydroperoxide group in the C9 or C13 position. The proportions of
C9 and C13 hydroperoxides formed may vary widely, from predominantly
C13 (soybean, watermelon), to predominantly C9 (tomato, potato), to intermediate mixtures (pear, bean, cucumber). Lyases then cleave these
hydroperoxides to the aldehyde and the corresponding ω-oxo acid. The
lyase prepared from cucumber is membranous, but may be solubilized with
Triton X-100 (Galliard *et al.* 1976A), while a similar enzyme from
watermelon seedling is apparently soluble (Vick and Zimmerman 1976).
The lyase reaction is anaerobic and shows a pH optimum in the 6.0–6.5
range. The watermelon lyase is inhibited by the thiol directed reagent
p-chloromercuribenzoate. With the cucumber lyase, the reaction rates
and Km values for the C9 and C13 hydroperoxides are equal. The partially
purified lyase from cucumber cleaves the C9 hydroperoxide to *cis*-3-
nonenal as the sole aldehyde product. In crude preparations, however, the
cis-3-nonenal formed is rapidly converted to *trans*-2-nonenal by a very
active isomerase. This isomerase is apparently present in many plant

tissues (Hatanaka *et al.* 1975; 1976A; Stone *et al.* 1975). A series of reactions analogous to those shown in Fig. 15.2, but starting with linolenic acid as the precursor, will produce *cis*-3-hexenal and *cis*-3,*cis*-6-nonadienal, which may subsequently be isomerized to *trans*-2-hexenal and *trans*-2,*cis*-6-nonadienal, respectively. Enzymatic reduction of the above aldehydes leads to the corresponding C6 and C9 alcohols, which are found in numerous fruits and vegetables (Salunkhe and Do 1977). Such aldehydes may also undergo oxidation, and subsequently give rise to esters such as *cis*-3-hexenyl-*trans*-2-hexenoate found in spring green tea (Takei *et al.* 1976).

In tomato, the hydroperoxide cleavage process differs significantly from that of cucumber (Galliard and Matthew 1977). Tomato homogenates produce primarily C9 hydroperoxide (95%) from linoleic and linolenic acids, but such preparations do not produce C_9 aldehydes; only C_6 aldehydes are formed. Thus, in spite of the fact that the C9 hydroperoxide is the major lipoxygenase product, the hydroperoxide cleavage process in tomato appears to be specific for the C13 hydroperoxide. Furthermore, whereas in cucumber the initial cleavage products (*cis*-3-enals) are rapidly converted to *trans*-2-isomers by endogenous isomerases, in tomato extracts *cis*-3-hexenal accumulates, indicating lower enal isomerase activity in this tissue. Thus, within the general framework of reactions shown in Fig. 15.2, differences in substrate fatty acid availability, as well as species variation in the lipoxygenase system, cleavage enzymes, isomerases, and reductases or dehydrogenases, may all contribute to the particular mixture of C_6 and C_9 aldehydes, alcohols, acids and esters produced by any given tissue.

It should also be pointed out here that fatty acid hydroperoxides can be formed nonenzymatically, and that, under both aerobic and anaerobic conditions, the fatty acid hydroperoxides may undergo further nonenzymatic attack to give products similar to those formed by enzymatic means (Gardner 1975; Grosch 1976; Coggon *et al.* 1977). These nonenzymatic reactions can also contribute to flavor formation.

Odd-chain (C_9–C_{17}) aldehydes are often present in the aroma complexes of fruits and vegetables including cucumber (Kemp *et al.* 1974; Kemp 1975), and an α-oxidation enzyme system capable of converting the common even-chain saturated and unsaturated fatty acids to such aldehydes has been obtained from the cucumber (Galliard and Matthew 1976). Similar α-oxidation systems appear to be widespread in plants (Galliard 1975), and these systems are probably responsible for the formation of the odd-chain aldehydes observed.

β-Oxidation of endogenous fatty acids provides a means of generating shorter-chain acids which in themselves, or after conversion to aldehydes, ketones, alcohols and esters, possess aroma bearing qualities. The aroma complex of pear, for example, contains a variety of methyl, ethyl, propyl, butyl and hexyl esters, the C_8–C_{14} acyl groups of which appear to be derived via β-oxidative chain shortening of endogenous saturated and unsaturated fatty acids (Creveling and Jennings 1970). A scheme for the formation of several such esters from the β-oxidation of linoleic acid is shown in Fig. 15.3. As some of the metabolites, such as cis-3,cis-6-dodecadienyl-CoA, that arise in this scheme are abnormal β-oxidation intermediates, 3-cis,2-trans-enoyl-CoA isomerase and 3-hydroxyacyl-CoA-3-epimerase are probably involved in converting these β-oxidation intermediates to the normal metabolites (Galliard 1975). Similar oxidation of oleic and linolenic acids could also generate a large group of esters. That β-oxidation is involved in the formation of such volatile aroma-bearing esters is strongly suggested by the observation that esters of the appropriate intermediate β-hydroxy- and β-keto-acids, such as ethyl β-hydroxy-hexanoate and methyl β-hydroxyoctanoate (Creveling et al. 1968; Tressl and Drawert 1973), also occur, though generally in small amounts. Exogenous C_6, C_8 and C_{10} fatty acids, when administered to banana slices, were converted largely to the corresponding esters, but some chain-shortened products were also detected (Tressl and Drawert 1973). Of particular interest, was the observation that labeled hexanoic and octanoic acid gave rise to 2-pentanone and 2-heptanone, respectively, in the banana slice system. Thus, β-oxidation may also be involved in methyl ketone formation, by a pathway similar to that observed in fungi (Kinsella and Hwang 1977) which involves decarboxylation of the intermediate β-ketoacids.

It is obvious from the foregoing discussion that β-oxidation can lead to the formation of a variety of short even-chain saturated, unsaturated, β-hydroxy and β-keto acids. Insertion of a single α-oxidation step into the β-oxidation sequence would allow the formation of odd-chain acyl derivatives, as well as α-hydroxy and α-keto acids. Thus, α- and β-oxidation of endogenous plant fatty acids can, theoretically, provide a potentially large number of short-chain acyl precursors of the volatile carbonyl compounds. Yet, very little direct evidence bearing on this possibility is presently available. In banana tissue slices, for example (Tressl and Drawert 1973), exogenous [U-^{14}C]palmitic acid is not significantly converted to volatile compounds, although this tissue actively converts shorter-chain acids (C_2–C_6) to the corresponding volatile alcohols and esters (see discussion below). Such a result may represent the inability of tissue slices to transport exogenous long-chain fatty acids to the appropriate site where they can undergo oxidative degradation to chain-lengths appropriate for volatile ester formation. Such transport difficulties might be overcome in cell-free preparations, and cell-free systems may be required to demonstrate the direct coupling of β-oxidation to the esterifying system.

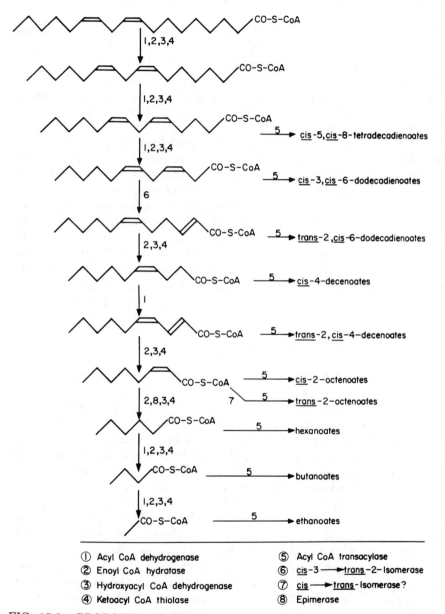

cis-5,cis-8-tetradecadienoates

cis-3,cis-6-dodecadienoates

trans-2,cis-6-dodecadienoates

cis-4-decenoates

trans-2,cis-4-decenoates

cis-2-octenoates

trans-2-octenoates

hexanoates

butanoates

ethanoates

① Acyl CoA dehydrogenase
② Enoyl CoA hydratase
③ Hydroxyacyl CoA dehydrogenase
④ Ketoacyl CoA thiolase
⑤ Acyl CoA transacylase
⑥ cis-3 ⟶ trans-2-Isomerase
⑦ cis ⟶ trans-Isomerase?
⑧ Epimerase

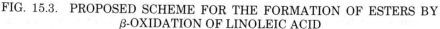

FIG. 15.3. PROPOSED SCHEME FOR THE FORMATION OF ESTERS BY
β-OXIDATION OF LINOLEIC ACID

VOLATILE CARBONYL COMPOUNDS DERIVED FROM AMINO ACIDS

Branched-chain esters, such as isovalerates and isobutyrates and esters of isovaleryl alcohol and isobutyl alcohol, are important components of banana aroma and of other fruit essences (Nursten 1970; Salunkhe and Do 1977). The formation of such branched esters with the concomitant accumulation of leucine and valine in ripening bananas prompted a number of investigators (Myers *et al.* 1969, 1970; Tressl and Drawert 1973) to examine the possible role of these amino acids as precursors of branch-chain esters. Labeling experiments using postclimacteric banana tissue slices did, in fact, show that [^{14}C]leucine was converted to isovalerate and isovaleryl esters, and that [^{14}C]valine was converted to isobutyrate and isobutyl esters. That these amino acids were incorporated into branched-chain esters of banana via transamination and subsequent oxidative-decarboxylation was strongly suggested by the fact that the appropriate labeled intermediates (*i.e.*, the α-ketoacids, 2-ketoisocaproate, and 2-ketoisovalerate) did accumulate (Tressl and Drawert 1973). Thus, the transamination and oxidative-decarboxylation of the common amino acids to the corresponding acyl-CoA derivatives (Table 15.2) would appear to provide another potential source of flavor precursors in fruits and vegetables.

FORMATION OF ESTERS

Volatile esters are common components of the aroma complexes of fruits and vegetables. A biosynthetic scheme for the formation of volatile esters from short-chain acids has been suggested by Tressl and Drawert (1973), based on an earlier proposal by Kolattukudy (1970A) for the conversion of fatty acids to long-chain wax esters. This generalized pathway is shown in Fig. 15.4. Acyl-CoA may arise directly from fatty acid oxidation or amino acid catabolism as indicated above, from glycolysis, or from any free acid by the action of thiokinase. The acyl-CoA derivatives may then be reduced to aldehydes by acyl-CoA reductase. The aldehydes are next reduced to the alcohols. At this stage, other aldehydes (from hydroperoxide cleavage, α-oxidation, etc.) and ketones (from β-ketoacids) may enter the cycle and be reduced to the corresponding alcohols. In any case, the final step is the esterification of an alcohol and an acyl-CoA by acyl-CoA transacylase. Acyl-CoA reductases and transacylases involved in the biosynthesis of short-chain compounds of flavor interest have not yet been isolated; however, long-chain acyl-CoA reductases and transacylases involved in wax ester and cutin biosynthesis (Kolattukudy 1967, 1970B, 1971; Croteau and Kolattukudy 1974), and aroyl-CoA reductases involved in lignin biosynthesis (Mansell *et al.* 1972; Rhodes and Wooltorton 1975; Wengenmayer *et*

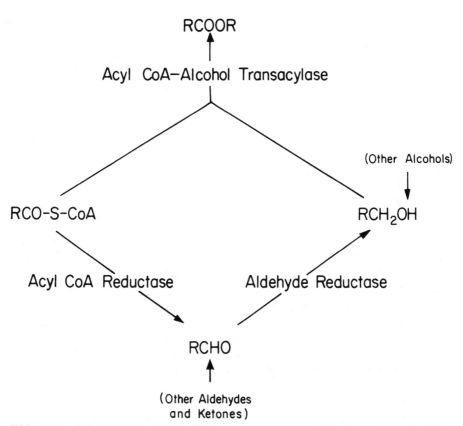

FIG. 15.4. PROPOSED PATHWAY FOR THE BIOSYNTHESIS OF SHORT-CHAIN ALDEHYDES, ALCOHOLS AND ESTERS FROM ACYL-CoA

al. 1976) have been demonstrated in plant extracts. Many aldehyde reductases and alcohol dehydrogenases have been isolated and characterized, and these enzymes probably supply the alcohols for ester synthesis, as well as play an important role establishing the aldehyde-alcohol balance encountered in the aroma complex of plant foodstuffs (Bruemmer 1975; Eriksson 1975; Eriksson *et al*. 1977).

Evidence consistent with the reaction scheme outlined in Fig. 15.4 has been obtained from a number of *in vivo* experiments. Acetate and butyrate esters account for about 70% of the aroma complex of banana, and post climacteric banana tissue slices readily incorporate exogenous [14C]acetate and [14C]butyrate into acetate and butyrate esters (and to a lesser extent into ethyl and butyl esters) (Tressl and Drawert 1973). Similarly, labeled leucine and valine (after catabolism as outlined above) and labeled

TABLE 15.2

CATABOLISM OF AMINO ACIDS TO FLAVOR PRECURSORS
BY TRANSAMINATION AND OXIDATIVE-DECARBOXYLATION

Amino Acid	Product
Glycine	Formyl CoA
Alanine	Acetyl CoA
Valine	2-Methylpropionyl CoA (Isobutyryl)
Leucine	3-Methylbutyryl CoA (Isovaleryl)
Isoleucine	2-Methylbutyryl CoA[1]
Phenylalanine	Phenylacetyl CoA

[1]β-Oxidation of 2-methylbutyryl CoA provides acetyl CoA and propionyl CoA.

caproic, caprylic, and capric acids are incorporated into the corresponding branched and straight-chain acyl and alkyl groups of esters in banana slices. Additionally, the C_6 and C_8 acid are converted to 2-pentanone and 2-heptanone respectively in banana slices (Tressl and Drawert 1973). Reduction of these ketones could provide the alcohol precursors of 2-pentyl and 2-heptyl esters that are occasionally found in aroma concentrates (Nursten 1970). Volatile carbonyl compounds accumulate in banana only after the climacteric rise in respiration, and significant incorporation of the above precursors into such volatile carbonyl compounds could be demonstrated only in post climacteric tissue (Tressl and Drawert 1973).

Recent studies with ripe strawberries are also consistent with the reaction scheme in Fig. 15.4. Thus, a variety of straight-chain and branched-chain alcohols and acids are incorporated into the appropriate esters of intact strawberry fruit, while both straight and branched-chain aldehydes are reduced to the corresponding alcohols and esterified by this tissue (Yamashita et al. 1975, 1976).

In addition to acyl esters, terpenyl and aroyl esters are often encountered in the aroma complex of plant foods, particularly fruits (Bruemmer 1975; Salunkhe and Do 1977). In cranberry essence, a number of benzoate and benzyl esters, as well as benzaldehyde and benzyl alcohol, are found (Anjou and von Sydow 1967, 1968; Croteau and Fagerson 1968). The cranberry also contains a rather high level of benzoic acid, the likely precursor of these aromatic volatiles. Recent studies (Croteau 1977C) have shown that [7-^{14}C]benzoic acid is converted to benzaldehyde, benzyl alcohol, benzyl benzoate, and minor amounts of other benzoate esters in tissue slices of ripe cranberry (Fig. 15.5). Chemical degradation studies of the products showed that [7-^{14}C]benzoic acid was incorporated directly, without scrambling of the label. Hydrolysis of benzyl benzoate indicated that label was about equally distributed between benzoic acid and benzyl

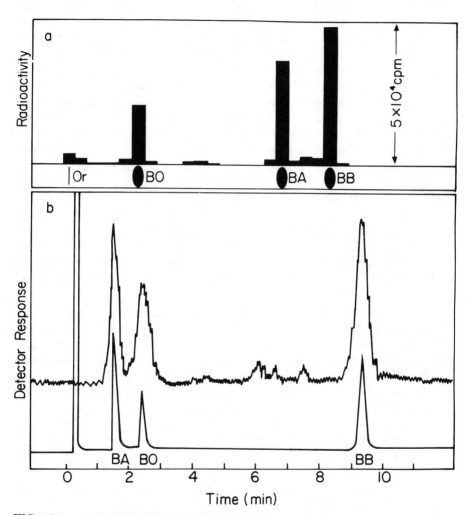

FIG. 15.5. (a) RADIO THIN-LAYER CHROMATOGRAM OF THE VOLATILE
PRODUCTS OBTAINED FROM CRANBERRY FRUIT SLICES THAT HAD
BEEN INCUBATED WITH [7-^{14}C]BENZOIC ACID FOR 4 HR AT 30°

The steam-distilled products (after removal of benzoic acid) were chromatographed
on silica gel G with hexane:ethyl acetate (60:40 v/v), and 0.5 cm sections of the gel
were taken for determination of radioactivity. Or, origin; BO, benzyl alcohol; BA,
benzaldehyde; BB, benzylbenzoate.

(b) RADIO GAS-LIQUID CHROMATOGRAM OF THE VOLATILE PRODUCTS
DESCRIBED ABOVE

The upper tracing shows radioactivity, and the smooth lower tracing is the flame
ionization detector response to the co-injected authentic standards.

alcohol. [7-14C]Benzaldehyde, on the other hand, was converted in tissue slices primarily to benzyl alcohol and benzyl benzoate (Table 15.3), and hydrolysis of the ester indicated that only the alcohol moiety was labeled. Incubation of tissue slices with [7-14C]benzyl alcohol yielded benzyl benzoate, and only the alcohol portion of the ester was labeled. These results clearly indicate that benzoic acid is the direct precursor of the benzoate and benzyl esters of cranberry, and they strongly suggest that the ester cycle (Fig. 15.4) operates primarily in the direction of ester synthesis and is not readily reversible (i.e., neither benzaldehyde nor benzyl alcohol gave rise to appreciable quantities of benzoic acid). None of the aforementioned conversions of [14C]benzoic acid and its derivatives could be demonstrated, in significant yield, in tissue slices prepared from unripe (green) cranberries. Unripe cranberries do contain an appreciable quantity of benzoic acid. This observation along with the tracer studies implies that, although the substrate for the synthesis of volatile compounds is present in immature fruit, the necessary enzymatic machinery is not yet available. Obviously, the development of flavor in cranberry, and in many other fruits and vegetables, is intimately related to the physiological and biochemical changes that accompany ripening.

Although in vivo studies can often reveal the pathway of formation of specific flavor components and provide other very useful insights into flavor biogenesis, more fundamental questions about the enzymes and mechanisms of flavor biosynthesis can only be answered by utilizing cell-free preparations. The involvement of acyl-CoA derivatives in short-chain ester synthesis, for example, is an important question that has yet to be resolved. Other very important questions relate to the regulation of flavor production and to the possibility of manipulating flavor formation. For example, does the composition of the aroma complex produced by a given tissue depend primarily on the types and amounts of precursors

TABLE 15.3

INCORPORATION OF LABELED PRECURSORS INTO
VOLATILE AROMATIC COMPOUNDS OF CRANBERRY[1]

Precursor	Products (cpm × 10⁻⁴)			
	Benzoic Acid	Benzal-dehyde	Benzyl Alcohol	Benzyl Benzoate
Benzoic Acid	—	3.9	2.1	5.3
Benzaldehyde	0.5	—	2.7	6.3
Benzyl alcohol	0.2	0.9	—	8.5

[1]In each experiment, 5 μCi of the [7-14C] labeled precursor were incubated with 3g of cranberry fruit slices for 4 hr. Products were isolated by steam distillation followed by thin-layer chromatography.

available to it, as suggested by Bruemmer (1975) for *Citrus*, or does it depend on the specificity of enzyme systems that can select starting materials from a diverse precursor pool (*e.g.*, the hydroperoxide cleavage enzyme of tomato that is specific for C13 hydroperoxides)? Our attempts to answer these and other questions by utilizing the cranberry system have been stymied, thus far, by our inability to obtain cell-free enzyme systems that will carry out, in significant yield, the conversions observed *in vivo*. Thus, the NADH-dependent reduction of benzaldehyde can be demonstrated *in vitro*, but the CoA dependent reduction or transacylation of benzoic acid cannot. Other investigators have encountered similar problems (Myers *et al.* 1969; Yamashita *et al.* 1975), which may be due in part to the presence of competing enzymes such as thioesterases. Until this present roadblock is overcome, it will remain the rate-limiting step in further investigations of this family of aroma compounds.

ORIGIN OF MONOTERPENES

The monoterpenes are the C10 representatives of the terpenoid family of natural products, and they comprise a very diverse group of acyclic and cyclic structural types and derivatives numbering several hundred individual compounds (Devon and Scott 1972). Several representative monoterpenes are shown in Fig. 15.6. The monoterpenes are almost ubiquitous as flavor substances, occurring as minor components in many fruit and vegetable essences, and as major metabolic products in *Citrus* and in the herbs and spices. In addition to their contribution to flavor *in situ*, the monoterpenes are major components of the extracted or distilled "essential oils", and are the raw chemicals for the synthesis and compounding of many artificial flavors (Erickson 1976).

Plants that accumulate large quantities of monoterpenes during their development generally contain specialized oil gland structures (Loomis and Croteau 1973), and these oil glands appear to be the primary site of monoterpene biosynthesis and storage (Croteau 1977A). The apparent compartmentation of biosynthetic sites within these oil glands prevents the ready penetration of exogenous precursors such as mevalonic acid, and this has somewhat hampered *in vivo* studies on the monoterpenes. In spite of this limitation, *in vivo* tracer studies have yielded information on monoterpene labeling patterns and metabolic turnover rates, and they have provided considerable insight into the physiology of monoterpene production. These studies have been reviewed (Banthorpe and Charlwood 1972; Banthorpe *et al.* 1972; Loomis and Croteau 1973; Croteau and Loomis 1975; Croteau 1975), and so the emphasis here will be on more recent work dealing with cell-free biosynthetic systems.

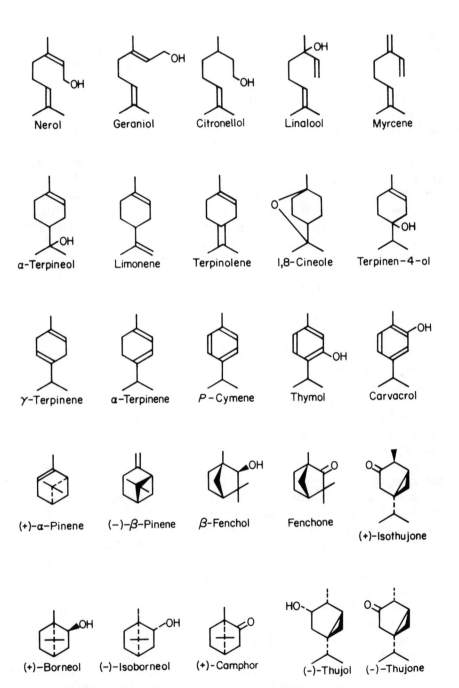

FIG. 15.6.   REPRESENTATIVE MONOTERPENES

Acyclic Precursors

The vast majority of the monoterpenes are constructed by the "head-to-tail" fusion of C_5 isoprene units, typified by the condensation of dimethylallyl pyrophosphate and isopentenyl pyrophosphate to form geranyl pyrophosphate (Fig. 15.7). As the first C_{10} compound to arise from mevalonic acid on the classical terpenoid pathway, geranyl pyrophosphate has been considered to be the direct precursor of the acyclic monoterpenes such as myrcene. Most of the monoterpenes, however, are cyclic (based on a cyclohexanoid ring), and so the *cis*-isomer of geranyl pyrophosphate (*i.e.*, neryl pyrophosphate) is generally considered to be the direct precursor of these compounds because the *cis* geometry at the Δ^2-position of neryl pyrophosphate readily permits cyclization. Consistent with this suggestion are numerous studies of the nonenzymatic solvolysis of neryl and geranyl pyrophosphates which show that the *cis*-derivative readily cyclizes while the *trans*-derivative does not (Cramer and Rittersdorf 1967; Valenzuela and Cori 1967). Linaloyl pyrophosphate also cyclizes under acid conditions, in a manner similar to neryl pyrophosphate (Rittersdorf and Cramer 1968). Linaloyl pyrophosphate has been suggested as the acyclic precursor of cyclohexanoid monoterpenes in *Citrus* (Attaway *et al.* 1966; Bruemmer 1975), but there is little biochemical evidence to support this proposal at present.

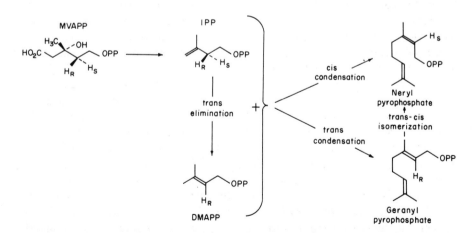

FIG. 15.7. BIOSYNTHESIS OF GERANYL PYROPHOSPHATE FROM MEVALONIC ACID PYROPHOSPHATE, AND TWO POSSIBLE BIOSYNTHETIC ROUTES TO NERYL PYROPHOSPHATE

Biosynthesis and Cyclization of Neryl Pyrophosphate

Although most investigators regard neryl pyrophosphate as the direct precursor of cyclohexanoid monoterpenes, and thus as a key intermediate in monoterpene biosynthesis, the origin of neryl pyrophosphate has not yet been clearly established. There are are two probable pathways for the biosynthesis of neryl pyrophosphate (Fig. 15.7). The first pathway is the *cis*-condensation of isopentenyl pyrophosphate and dimethylallyl pyrophosphate to yield neryl pyrophosphate directly. This mechanism is analogous to the *trans*-condensation involved in geranyl pyrophosphate biosynthesis, but of opposite stereospecificity. The second route is the classical *trans*-condensation of isopentenyl pyrophosphate and dimethylallyl pyrophosphate to geranyl pyrophosphate, followed by isomerization of geranyl pyrophosphate, or a derivative of geranyl pyrophosphate, to the *cis*-isomer. Some support for both pathways is available, and this evidence has been reviewed (Croteau and Loomis 1975). The recent report of a cell-free preparation from *Tanacetum*, capable of synthesizing geraniol and nerol in high yield from mevalonic acid (Banthorpe *et al.* 1976), may provide the necessary tool to solve this problem.

The first unified scheme for the biosynthesis of cyclic monoterpenes was proposed by Ruzicka and associates (1953), and was later modified slightly by Loomis (1967). According to this generally accepted hypothesis (Fig. 15.8), neryl pyrophosphate undergoes cyclization to an α-terpinyl cation intermediate, which by subsequent internal additions, hydride shifts, and rearrangements, gives rise to the various monoterpene skeletal types. Such derivatives are then thought to undergo a variety of metabolic interconversions, including hydroxylations, dehydrogenations, and dehydrations, to yield the great variety of monoterpenes encountered in plants. Thus, it is thought that a relatively small number of cyclases must determine the basic structural character of the monoterpenes produced by a given organism, but that numerous enzymes may be involved in the subsequent metabolic interconversions of the parent cyclic compounds. *In vivo* tracer studies have shown that the labeling patterns of many cyclic monoterpenes derived from exogenous precursors, such as mevalonic acid, are consistent with this basic proposal (Banthorpe *et al.* 1972), and several enzymes (dehydrogenases and reductases) involved in metabolic interconversions of various acyclic and monocyclic monoterpenes have been demonstrated (Battaile *et al.* 1968; Potty and Bruemmer 1970A, B; Dunphy and Allcock 1972). Yet, very few cell-free biosynthetic systems capable of converting acyclic precursors to cyclic monoterpenes have been reported. Thus, this key step in the biosynthesis of most cyclic monoterpenes remains uncertain, and very little is known about the enzymes involved in monoterpene cyclizations.

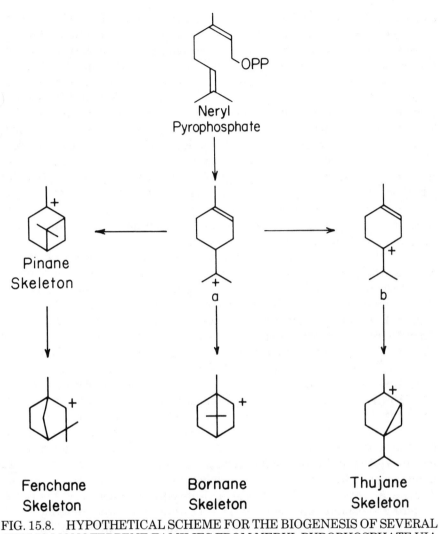

Neryl
Pyrophosphate

Pinane
Skeleton

a

b

Fenchane
Skeleton

Bornane
Skeleton

Thujane
Skeleton

FIG. 15.8.   HYPOTHETICAL SCHEME FOR THE BIOGENESIS OF SEVERAL
CYCLIC MONOTERPENE FAMILIES FROM NERYL PYROPHOSPHATE VIA
THE α-TERPINYL CATION (a) AND THE TERPINEN-4-YL CATION (b)

Monocyclic and Heterocyclic Monoterpenes

Cell-free extracts of mint convert neryl pyrophosphate to α-terpineol as
the major cyclic product (Croteau et al. 1973). With substrate concentrations of α-terpineol these same extracts can produce small amounts of
limonene and terpinolene (Burbott et al. 1973) (Fig. 15.6), suggesting that
these monocyclic hydrocarbons might be derived from α-terpineol by de-

hydration. On the other hand, cell-free preparations from *Citrus* were reported to convert neryl pyrophosphate and geranyl pyrophosphate to limonene, while α-terpineol was not reported as a biosynthetic product in this system (George-Nascimento and Cori 1971; Chayet *et al.* 1977). Limonene, terpinolene, α-terpineol, and the heterocyclic 1,8-cineole (Fig. 15.6), which could arise from α-terpineol by internal alkoxylation, are very common monoterpenes often found together in aroma complexes. The co-occurrence of such closely related monoterpenes, coupled to the apparent ease by which such compounds might be metabolically interconverted, raises an important question as to whether these compounds arise independently from an acyclic precursor, such as neryl pyrophosphate, or whether such monoterpenes may be formed by modification of a single monocyclic intermediate, such as α-terpineol. Until a cell-free preparation capable of synthesizing several related compounds was obtained, this question could not be adequately answered.

Recently, a soluble enzyme preparation from sage leaves was shown to catalyze the conversion of [³H]neryl pyrophosphate to a number of cyclic monoterpenes, including limonene, terpinolene, α-terpineol, and 1,8-cineole, which were identified by the preparation of derivatives, chemical degradation studies, and radiochromatographic analyses (Croteau and Karp 1976B). The formation of all the cyclic products increased linearly with protein concentration and time, was maximum at about pH 6.5, and was stimulated by the presence of $MnCl_2$ or $MgCl_2$. Addition of NaF to the incubation medium also stimulated cyclic monoterpene formation by inhibiting the hydrolysis of neryl pyrophosphate by endogenous pyrophosphatases. Inclusion of 1 mM ($\pm$)-α-terpineol in the incubation medium suppressed 1,8-cineole and cyclic hydrocarbon formation from neryl pyrophosphate, suggesting that this alcohol could be on intermediate. However, the inhibition appeared to be nonspecific as several other monoterpene alcohols were similarly effective. In order to provide a more direct examination of the pathway, ($\pm$)-[3-³H]-α-terpineol, and the corresponding phosphate and pyrophosphate esters, were prepared and tested as substrates with the enzyme system. Neither α-terpineol, nor its phosphorylated derivatives, was converted to 1,8-cineole or to cyclic hydrocarbons, indicating that these substances were not intermediates in the biosynthesis of the structurally related monoterpenes. Furthermore, labeled α-terpineol was not incorporated into other monoterpenes in sage leaf slices, whereas the acyclic alcohols, nerol, and geraniol, were readily incorporated into the characteristic cyclic monoterpenes of sage leaf. Similar lines of evidence excluded 1,8-cineole, limonene, and terpinolene as intermediates in the formation of cyclic products from neryl pyrophosphate. Thus, all of these results suggest that 1,8-cineole, α-terpineol, and the cyclic olefins are derived independently from neryl pyrophosphate, rather than as free intermediates of a common reaction sequence.

Fractionation of the crude soluble enzyme system by chromatography on hydroxylapatite allowed the separation of 1,8-cineole synthetase from competing phosphatases and from α-terpineol synthetase and limonene synthetase activities (Croteau and Karp 1977A). Subsequent studies (Croteau and Karp unpublished) revealed that α-terpineol synthetase and limonene synthetase activities were readily resolved by chromatography on Sephadex G-150. These results proved conclusively that distinct, though perhaps related, synthetases were involved in the biosynthesis of these structurally related monoterpenes.

Preliminary characterization of the partially purified 1,8-cineole synthetase indicated that the enzyme had a pH optimum at pH 6.5 and that a divalent cation was required for catalysis (Croteau and Karp 1977A). $Mn_{++}$was preferred, and Mg_{++}could substitute only at tenfold higher concentrations. The conversion of neryl pyrophosphate to 1,8-cineole was an anaerobic process, implying that the source of oxygen in the heterocyclic product was H_2O and not O_2. The apparent Km for neryl pyrophosphate was $10^{-5}M$, and inhibition of activity was observed at substrate concentrations above $2 \times 10^{-4}M$. 1,8-Cineole did not significantly inhibit activity, but the other product of the reaction, inorganic pyrophosphate, was mildly inhibitory (33% inhibition at $2 \times 10^{-4}M$). The thiol-directed reagent, p-hydroxymercuribenzoate, strongly inhibited the conversion of neryl pyrophosphate to 1,8-cineole (90% inhibition at $4 \times 10^{-5}M$). Further purification of this synthetase by ion exchange chromatography and chromatography on Sephadex G-150 has suggested that a single protein with a molecular weight of about 100,000 daltons is involved in converting neryl pyrophosphate to 1,8-cineole, and that no free intermediates are formed in the reaction. This observation, coupled to the inhibition of cineole synthetase activity by p-hydroxymercuribenzoate, suggests the possible involvement of an enzyme thiol-bound intermediate in the reaction. Substrate specificity studies indicated that the cineole synthetase was highly specific for neryl pyrophosphate, unlike γ-terpinene synthetase (see below) and hydrocarbon synthetases from *Citrus* (Chayet *et al.* 1977) which can utilize both neryl pyrophosphate and geranyl pyrophosphate as substrates. The utilization of both geranyl pyrophosphate and neryl pyrophosphate by such enzymes has been interpreted as involving a multistep cyclization mechanism in which the first step is the loss of the pyrophosphate group and the conversion of geranyl pyrophosphate or neryl pyrophosphate to an equivalent cation of the *cis*-configuration (Croteau 1975; Chayet *et al.* 1977). The strict requirement for neryl pyrophosphate in the case of cineole synthetase might, thus, imply a concerted cyclization mechanism. One feasible mechanism to account for the biosynthesis of 1,8-cineole from neryl pyrophosphate is presented in Fig. 15.9. Cyclization is initiated by enzyme attack at the isopropylidene double bond with consequent ring closure and displacement of the pyrophosphate group. Whether attack is initiated from above or below the plane of the double bond is inconsequential, mechanistically, so long as the remaining double bond is hydrated in such a way that the hydroxyl and isopropyl are *cis*, enabling the heterocyclization step to proceed with displacement of the enzyme. Other interpretations are possible, and many questions about this enzyme, and about the properties and mechanisms of the α-terpineol synthetase and olefin synthetases remain to be answered.

Aromatic Monoterpenes

p-Cymene, carvacrol, and thymol (Fig. 15.6) are the most common aromatic monoterpenes encountered in plant tissue. The proportion of carvacrol and thymol vary widely from species to species, and in some cases only one of these isomeric alcohols may be present along with *p*-cymene. These aromatic monoterpenes generally, if not always, co-occur with terpinen-4-ol, γ-terpinene, and α-terpinene (Fig. 15.6). The co-occurrence of these monoterpenes, coupled to an obvious structural relationship, suggest that these compounds are biogenetically related, and originate by way of the hypothetical terpinen-4-yl cation intermediate shown in Fig. 15.8.

Common thyme (*Thymus vulgarus*) produces a volatile oil that contains thymol (54%), γ-terpinene (30%), *p*-cymene (8%), and a number of other monoterpenes, including α-terpinene and terpinen-4-ol, at much lower levels. The time-course of incorporation of $^{14}CO_2$ into the monoterpenes of thyme shoot tips revealed that thymol, γ-terpinene, and *p*-cymene were heavily labeled with tracer (Poulose and Croteau 1977). The change in specific activity with time of these three major components suggested a

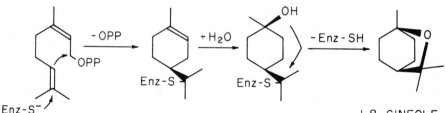

FIG. 15.9.   A POSSIBLE MECHANISM FOR THE CONVERSION OF NERYL PYROPHOSPHATE TO 1,8-CINEOLE

biosynthetic sequence by which γ-terpinene gave rise to p-cymene, which in turn yielded thymol (Fig. 15.10). Consistent with this proposed biosynthetic scheme, [G-³H]γ-terpinene was incorporated into p-cymene and thymol in thyme leaf slices, whereas [G-³H]p-cymene gave rise only to thymol in leaf slices. None of the minor components of thyme (i.e., α-terpinene and terpinen-4-ol) accumulated significant label when leaf slices were incubated with the above radioactive precursors, and so the role of these compounds in the proposed biosynthetic scheme cannot presently be assessed. In any case, the tracer studies did suggest a key role for γ-terpinene in the biosynthesis of the aromatic monoterpenes.

A soluble enzyme preparation from young thyme leaves was shown to catalyze the conversion of neryl pyrophosphate to γ-terpinene (Poulose and Croteau 1977). This crude preparation also contained a very active neryl pyrophosphate pyrophosphohydrolase activity. The possible function of this latter enzyme(s) in monoterpene biosynthesis is not known, but it could be involved in some regulatory role. Subsequent fractionation of the crude preparation by chromatography on Sephadex G-150 and hy-

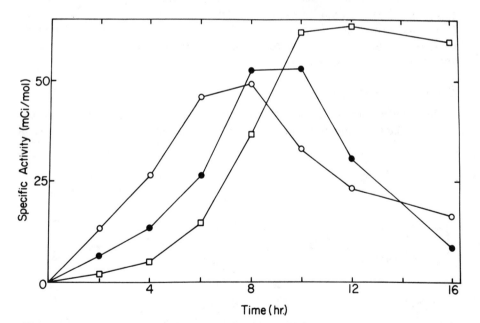

FIG. 15.10.   TIME-COURSE OF LABELING OF THE MAJOR MONOTER-PENES IN *THYMUS VULGARIS* SHOOT TIPS AFTER 1 HR EXPOSURE TO ¹⁴CO₂ IN CONTINUOUS LIGHT

The monoterpenes are: γ-terpinene, o; p-cymene, •; and thymol, □.

droxylapatite, followed by ion exchange chromatography and hydrophobic chromatography on neroic acid substituted Sepharose, allowed the resolution of the γ-terpinene synthetase from the pyrophosphohydrolase. The partially purified synthetase had a pH optimum in the 6.5–7.0 range, showed an absolute requirement for a divalent cation (Mg^{++}), and was strongly inhibited by thiol-directed reagents. All evidence to date indicates that a single protein with a molecular weight of about 100,000 daltons is involved in the reaction, and that no free intermediates are formed. Subsequent substrate specificity studies showed that the enzyme utilized geranyl pyrophosphate with nearly the same efficiency as neryl pyrophosphate, suggestive of the ionization-cyclization mechanism discussed above. A possible mechanism for the conversion of neryl pyrophosphate to γ-terpinene, in which the enzyme stabilizes the intermediate carbonium ion and is then displaced by the hydride shift, is shown in Fig. 15.11.

The above results are preliminary, and much more remains to be done before the pathway of biosynthesis of the aromatic monoterpenes can be confirmed. Thus far, the evidence suggests that thymol (or carvacrol) can be derived by the cyclization of neryl pyrophosphate or geranyl pyrophosphate to γ-terpinene, followed by desaturation to p-cymene, and hydroxylation of this aromatic hydrocarbon. The aromatization process involved in p-cymene formation appears to be different from that involved in either the shikimate pathway or steroidal aromatization.

Bicyclic Monoterpenes

Relatively little is known about the biosynthesis of bicyclic monoterpenes of the bornane, fenchane, pinane, and thujane families. However, as

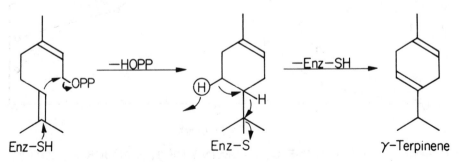

FIG. 15.11.   A POSSIBLE MECHANISM FOR THE CONVERSION OF NERYL PYROPHOSPHATE TO γ-TERPINENE

indicated previously, the labeling patterns of several bicyclic monoterpenes, derived from exogenous specifically labeled precursors, are consistent with the hypothetical biogenetic pathway shown in Fig. 15.8 (Francis
1971; Banthorpe et al. 1972). Preliminary reports on the cell-free biosynthesis of α-pinene, β-pinene, and sabinene (Cori 1969; George-Nascimento
and Cori 1971; Croteau and Karp 1976B; Chayet et al. 1977) and
isothujone (Banthorpe et al. 1976) are available, but almost nothing is
known about the enzymes involved in these transformations.

Recently, a soluble enzyme preparation from sage was shown to catalyze
the anaerobic conversion of [1-^{3}H]neryl pyrophosphate to the bicyclic
alcohol borneol (Fig. 15.6) (Croteau and Karp 1976A). The pH optimum for
this reaction was near 8.0, and high yields of borneol were obtained only in
the presence of $MgCl_2$. Although neryl pyrophosphate was readily converted to borneol, a specific search of the enzymatic products for the
diastereomeric isoborneol revealed that this alcohol was not formed in
detectable yield. A small amount of α-terpineol was formed by the cell-free
preparation. However, synthetic $(\pm)$-[3-^{3}H]-α-terpineol was not converted
to borneol by the enzyme preparation, indicating that α-terpineol was not
an intermediate in borneol biosynthesis. In addition to α-terpineol,
another minor radioactive component was formed by the enzyme extract
when incubated with [1-^{3}H]neryl pyrophosphate. This compound had the
properties of camphor, suggesting that a dehydrogenase was present in the
preparation that converted borneol to camphor. Addition of oxidized
pyridine nucleotides to the incubation mixture greatly stimulated the
oxidation of borneol (derived from neryl pyrophosphate) to camphor. The
biosynthetic products were identified via the synthesis of derivatives,
which were analyzed by radio-chromatography or were crystallized to
constant specific activity. The direct conversion of [1-^{3}H]neryl pyrophosphate to [3-^{3}H]borneol and [3-^{3}H]camphor was demonstrated by chemical
degradation studies and, thus, shown to be consistent with earlier in vivo
studies (Banthorpe and Baxendale 1970; Battersby et al. 1972). Furthermore, resolution of the biosynthetic products proved that they were the
(+)-isomers, the same as the natural constituents of sage oil. These results
strongly suggested the (+)-camphor was synthesized by dehydrogenation
of (+)-borneol formed by the cyclization of neryl pyrophosphate.

In order to examine the enzymes involved in camphor biosynthesis in
greater detail, and, in particular, to determine if any free intermediates
were involved in the cyclization of neryl pyrophosphate to borneol, an
attempt was made to purify the borneol synthetase activity by gel permea-
tion chromatography. This procedure resulted in the apparent loss of
catalytic capability; however, subsequent detailed recombination of column fractions demonstrated that two separable enzymatic activities were

required for the conversion of neryl pyrophosphate to borneol (Fig. 15.12) (Croteau and Karp 1977B). Incubation of the two enzymes, while separated by a dialysis membrane, still allowed borneol formation from neryl pyrophosphate, indicating that a free, dialyzable intermediate was involved in this transformation. Subsequent studies showed that the higher molecular weight enzyme component (F_1) converted neryl pyrophosphate to the water-soluble intermediate, while the lower molecular weight component (F_2) converted the intermediate to borneol. The intermediate was isolated by o-(diethylaminoethyl)cellulose column chromatography and subsequently identified as bornyl pyrophosphate by direct chromatographic analysis, and by the preparation of derivatives and chromatographic anlysis of both the hydrogenolysis and enzymatic hydrolysis products of bornyl pyrophosphate. Preliminary studies with this bornyl pyrophosphate synthetase indicated that the enzyme had a molecular weight in the 130,000 dalton range, had a pH optimum near 8.0, and required a divalent cation for catalysis.

The demonstration of the bornyl pyrophosphate synthetase strongly suggested that the second enzyme component (F_2) was functioning in the hydrolysis of bornyl pyrophosphate to borneol. This suggestion was confirmed directly using synthetic (+)-[G-^{3}H]bornyl pyrophosphate as a substrate. Preliminary studies indicated that the enzyme had a pH optimum

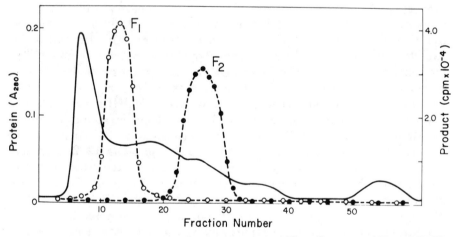

From Croteau and Karp (1977B)

FIG. 15.12. SEPHADEX G-150 GEL FILTRATION OF THE 105,000g SUPERNATANT PREPARATION FROM SAGE LEAVES SHOWING THE SEPARATION OF THE TWO ENZYME COMPONENTS REQUIRED FOR THE CONVERSION OF NERYL PYROPHOSPHATE TO BORNEOL

Absorbance at 280 nm (—), bornyl pyrophosphate synthetase (F_1) activity (o—o) and bornyl pyrophosphate pyrophosphohydrolase (F_2) activity (•—•) are plotted.

in the 5.0–6.0 range, and possessed the ability to hydrolyze both bornyl pyrophosphate and bornyl phosphate to borneol. Thus, a two-step hydrolysis seemed probable.

Recently, these two activities (hydrolysis of the pyrophosphate to the monophosphate and hydrolysis of the monophosphate to borneol) were resolved by chromatography on hydroxylapatite (Croteau and Karp, in preparation). A detailed search for a pyrophosphohydrolase activity with an alkaline pH optimum (as with the synthetase) revealed the presence of such an activity. This alkaline pyrophosphate phosphatase was readily resolved from the acid pyrophosphate phosphatase on Sephadex G-150 column chromatography, and was much less active than its acid pH optimum counterpart. The alkaline activity has been purified more than 200-fold by a five-step fractionation procedure. The pyrophosphate and monophosphate hydrolase activities are coincident at each step of the fractionation, and the ratio of specific activities remains constant throughout purification. Therefore, it seems likely that the hydrolysis of bornyl pyrophosphate may be carried out by two distinct types of activities, one with acid pH optimum and the other with alkaline optimum. The activity at acid pH consists of two distinct steps catalyzed by separate enzymes, while with the alkaline activity both steps may be catalyzed by the same enzyme.

Although the bornyl pyrophosphate synthetase and the phosphatases have not been examined in detail, it is quite clear that borneol is synthesized by the initial cyclization of neryl pyrophosphate to bornyl pyrophosphate followed by hydrolysis of this intermediate to the alcohol, which may then be oxidized to camphor (Fig. 15.13). The intermediate role of bornyl pyrophosphate in the biosynthesis of borneol has also been demonstrated in soluble preparations from rosemary (*Rosmarinus officinalis*) leaves (Croteau and Karp 1977B). Unlike the oil of sage, which contains (+)-borneol, the oil of rosemary contains (−)-borneol. Thus, the

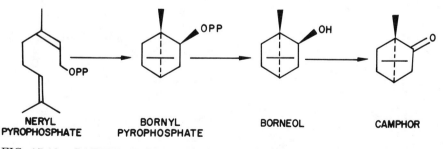

NERYL
PYROPHOSPHATE

BORNYL
PYROPHOSPHATE

BORNEOL

CAMPHOR

FIG. 15.13. PATHWAY FOR THE BIOSYNTHESIS OF (+)-BORNEOL AND (+)-CAMPHOR FROM NERYL PYROPHOSPHATE

proposed pathway for borneol formation is probably widespread in the plant kingdom. Cyclic pyrophosphates, such as pre-squalene pyrophosphate, are involved in the formation of higher terpenes; however, bornyl pyrophosphate is the first cyclic pyrophosphate to be reported as an intermediate in monoterpene biosynthesis. Unlike other cyclic pyrophosphates, bornyl pyrophosphate is unique in that the pyrophosphate moiety of its acyclic precursor appears to have migrated during cyclization, and to have participated in the formation of the 1,4-bridge. A possible mechanism for the conversion of neryl pyrophosphate to (+)-bornyl pyrophosphate is shown in Fig. 15.14. The enzyme may be involved in stabilizing the transient cation. It seems possible that cyclic pyrophosphate intermediates may also be involved in the biosynthesis of other cyclic monoterpene alcohols, such as thujol and fenchol, although no information bearing on this possibility is presently available.

The final step in camphor biosynthesis is the oxidation of borneol. The dehydrogenase involved in this conversion has been isolated from sage leaf, and partially purified and characterized (Croteau and Hooper 1977). Fractionation of the crude soluble preparation by $(NH_4)_2SO_4$ precipitation, chromatography on o-(diethylaminoethyl)cellulose and Sephadex G-150, and affinity chromatography on DL-camphor-3-carboxy substituted cellulose afforded about 200-fold purification of the enzyme, and borneol dehydrogenase was completely resolved from alcohol dehydrogenase. The borneol dehydrogenase had an apparent molecular weight of 91,000 daltons and a pH optimum at 8.0. The enzyme was specific for NAD, and had a Km for this cofactor of 7×10^{-5}M. The apparent Km for (+)-borneol was 3×10^{-5}M. Thiol directed reagents strongly inhibited the enzyme. The stereospecificity of the hydride transfer from (+)-[2-^{3}H]borneol to NAD was examined, and the borneol dehydrogenase was shown to possess B-type

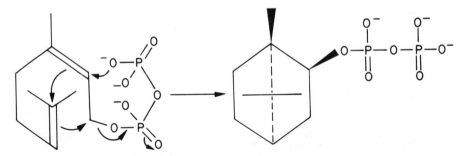

FIG. 15.14.   A POSSIBLE MECHANISM FOR THE CONVERSION OF NERYL
PYROPHOSPHATE TO (+)-BORNYL PYROPHOSPHATE

stereospecificity, unlike alcohol dehydrogenase. Substrate specificity studies showed the enzyme to be rather specific for (+)-borneol, although (−)-thujol was also oxidized (to (−)-thujone) at high rates. The ratio of specific activities for borneol and thujol oxidation remained constant throughout enzyme purification, suggesting that a single dehydrogenase was involved. (−)-Thujone is the major component of sage oil, and it seems likely that this ketone is derived, *in vivo*, by dehydrogenation of (−)-thujol. Thus, the present dehydrogenase might be involved in the biosynthesis of both camphor and thujone. The content of camphor and thujone in the volatile oil of sage increases with the age of the plant, and the development of dehydrogenase activity in sage leaves parallels this increase.

All of the enzymatic steps in the pathway from neryl pyrophosphate to camphor now appear to be elucidated. Thus, camphor is the only bicyclic monoterpene about which we know even this much. Although very little is understood, as yet, about the enzymes involved in camphor formation, the present studies could provide a model for the examination of other bicyclic monoterpenes.

CONCLUDING REMARKS

The enzymatic formation of volatile flavor compounds during growth, ripening, and physical disruption of plant tissues is now well appreciated, and over the last few years considerable research has been directed toward an attempt to understand the basic principles underlying this phenomenon. In many instances, the qualitative and quantitative changes that occur in the aroma complex during flavor development have been cataloged, and inferences have been drawn from this data about the possible origins of certain flavor constituents. In other cases, research has progressed to tracer studies that have allowed the mapping of biosynthetic pathways, and the demonstration of expected intermediates and precursors. In several instances, cell-free systems have been developed, and the enzymes involved in flavor biosynthesis are being purified and characterized and their mechanisms of action explored. In a few cases, a beginning has been made in understanding the integration of flavor biogenesis into the overall metabolic or physiological context in which flavor development occurs (e.g., ripening, maceration, stress, etc.). Thus, progress is being made, but much more remains to be done. In the next few years we may look forward to a greater understanding of the pathways and enzymes of flavor biosynthesis, and to a deeper understanding of how flavor biogenesis is related to other metabolic processes in the plant. Only with this knowledge can we hope to understand the mechanisms of control and regulation that will allow us to devise effective means of manipulating flavor production. Before this ultimate goal is reached, however, a number of shorter-term benefits may accrue from studies on flavor biogenesis including: the ability to predict the identity of as yet unknown flavor components; the ability to biosynthesize, *in vitro,* a particularly desirable flavor component or unusual stereoisomer that is difficult to obtain synthetically; and the ability to gain insight into a relatively unexplored area of science that touches on many disciplines including food science, horticulture, ecology, biochemistry, and physiology.

Although the basic biochemical approach to flavor formation has been stressed in this review, it must be emphasized that while flavor is the end result of a series of enzymatic reactions, this process is greatly influenced by various environmental factors and agricultural practices, and is determined, ultimately, by the genetic make-up of the plant. Studies on the effects of environmental factors (Burbott and Loomis 1967; Scora and Adams 1973) and agricultural practices (Ghosh and Chatterjee 1976; Croteau 1977B), and studies directed toward genetic manipulation (Hefendehl and Murray 1976; Shaw and Wilson 1976) should also contribute greatly to our understanding of, and our ability to control, flavor production in fruits and vegetables.

ACKNOWLEDGMENTS

This work was supported in part by a National Science Foundation Grant (PCM76-23632) and by a Cottrell Research Grant from the Research Corporation. Scientific Paper No. 4847, Project 0268, College of Agriculture Research Center, Washington State University, Pullman, WA 99164.

BIBLIOGRAPHY

ANJOU, K. and von SYDOW, E. 1967. The aroma of cranberries. 2. *Vaccinium macrocarpon* Ait. Acta Chem. Scand. *21*, 2076–2082.

ANJOU, K. and von SYDOW, E. 1968. The aroma of cranberries. 4. *Vaccinium macrocarpon* Ait. Arkiv. Kemi *30*, 9–14.

ATTAWAY, J. A., PIERINGER, A. P. and BARABAS, L. J. 1966. The origin of citrus flavor components. 1. The analysis of citrus leaf oil using gas liquid chromatography, thin layer chromatography and mass spectrometry. Phytochemistry *5*, 141–151.

BANTHORPE, D. V. and BAXENDALE, D. 1970. Terpene biosynthesis. Part 3. Biosynthesis of (+)- and (−)-camphor in *Artemisia, Salvia,* and *Chrysanthemum* species. J. Chem. Soc. (C), 2694–2696.

BANTHORPE, D. V., BUCKNALL, G. A., DOONAN, H. J., DOONAN, S. and ROWAN, M. G. 1976. Biosynthesis of geraniol and nerol in cell-free extracts of *Tanacetum vulgare.* Phytochemistry *15*, 91–100.

BANTHORPE, D. V. and CHARLWOOD, B. V. 1972. Biogenesis of terpenes. *In*

Chemistry of Terpenes and Terpenoids, A. A. Newman (Editor). Academic Press, New York.

BANTHORPE, D. V., CHARLWOOD, B. V. and FRANCIS, M. J. O. 1972. The biosynthesis of monoterpenes. Chem. Rev. 72, 115–155.

BATTAILE, J., BURBOTT, A. J. and LOOMIS, W. D. 1968. Monoterpene interconversions: Metabolism of pulegone by a cell-free system from Mentha piperita. Phytochemistry 7, 1159–1163.

BATTERSBY, A. R., LAING, D. G. and RAMAGE, R. 1972. Biosynthesis. Part 19. Concerning the biosynthesis of (−)-camphor and (−)-borneol in Salvia officinalis. J. Chem. Soc. (Perkin I), 2743–2748.

BRUEMMER, J. H. 1975. Aroma substances of citrus fruits and their biogenesis. In Odor and Taste Substances, F. Drawert (Editor). Verlag Hans Carl, Nurnberg.

BURBOTT, A. J., CROTEAU, R., SHINE, W. E. and LOOMIS, W. D. 1973. Biosynthesis of cyclic monoterpenes by cell-free extracts of Mentha piperita L. Plant Physiol. 51 (suppl.), 49.

BURBOTT, A. J. and LOOMIS, W. D. 1967. Effects of light and temperature on the monoterpenes of peppermint. Plant Physiol. 42, 20–28.

CHAYET, L., ROJAS, C., CARDEMILL, E., JABALQUINTO, A. M., VICUNA, R. and CORI, O. 1977. Biosynthesis of monoterpene hydrocarbons from [1-^{3}H]neryl pyrophosphate and [1-^{3}H]geranyl pyrophosphate by soluble enzymes from Citrus limonum. Arch. Biochem. Biophys. 180, 318–327.

COGGON, P., ROMANCZYK, L. J. JR. and SANDERSON, G. W. 1977. Extraction, purification and partial characterization of a tea metalloprotein and its role in the formation of black tea aroma constituents. J. Agr. Food Chem. 25, 278–283.

CORI, O. 1969. Terpene biosynthesis: Utilization of neryl pyrophosphate by an enzyme system from Pinus radiata seedlings. Arch. Biochem. Biophys. 135, 416–418.

CRAMER, F. and RITTERSDORF. 1967. The hydrolysis of certain monoterpene alcohol phosphates and pyrophosphates. Model reactions for the biosynthesis of monoterpenes. Tetrahedron 23, 3015–3022. (German).

CREVELING, R. K. and JENNINGS, W. G. 1970. Volatile components of Bartlett pear. Higher boiling esters. J. Agr. Food Chem. 18, 19–24.

CREVELING, R. K., SILVERSTEIN, R. M. and JENNINGS, W. G. 1968. Volatile components of pineapple. J. Food Sci. 33, 284–289.

CROTEAU, R. 1975. Biosynthesis of monoterpenes and sesquiterpenes. In Odor and Taste Substances, F. Drawert (Editor). Verlag Hans Carl, Nurnberg.

CROTEAU, R. 1977A. Site of monoterpene biosynthesis in Marjorana hortensis leaves. Plant Physiol. 59, 519–520.

CROTEAU, R. 1977B. Effect of irrigation method on essential oil yield and rate of oil evaporation in mint. HortScience 12, 563–565.

CROTEAU, R. 1977C. Biosynthesis of benzaldehyde, benzyl alcohol, and benzyl benzoate from benzoic acid in cranberry (Vaccinium macrocarpon). J. Food Biochem. 1, 317–326.

CROTEAU, R., BURBOTT, A. J. and LOOMIS, W. D. 1973. Enzymatic cyclization of neryl pyrophosphate to α-terpineol by cell-free extracts from peppermint. Biochem. Biophys. Res. Commun. 50, 1006–1012.

CROTEAU, R. and FAGERSON, I. S. 1968. Major volatile components of the juice of American cranberry. J. Food Sci. 33, 386–389.

CROTEAU, R. and HOOPER, C. L. 1978. Metabolism of monoterpenes: Partial purification and characterization of a bicyclic monoterpenol dehydrogenase from sage (Salvia officinalis). Arch. Biochem. Biophys., in press.

CROTEAU, R. and KARP, F. 1976A. Enzymatic synthesis of camphor from neryl pyrophosphate by a soluble preparation from sage (Salvia officinalis). Biochem. Biophys. Res. Commun. 72, 440–447.

CROTEAU, R. and KARP, F. 1976B. Biosynthesis of monoterpenes: Enzymatic conversion of neryl pyrophosphate to 1,8-cineole, α-terpineol, and cyclic monoterpene hydrocarbons by a cell-free preparation from sage (Salvia officinalis). Arch. Biochem. Biophys. 176, 734–746.

CROTEAU, R. and KARP, F. 1977A. Biosynthesis of monoterpenes: Partial purification and characterization of 1,8-cineole synthetase from Salvia officinalis. Arch. Biochem. Biophys. 179, 257–265.

CROTEAU, R. and KARP, F. 1977B. Demonstration of a cyclic pyrophosphate intermediate in the enzymatic conversion of neryl pyrophosphate to borneol. Arch. Biochem. Biophys. 184, 77–86.

CROTEAU, R: and KOLATTUKUDY, P. E. 1974. Biosynthesis of hydroxyfatty acid polymers. Enzymatic synthesis of cutin from monomer acids by cell-free preparations from the epidermis of Vicia faba leaves. Biochemistry 13, 3193—3202.

CROTEAU, R. and LOOMIS, W. D. 1975. Biosynthesis and metabolism of monoterpenes. Int. Flavours, 292–296.

DEVON, T. K. and SCOTT, A. I. 1972. Handbook of naturally occurring compounds. Vol. 2. Terpenes. Academic Press, New York.

DUNPHY, P. J. and ALLCOCK, C. 1972. Isolation and properties of a monoterpene reductase from rose petals. Phytochemistry 11, 1887–1891.

ERICKSON, R. E. 1976. The industrial importance of monoterpenes and essential oils. Lloydia 39, 8–19.

ERIKSSON, C. 1975. Aroma compounds derived from oxidized lipids. Some biochemical and analytical aspects. J. Agr. Food Chem. 23, 126–128.

ERIKSSON, C., QVIST, I. and VALLENTIN, K. 1977. Conversion of aldehydes to alcohols in liquid foods by alcohol dehydrogenase. In Enzymes in Food and Beverage Processing, Symposium Series No. 47, R. L. Ory and A. J. St. Angelo (Editors). American Chemical Society, Washington, D.C.

FRANCIS, M. J. O. 1971. Monoterpene biosynthesis. In Aspects of Terpenoid Chemistry and Biochemistry, T. W. Goodwin (Editor). Academic Press, New York.

GALLIARD, T. 1975. Degradation of plant lipids by hydrolytic and oxidative enzymes, In Recent Advances in the Chemistry and Biochemistry of Plant Lipids, T. Galliard and E. I. Mercer (Editors). Academic Press, London.

GALLIARD, T. and MATTHEW, J. A. 1976. The enzymic formation of long chain aldehydes and alcohols by α-oxidation of fatty acids in extracts of cucumber fruit (Cucumis sativus). Biochim. Biophys. Acta 424, 26–35.

GALLIARD, T., MATTHEW, J. A., FISHWICK, M. J. and WRIGHT, A. J. 1976B. The enzymatic degradation of lipids resulting from physical disruption of cucumber (Cucumus sativus) fruit. Phytochemistry 15, 1647–1650.

GALLIARD, T. and MATTHEW, J. A. 1977. Lipoxygenase-medicated cleavage of fatty acids to carbonyl fragments in tomato fruits. Phytochemistry 16, 339–343.

GALLIARD, T. and PHILLIPS, D. R. 1976. The enzymic cleavage of linoleic acid to C9 carbonyl fragments in extracts of cucumber (Cucumus sativus) fruit and the possible role of lipoxygenase. Biochim. Biophys. Acta 431, 278–287.

GALLIARD, T., PHILLIPS, D. R. and REYNOLDS, J. 1976A. The formation of cis-3-nonenal, trans-2-nonenal, and hexanal from linoleic acid hydroperoxide isomers by a hydroperoxide cleavage enzyme system in cucumber (*Cucumis sativus*) fruits. Biochim. Biophys. Acta *441*, 181–192.

GARDNER, H. W. 1975. Decomposition of linoleic hydroperoxide. Enzymic reactions compared with nonenzymic. J. Agr. Food Chem. *23*, 129–136.

GEORGE-NASCIMENTO, C. and CORI, O. 1971. Terpene biosynthesis from geranyl and neryl pyrophosphates by enzymes from orange flavedo. Phytochemistry *10*, 1803–1810.

GHOSH, M. L. and CHATTERJEE, S. K. 1976. Effect of N:P:K fertilizers on growth, development, and essential oil content of *Mentha* spp. Ind. J. Expt. Biol. *14*, 366–368.

GROSCH, W. 1976. Breakdown of linoleic acid hydroperoxides. Formation of volatile carbonyl compounds. Z. Lebensm. Unters.-Forsch. *160*, 371–375. (German).

HATANAKA, A., KAJIWARA, T. and HARADA, T. 1975. Biosynthetic pathway of cucumber alcohol: *trans*-2,*cis*-6-nonadienol via *cis*-3,*cis*-6-nonadienal. Phytochemistry *14*, 2589–2592.

HATANAKA, A., KAJIWARA, T. and SEKIYA, J. 1976A. Biosynthesis of *trans*-2-hexenal in chloroplasts from *Thea sinensis*. Phytochemistry *15*, 1125–1126.

HATANAKA, A., KAJIWARA, T. and SEKIYA, J. 1976B. Seasonal variations in *trans*-2-hexenal and linolenic acid in homogenates of *Thea sinensis* leaves. Phytochemistry *15*, 1889–1891.

HATANAKA, A., KAJIWARA, T., SEKIYA, J. and HIRATA, H. 1976C. Biosynthetic pathway of leaf aldehyde in *Farfugium japonicum* Kitamura leaves. Agr. Biol. Chem. *40*, 2177–2180.

HEFENDEHL, F. W. and MURRAY, M. J. 1976. Genetic aspects of the biosynthesis of natural odors. Lloydia *39*, 39–52.

KEMP, T. R. 1975. Characterization of some new C_{16} and C_{17} unsaturated fatty aldehydes. J. Am. Oil Chem. Soc. *52*, 300–302.

KEMP, T. R., KNAVEL, D. E. and STOLTZ, L. P. 1974. Identification of some volatile compounds from cucumber. J. Agr. Food Chem. *22*, 717–718.

KINSELLA, J. E. and HWANG, D. 1977. Biosynthesis of flavors by *Penicillium roqueforti*. Biotechnol. Bioeng. *18*, 927–938.

KOLATTUKUDY, P. E. 1967. Mechanisms of synthesis of waxy esters in broccoli (*Brassica oleracea*). Biochemistry *6*, 2705–2717.

KOLATTUKUDY, P. E. 1970A. Biosynthesis of cuticular lipids. Ann. Rev. Plant Physiol. *21*, 163–192.

KOLATTUKUDY, P. E. 1970B. Reduction of fatty acids to alcohols by cell-free preparations of *Euglena gracilis*. Biochemistry *9*, 1095–1102.

KOLATTUKUDY, P. E. 1971. Enzymatic synthesis of fatty alcohols in *Brassica oleracea*. Arch. Biochem. Biophys. *142*, 701–709.

LOOMIS, W. D. 1967. Biosynthesis and metabolism of monoterpenes. *In* Terpenoids in Plants, J. B. Pridham (Editor). Academic Press, New York.

LOOMIS, W. D. and CROTEAU, R. 1973. Biochemistry and physiology of lower terpenoids. *In* Terpenoids: Structure, Biogenesis, and Distribution, Recent Advances in Phytochemistry, Vol. 6, V. C. Runeckles and T. J. Mabry (Editors). Academic Press, New York.

MANSELL, R. L., STOCKIGT, T. and ZENK, M. H. 1972. Reduction of ferulic acid to coniferyl alcohol in a cell-free system from a higher plant. Z. Planzenphysiol. *68*, 286–288.

MYERS, M. J., ISSENBERG, P. and WICK, E. L. 1969. Vapor analysis of the production by banana fruit of certain volatile constituents. J. Food Sci. *34*, 504–508.

MYERS, M. J., ISSENBERG, P. and WICK, E. L. 1970. L-Leucine as a precursor of isoamyl alcohol and isoamyl acetate, volatile aroma constituents of banana fruit discs. Phytochemistry *9*, 1693–1700.

NURSTEN, H. E. 1970. Volatile compounds: The aroma of fruits. *In* The Biochemistry of Fruits and Their Products, Vol. 1., A. C. Hulme (Editor). Academic Press, New York.

POTTY, V. H. and BRUEMMER, J. H. 1970A. Oxidation of geraniol by an enzyme system from orange. Phytochemistry *9*, 1001–1007.

POTTY, V. H. and BRUEMMER, J. H. 1970B. Limonene reductase system in the orange. Phytochemistry *9*, 2319–2321.

POULOSE, A. J. and CROTEAU, R. 1978. Biosynthesis of aromatic monoterpenes: Conversion of γ-terpinene to *p*-cymene and thymol in *Thymus vulgaris*. Arch. Biochem. Biophys. *187*, 307–314.

RHODES, M. J. and WOOLTORTON, L. S. C. 1975. Enzymes involved in the reduction of ferulic acid to coniferyl alcohol during the aging of disks of swede root tissue. Phytochemistry *14*, 1235–1240.

RITTERSDORF, W. and CRAMER, F. 1968. Cyclization of nerol and linalöol on solvolysis of their phosphate esters. Tetrahedron *24*, 43–52.

RUZICKA, L. A., ESCHENMOSER, A. and HEUSSER, H. 1953. The isoprene rule and the biogenesis of terpenic compounds. Experientia *9*, 357–367.

SALUNKHE, D. K. and DO, J. Y. 1977. Biogenesis of aroma constituents of fruits and vegetables. CRC Crit. Rev. Food Sci. Nutr. *8*, 161–190.

SCORA, R. W. and ADAMS, C. 1973. Effect of oleocellosis, desiccation, and fungal infection upon the terpenes of individual oil glands of *Citrus latipes*. Phytochemistry *12*, 2347–2350.

SHAW, P. E. and WILSON, C. W. 1976. Comparison of extracted peel oil composition and juice flavor for rough lemon, persian lime, and a lemon-lime cross. J. Agr. Food Chem. *24*, 664–666.

STONE, E. J., HALL, R. M. and KAZENIAC, S. J. 1975. Formation of aldehydes and alcohols in tomato fruit from U-^{14}C-labeled linolenic and linoleic acids. J. Food Sci. *40*, 1138–1141.

TAKEI, Y., ISHIWATA, K. and YAMANISHI, T. 1976. Aroma components characteristic of spring green tea. Agr. Biol. Chem. *40*, 2151–2157.

TRESSL, R. and DRAWERT, F. 1973. Biogenesis of banana volatiles. J. Agr. Food Chem. *21*, 560–565.

VALENZUELA, P. and CORI, O. 1967. Acid catalyzed hydrolysis of neryl pyrophosphate and geranyl pyrophosphate. Tetrahedron Lett. 3089–3094.

VICK, B. A. and ZIMMERMAN, D. C. 1976. Lipoxygenase and hydroperoxide lyase in germinating watermelon seedlings. Plant Physiol. *57*, 780–788.

WENGENMAYER, H., EBEL, J. and GRISEBACH, H. 1976. Enzymic synthesis of lignin precursors. Purification and properties of a cinnamoyl-CoA: NADPH reductase from cell suspension cultures of soybean (*Glycine max*). Eur. J. Biochem. *65*, 529–536.

YAMASHITA, I., NEMOTO, Y. and YOSHIKAWA, S. 1975. Formation of volatile esters in strawberries. Agric. Biol. Chem. *39*, 2303–2307.

YAMASHITA, I., NEMOTO, Y. and YOSHIKAWA, S. 1976. Formation of volatile alcohols and esters from aldehydes in strawberries. Phytochemistry *15*, 1633—1637.

ROLE OF HYDROPEROXIDES IN THE ONSET OF SENESCENCE PROCESSES IN PLANT TISSUES

CHAIM FRENKEL

Department of Horticulture and Forestry
Rutgers University
New Brunswick, NJ 08903

ABSTRACT

The onset of the senescence process in plant storage tissues, including fruit and potato tubers, as occurring normally or as induced by ethylene or other senescence factors, is accompanied by a pronounced increase in the level of H_2O_2. The peroxide is viewed as an active oxygen species which can attack metabolites which otherwise resist the action of molecular oxygen, thereby initiating oxidative reactions leading to the onset of senescence processes.

INTRODUCTION

Van Fleet (1954) and other workers (Pilet and Zryd 1965; Siegel and Porto 1961) proposed that development in plants is accompanied by a progressive shift from a reduced toward an oxidative state in tissues. By reason of this view the onset of senescence, the last phase in plant development (Leopold 1961; Varner 1965), marks the endpoint in the transition toward an oxidative state in tissues. Moreover, oxygen utilization may accelerate the transitions in tissues toward an oxidative state, and thereby the initiation of senescence processes.

It is important to elucidate the nature of the active oxygen species involved in oxidative processes in aging tissues since metabolites commonly resist the direct oxidation by molecular oxygen (O_2) (Hamilton 1974). Apparently other oxygen species serve as the active form in oxidative reactions leading to senescence. Hamilton (1974) reasoned that the single most important reason for the sluggish action of molecular oxygen with organic compounds is that O_2 exists in the stable triplet ground state. The direct reaction of a triplet molecule (O_2) with a singlet (organic metabolite) is a spin forbidden process and thus will not readily occur. This is so because the time required for the electron spin inversion of a triplet to

433

a singlet (10^{-9} sec) is much longer relative to the time period which is required for molecular collisions (10^{-13} sec) leading to chemical reactions.

The resistance of metabolites to oxidation can be overcome by active oxygen species represented by reduced oxygen forms as shown below. The electron transfer to O_2, as catalyzed by terminal oxidases, involves the overall tetravalent reduction of O_2 (formation of water). However, O_2 exhibits preference for a univalent reduction pathway (Fridovich 1975B) and subsequently the formation of oxygen intermediates in a reduced state as outlined below (Hamilton 1974):

$$O_2 \xrightarrow[H^+]{e^-} HO^2 \xrightarrow[H^+]{e^-} H_2O_2 \xrightarrow[H^+ \quad -H_2O]{e^-} HO^{\cdot} \xrightarrow[H^+]{e^-} H_2O$$

These intermediates are formed in organisms as follows:

1. a. $RH + O_2 \longrightarrow R^{\cdot} + HO_2^{\cdot}$ (perhydroxyl radical)
    b. $RH + O_2 \longrightarrow RH^+ + O_2^{\cdot}$ (superoxide)
The dissociation pK of the perhydroxyl radical is 4.8 and is therefore commonly found in metabolic media in the dissociated form ($H^+ + O_2^{-}$)
2. Hydrogen peroxide results from the transfer of an electron pair to oxygen:
    a. $RH_2 + O_2 \longrightarrow R + H_2O_2$
H_2O_2 is also a product of the dismutation of superoxide radicals. The reaction occurring spontaneously or as catalyzed by superoxide dismutase (SOD) (Fridovich 1974, 1975A, B):
    b. $2 O_2^{\cdot} + 2H^+ \longrightarrow H_2O_2 + O_2$
3. Formation of the hydroxyl radical ($HO^{\cdot}$) is attributed to the Haber-Weiss reaction (Haber and Weiss 1934) involving the interaction of the superoxide radical and H_2O_2.
    a. $O_2^{\cdot} + H_2O_2 \longrightarrow HO^- + HO^{\cdot} + O_2$
    b. HO can also be formed from H_2O_2 in the presence of low valance transition metals (Hamilton 1974).
4. Singlet oxygen ($O_2{}^*$) is another active oxygen species. It can be apparently formed by the Haber-Weiss reaction (Eq 3a) and by several other mechanisms (Krimsky 1977).

Involvement of Reduced Oxygen Intermediates in the Onset of Senescence Process in Fruit

Singlet and hydroxy radicals were implicated in the oxidative degradation of metabolites *in vitro* (Kellog and Fridovich 1975). However, the formation and the action of these oxygen intermediates *in vivo* awaits confirmation.

Superoxides are proven metabolic products in a wide range of organisms (Fridovich 1974; 1975A, B). The activity of superoxide dismutase (SOD), an enzyme scavenger of superoxide, was used as a reciprocal index to obtain estimates of the changes in the level of superoxide in tissues. With the use of this criterion, it was shown that superoxide levels show little or no change in senescing fruit tissues (Baker 1976). Essentially the same results were obtained in aging tissues of rat and mouse (Kellog and Fridovich 1976). Additional studies are needed to ascertain the involvement of the superoxide radical in oxidative reactions in aging tissues.

In our studies we examined the relationship between the metabolism and the action of H_2O_2 as occurring *in vivo* to the onset of senescence processes in fruit and other storage organs such as potato tubers. Hydroperoxides, including H_2O_2 and lipid peroxides, were determined using a titanium assay method (Brennan and Frenkel 1977). Test compounds were applied in mannitol solutions as described previously (Frenkel 1974; Frenkel *et al.* 1969).

In tomato the formation of lycopene, the red tomato pigment, and ethylene evolution are accompanied by a rise in H_2O_2 and lipid peroxides (Fig. 16.1). Likewise, the onset of ripening in Bartlett pear, as measured by changes in fruit firmness and ethylene evolution, is accompanied by an increase in peroxide levels (Fig. 16.2a). The relationship between the onset of senescence processes and the rise in peroxides is apparently not casual since the manipulation of H_2O_2 levels in tissues influence the rate of senescence processes. Application of glycolate which serves as an electron donor for the reduction of O_2 to H_2O_2, resulted in the stimulation of H_2O_2 formation in fruit (Fig. 16.2a) and consequently the enhancement in softening (16.2b) and ethylene formation (16.2c). Similar results were obtained by the application of xanthine which also serves as a substrate for the formation of peroxides (Brennan and Frenkel 1977). The level of H_2O_2 in tissues can also be promoted if the peroxide breakdown by catalase is inhibited. Application of catalase inhibitors, including hydroxylamine and cyanide (Brennan and Frenkel 1977) led to the increase in H_2O_2 in pears (Fig. 16.3a) and consequently the enhancement in softening (Fig. 16.3b), although ethylene synthesis was retarded (Fig. 16.3c). The employed enzyme inhibitors apparently bind metal ions which are required in the pathway of ethylene synthesis (Lieberman *et al.* 1965). While the promotion of H_2O_2 levels is stimulatory to ripening, the inhibition of H_2O_2 formation delayed the onset of ripening. The activity of glycolate oxidase, which catalyzes the electron transfer for glycolate to O_2 (Tolbert 1971) is inhibited by alpha-hydroxy-2-pyridine methane-sulfonic acid

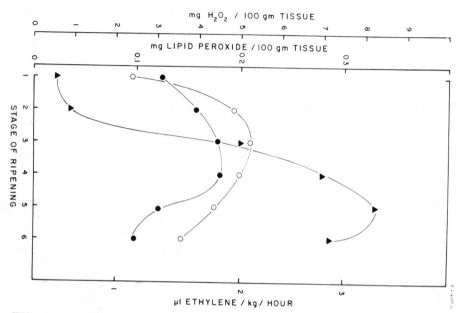

FIG. 16.1 .EVOLUTION OF ETHYLENE (▲) AS RELATED TO THE CHANGES IN H_2O_2 (○) AND LIPID PEROXIDES (●) IN TOMATO (VAR. AZES) AT DIFFERENT STAGES OF RIPENING
The presented ripening stages are: 1. green; 2. white; 3. breaking; 4. pink; 5. red; 6. overripe.

(HPMS) (Zelitch 1972). The application of HPMS to pears resulted in the decrease in H_2O_2 concentrations as compared to the untreated fruit while glycolate applications stimulated the peroxide concentrations (Fig. 16.4a). The obtained ripening processes corresponded to the changes in H_2O_2. The stimulation in peroxide formation enhanced softening and ethylene evolution whereas the retardation in the formation of H_2O_2 leads to the inhibition in these ripening processes (Fig. 16.4b, and c, respectively). These results indicate that the onset of senescence in fruit correlates with the concentrations of H_2O_2 in tissues as occurring normally or as influenced by the treatments.

The regulation of the turnover in H_2O_2 in tissues may be subject to the action of ethylene. In pears application of ethylene led to the stimulation in H_2O_2 levels (Fig. 16.5). However, in climacteric fruit such as tomato or pear both ethylene and peroxide levels increase concomitantly (Fig. 16.1 and 16.2, respectively) and for that reason it is difficult to discern the interdependency of these processes. To test the role of ethylene we employed potato tubers, since the initiation of ethylene dependent responses, such as a rise in respiration, clearly depends on the action of

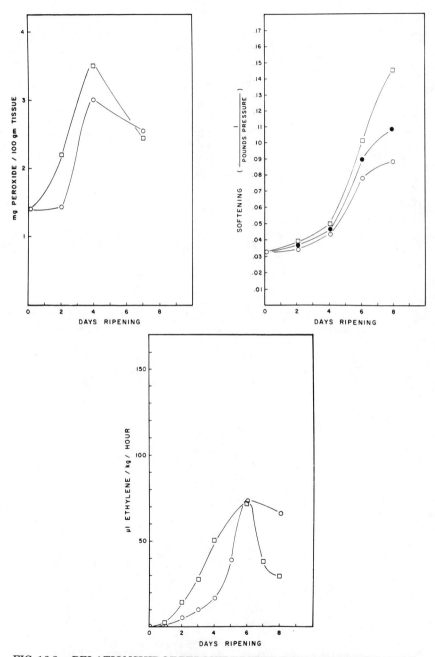

FIG. 16.2.   RELATIONSHIP OF PEROXIDE LEVELS IN BARTLETT PEAR, AS
OCCURRING NORMALLY AND AS INFLUENCED BY THE APPLICATION
OF GLYCOLATE (LEFT) TO RATE OF SOFTENING (RIGHT) AND
ETHYLENE EVOLUTION (CENTER)
The glycolate concentrations used were zero (mannitol control) (○) and mannitol
plus 1 μM glycolate (□). Effect of ethylene (10 μ1/L) on softening (●) is shown for
comparison.

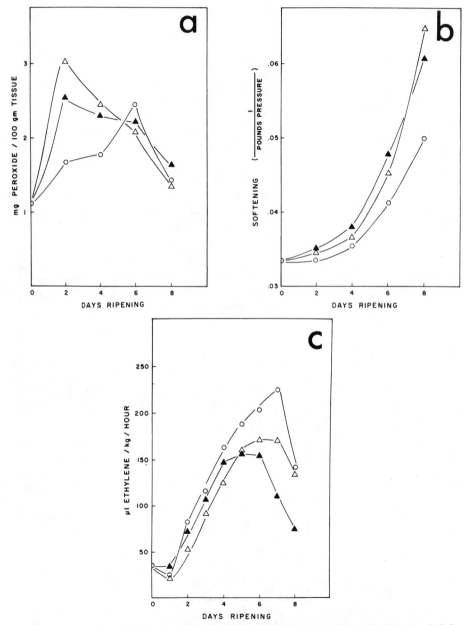

FIG. 16.3.   RELATION OF PEROXIDE LEVELS IN BARTLETT PEAR AS OC-
CURRING NORMALLY AND (a) AS INFLUENCED BY THE APPLICATION OF
CATALASE INHIBITORS (b) TO RATE OF SOFTENING AND (c) ETHYLENE
EVOLUTION
The concentrations of inhibitors used were zero (mannitol control) (○) and of
mannitol plus 10 mM of KCN (△) or hydroxylamine HCl (▲).

exogenous ethylene (Reid and Pratt 1972). The effect of ethylene on the formation of peroxides in potato tubers is shown in Fig. 16.6.   Fig. 16.6A shows that potato tubers normally exhibit a steady state in respiration and that the application of ethylene induces a respiratory upsurge as previously observed (Reid and Pratt 1972). The rise in respiration after a lag period reached a peak 2- to 3-fold the initial rate, followed by a decline. Moreover, the upsurge in respiration was accompanied by corresponding changes in peroxides (Fig. 16.6C), indicating that as with respiration, ethylene triggered an upsurge in the formation of H_2O_2. The application of ethylene in 100% O_2 resulted in a marked increase in respiration (DB) and likewise the onset in peroxide formation (Fig. 16.6D). While in air, the ethylene induced respiration was two to three times the initial rate, in O_2 the respiration was 10-fold greater. Similarly, ethylene in O_2 induced a higher level of H_2O_2 as compared with ethylene in air. The results indicate that the ethylene induced respiration and peroxide formation are augmented by high O_2 tensions suggesting that both processes are apparently catalyzed by enzyme(s) with high Km (O_2).

The results showing the stimulatory effects of high O_2 tensions on senescence processes suggest that oxygen is rate limiting. However, it appears that O_2 by itself had little or no effect. For example, O_2 even at the highest concentrations (100%) could not initiate the respiratory upsurge or peroxide formation in potatoes (Fig. 16.6B and D). Ethylene was clearly required for the initiation of these processes. Likewise, we show that lycopene formation in the nonripening rin tomato mutant is induced by the synergistic effect of ethylene and oxygen. Although ethylene induces a respiratory upsurge in the rin mutant, other ripening processes are not initiated. As in potatoes oxygen alone, even at high concentrations could not induce pigment formation (Fig. 16.7). However, in the presence of oxygen and ethylene lycopene formation ensued and was proportional to the employed O_2 concentrations. In climacteric fruit which synthesize adequate ethylene, such as pear, high O_2 alone accelerated ripening processes including softening or chlorophyll degradation (Frenkel 1975). The results showing that oxygen utilization is stimulatory to senescence process, as observed for example in fruit, support the concept that senescence is an oxidative phenomenon. Peroxides, but possibly other oxygen forms, represent oxygen species which apparently initiate oxidative processes. In this way the formation of active oxygen species as regulated by ethylene, or other senescence factors, represent the loss of the relative immunity of tissues toward molecular oxygen and the onset of senescence processes.

The mode by which ethylene stimulates the formation of peroxides is not certain. Solomos and Laties (1976) proposed that ethylene stimulates an alternate respiratory pathway in mitochondia, which apparently can lead to the formation of H_2O_2 (Rich et al. 1976). However, other peroxide producing enzyme systems are present in plant tissues (Halliwell 1974) and may also be subject to the action of ethylene. It is also difficult to ascertain, based on the present information, whether the stimulation of

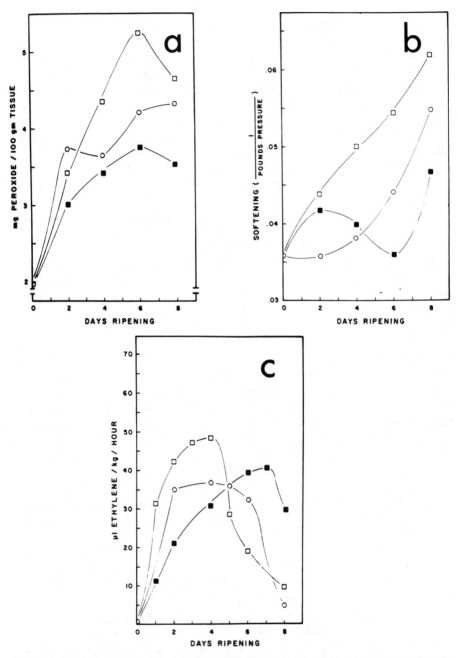

FIG. 16.4. RELATION OF PEROXIDE LEVELS IN PEAR AS OCCURRING NORMALLY AND (a) AS INFLUENCED BY GLYCOLATE OR HPMS (b) TO RATE OF SOFTENING AND (c) ETHYLENE EVOLUTION

The employed concentrations of the applied substances were zero (mannitol control) (○) and 10 μM glycolate (□) and 10 mM HPMS (■).

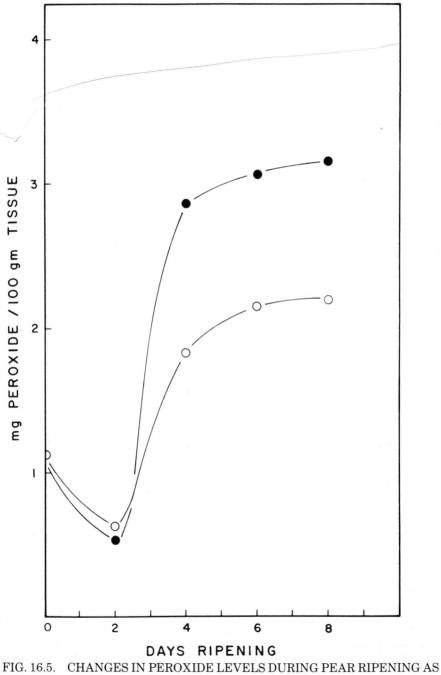

FIG. 16.5.   CHANGES IN PEROXIDE LEVELS DURING PEAR RIPENING AS OCCURRING NORMALLY (○) AND AS INFLUENCED BY ETHYLENE (10 μl/L) (●)

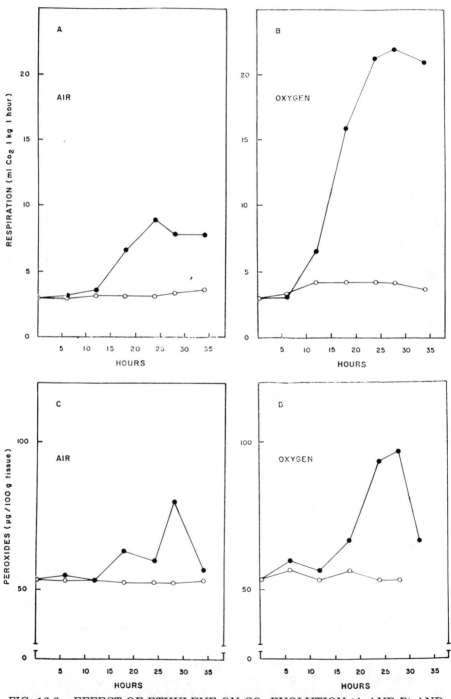

FIG. 16.6.   EFFECT OF ETHYLENE ON CO_2 EVOLUTION (A AND B) AND
PEROXIDE FORMATION (C AND D) IN AIR (21% O_2) AND 100%
The employed ethylene concentrations were zero (○) and 10 μl/L (●).

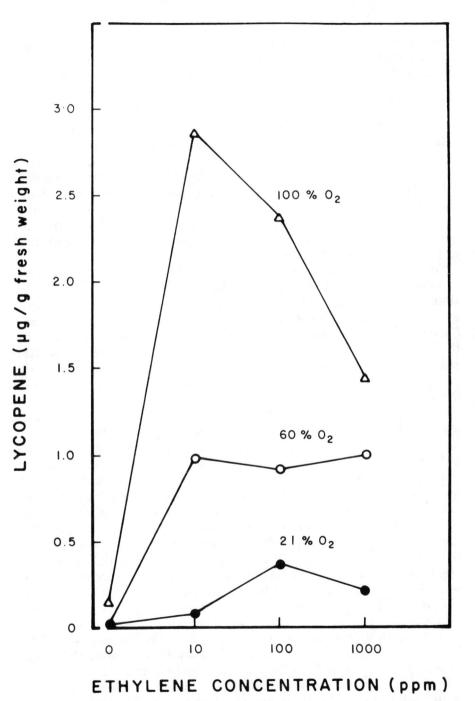

FIG. 16.7.   LYCOPENE FORMATION IN *rin* TOMATO AS INFLUENCED BY
DIFFERENT ETHYLENE AND OXYGEN COMBINATIONS

peroxide formation by ethylene is direct or indirect. Also the possible action of other senescence factors capable of stimulating peroxide formation and, thereby, the onset of senescence must not be overlooked.

Metabolic Action of Peroxides in Tissues

Metabolic changes in tissues may result from both the formation or the action of peroxides. Unlike respiration the electron transfer to oxygen leading to the formation of reduced oxygen intermediates is not coupled to the formation of useful energy such as ATP or reduced pyridine nucleotides, therefore representing an incinerating process in which the potential bond energy may dissipate as heat. From the standpoint of metabolic efficiency the process is, therefore, wasteful as shown by the loss of growth efficiency in plants employing peroxide producing systems during photorespiration (Tolbert 1971; Zelitch 1972).

Specific metabolic changes may be induced following the formation of peroxides since the latter can lead to the onset of oxidative processes. Glutathione-peroxidase catalyzes the oxidation of sulfhydryl (SH) groups by H_2O_2 in plants (Stonier 1973) and in fungal systems (Munkres and Colvin 1976). The diminution of SH gradients in fruit may be important in the onset of ripening. Reduction of SH gradients with the use of SH modifying reagents markedly hastened ripening, while protecting SH groups in fruit with SH compounds led to the inhibition of the process (Frenkel 1976).

Senescence in fruit is accompanied by a decline in unsaturated fatty acids (Meigh et al. 1976) which is attributed to lipid peroxidation (Wooltorton et al. 1965) as was also suggested in aging mammalian tissues (Tappel 1973; Tappel et al. 1973). In our studies we show that there is an increase in lipid peroxidation with the rise in H_2O_2 (Fig. 16.1). H_2O_2 can serve as a source of oxygen radicals in lipid peroxidation (Kellog and Fridovich 1975) but may also be utilized by peroxidase in lipid peroxidation (Tappel 1973), although it is also possible that in plants the process is catalyzed by lipoxygenase (Frenkel et al. 1976; Wooltorton et al. 1965). Lipid autooxidation can proceed as a chain reaction (Tappel 1973). Malondialdehyde, the breakdown product of lipid oxidation, attacks free amino groups in macromolecules, including proteins, forming 1-amino-3-imino-propene bonds, and resulting in the formation of fluorescent pigment (lipofuscin) (Munkres and Colvin 1976). The latter has been shown to be associated with aging in animal systems (Epstein et al. 1972; Miguel et al. 1974; Reichel 1968; Strehler et al. 1959), fungi (Munkres and Colvin 1976) and increases during ripening in fruit (Maguire and Haard 1975).

Oxidative reactions are apparently important in the turnover of hormonal factors involved in the regulation of senescence. The plant hormone indoleacetic acid (IAA) is normally associated with a state of juvenility in plants (Sachar 1973) and functions as a senescence retardant in abscissing leaves (Osborne 1967), in fruit tissue slices (Vendrell 1969), and intact fruit (Frenkel 1974; Frenkel and Dyck 1973; Frenkel et al. 1976; Tignwa and Young 1975). The decline in IAA apparently leads to the release of ripening from the inhibition of IAA and may occur in part by hormonal oxidation as catalyzed by an oxidase/peroxidase system (Yamazaki 1974). The latter has been shown to increase in activity at the onset of senescence in fruit (Frenkel 1972). Moreover, IAA oxidation products obtained in vitro by the action of peroxidase and H_2O_2, or by synthesis, can actually function as senescence promoting factors in fruit (Frenkel 1975; Frenkel et al. 1975). IAA oxidation may serve therefore to deplete senescence retardants and furthermore to lead to the formation of senescence promoting factors in tissues. Further work may reveal whether other hormonal factors undergo a similar metabolic fate.

The evolution of ethylene accompanies senescence in many plant systems, notably in fruit (Burg and Burg 1965). In vitro studies show that hydroperoxides, including H_2O_2 and lipid peroxides, stimulate the formation of ethylene (Lieberman and Mapson 1964). We show that the rise in hydroperoxides in vivo precedes the onset of ethylene evolution during fruit senescence (Fig. 16.1). Generally, the changes in ethylene evolution during fruit senescence correspond to the changes in peroxides in fruit tissues (Fig. 16.2 and 16.4). H_2O_2 by itself or together with superoxides may serve as a source of oxygen radicals leading to the release of ethylene from precursor substrates (Elstner and Konze 1974).

The possibility that the action of H_2O_2 consists, in part, of shifting the redox state of tissues cannot be precluded. Stonier (1972) showed that H_2O_2 reacts with a "protector" macromolecule in plant tissues. When the reducing capacity of the protector is exhausted, the oxidation of autooxidizable metabolites in plants may ensue. Further studies are required to show to what extent peroxides selectively attack specific target metabolites or generally cause a shift in the redox state of tissues.

ACKNOWLEDGEMENTS

The author acknowledges the contributions of Dr. Thomas Brennan, Dr. Chee-kok Chin, and Dr. N. A. Michael Eskin.

BIBLIOGRAPHY

BAKER, J. E. 1976. Superoxide dismutase in ripening fruits. Plant Physiol. 58, 644–647.

BRENNAN, T. and FRENKEL, C. 1977. Involvement of hydrogen peroxide in the regulation of senescence in pear. Plant Physiol. *49*, 411–416.

BURG, S. P. and BURG, E. A. 1965. Ethylene action and the ripening of fruits. Science *148*, 1190–1196.

ELSTNER, E. F. and KONZE, J. R. 1974. Ethylene production by isolated chloroplasts. Z. Naturforsch. *29*, 710–716.

EPSTEIN, J., HIMMELHOCH, S. and GERSHON, D. 1972. Studies on aging in nematodes. 3. Electron microscopic studies on age-associated cellular damage. Mech. Aging Devel. *1*, 245–255.

FRENKEL, C. 1972. Involvement of peroxidase and indole-3-acetic acid oxidase isozymes from pear, tomato and blueberry fruit in ripening. Plant Physiol. *49*, 757–763.

FRENKEL, C. 1974. Role of oxidative metabolism in the regulation of fruit ripening. *In* Facteurs et Regulation de la Maturation des Fruits. Colloques Internationaux du Center National de la Researche Scientifique, Paris, France. No. 231, 201–209.

FRENKEL, C. 1975. Oxidative turnover of auxins in relation to the onset of ripening in Bartlett pear. Plant Physiol. *55*, 480–484.

FRENKEL, C. 1976. Regulation of ripening in Bartlett pears with sulfhydryl reagents. Bot. Gaz. *137*, 154–159.

FRENKEL, C., KLEIN, I. and DILLEY, D. R. 1968. Protein synthesis in relation to ripening of pome fruits. Plant Physiol. *43*(7), 1143–1146.

FRENKEL, C., KLEIN, I. and DILLEY, D. R. 1969. Methods for the study of ripening and protein synthesis in intact pome fruits. Phytochem. *8*, 945–955.

FRENKEL, C. and GARRISON, S. A. 1976. Initiation of lycopene synthesis in the tomato mutant *rin* as influenced by oxygen and ethylene interactions. HortScience *11*, 20–21.

FRENKEL, C. and DYCK, R. 1973. Auxin inhibition of ripening. Plant Physiol. *51*, 6–9.

FRENKEL, C., HADDON, V. R. and SMALLHEER, J. 1975. Promotion of softening and ethylene synthesis in Bartlett pear by 3-methyleneoxindole. Plant Physiol. *56*, 647–649.

FRENKEL, C., DYCK, R. and HAARD, N. F. 1976. Role of auxins in the regulation of fruit ripening. *In* Symposium: Post-harvest Physiol., Biochem. Pathology and Handling of Fruits and Vegetables. N. F. Haard and D. K. Salunkhe (Editors). The Avi Publ. Co., Westport, Conn.

FRENKEL, C., BRENNAN, T. and ESKIN, M. 1976. Relation of ethylene metabolism to peroxide formation in fruit and potato tubers. Plant Physiol. *57*, S–507.

FRIDOVICH, I. 1974. Superoxide dismutase. *In* Molecular Mechanisms of Oxygen Activation. Acad. Press, N.Y.

FRIDOVICH, I. 1975A. Superoxide dismutase. Annu. Rev. Biochem. *44*, 147–159.

FRIDOVICH, I. 1975B. Oxygen: Boon and Bane. American Scientist *63*, 54–59.

HABER, F. and WEISS, J. 1934. The catalytic decomposition of hydrogen peroxide by iron salts. Proc. Roy. Soc. London, *A147*, 332.

HALLIWELL, B. 1974. Superoxide dismutase, catalase and glutathion peroxidase: Solutions to the problems of living with oxygen. New Phytol. *73*, 1075–1086.

HAMILTON, G. A. 1974. Chemical models and mechanisms for oxygenases. *In* Molecular Models of Oxygen Activation. Academic Press, N.Y.

KELLOG, E. W. and FRIDOVICH, I. 1975. Superoxide, hydrogen peroxides, and singlet oxygen in lipid peroxidation by xanthine oxidase system. J. Biol. Chem. *250*, 8812–8817.

KELLOG, E. W. and FRIDOVICH, I. 1976. Superoxide dismutase in the rat and mouse as a function of age and longevity. J. Gerontol. *4*, 405–408.

KRIMSKY, N. I. 1977. Singlet oxygen in biological systems. TIBS 35–38.

LEOPOLD, A. C. 1961. Senescence in plant development. Science *134*, 1727–1732.

LIEBERMAN, M. and MAPSON, L. W. 1964. Genesis and biogenesis of ethylene. Nature *204*, 343–344.

LIEBERMAN, M., KUNISHI, A. T. and WARDALE, D. A. 1965. Ethylene production from methionine. Biochem. *97*, 449–489.

MAGUIRE, Y. P. and HAARD, N. F. 1975. Fluorescent product accumulation in ripening fruit. Nature *258*, 599.

MEIGH, D. F., JONES, J. D. and HULME, A. C. 1976. The respiration climacteric in apple. Phytochem. *6*, 1507–1515.

MIGUEL, J., TAPPEL, A. L., DELLARD, C. J., HERMAN, M. M. and BENSCH, K. G. 1974. Fluorescent products and lysosomal components in aging *Drosophila melanogaster*. J. Gerontol. *29*, 622–637.

MUNKRES, K. D. and COLVIN, J. J. 1976. Aging of *Neurospora crassa*. 2. Organic hydroperoxides toxicity and the protective role of antioxidant and the antioxygenic enzymes. Mech. Aging Dev. *5*, 99–107.

MUNKRES, D. D. and MINSSEN, M. 1976. Aging of *Neurospora crassa*. 1. Evidence for the free radical theory of aging from studies of a natural-death mutant. Mech. Aging Dev. *8*, 79–98.

OSBORNE, D. J. 1967. Hormonal regulation of leaf senescence. Symp. Soc. Exp. Biol. *21*, 699–304.

PILET, P. E. and ZRYD, J. P. 1965. Distribution des Composes Sulfhydrilees dans les Racimes. Ann. Physio. Vegetales *7*, 243–350.

REICHEL, W. 1968. Lipofuscin pigment accumulation and distribution in five rat organs as a function of age. J. Gerontol. *23*, 145–153.

REID, S. M. and PRATT, H. R. 1972. Effect of ethylene on potato tuber respiration. Plant Physiol. *49*, 252–255.

RICH, P. R., BOVERIS, A., BONNER, W. D. and MOORE, A. L. 1976. Hydrogen peroxide generated by the alternate oxidase of higher plants. Biophys. Biochem. Res. Comm. *3*, 695–703.

SACHAR, J. 1973. Senescence and post-harvest physiology. Ann. Rev. Plant Physiol. *25*, 197–224.

SEIGEL, S. M. and PORTO, F. 1961. Oxidants, antioxidants, and growth regulation. *In* Plant Growth Regulation. R. M. Klein (Editor). Iowa State University Press, Ames, IA.

SOLOMOS, T. and LATIES, G. G. 1976. Induction by ethylene of cyanide-resistant respiration. Biochem. Biophys. Res. Comm. *70*, 663–671.

STREHLER, B. L., MARK, D. D., MILDVAN, A. S. and GEE, M. V. 1959. Rate and magnitude of age pigment accumulation in the human myocardium. J. Gerontol. *14*, 430–439.

STONIER, T. 1972. The role of auxin protectors in autonomous growth. *In* Les Cultures de Tissues de Plants, M. L. Hirth and G. Morel (Editors). Colloques Internationaux, CNRS, Paris, No. 193.

STONIER, T. 1973. Studies on auxin protectors 11. Inhibition of peroxidase-catalyzed oxidation of glutathione by auxin protectors and o-dihydroxyphenols. Plant Physiol. *51*, 391–395.

TAPPEL, A. L. 1973. Lipid peroxidation damage to cell components. Fed. Proc. *32*, 1870–1874.

TAPPEL, A., FLETCHER, B. and DEAMER, D. 1973. Effects of antioxidants and nutrients on lipid peroxidation fluorescent products and aging parameters in the mouse. J. Gerontol. *28*, 145–424.

TIGNWA, P. O. and YOUNG, R. E. 1975. The effect of indole-3-acetic acid and other growth regulators on the ripening of avocado fruit. Plant Physiol. *55*, 937–940.

TOLBERT, N. W. 1971. Microbodies, peroxisomes, glyoxysomes. Annu. Rev. Plant Physiol. *22*, 45–74.

VAN FLEET, D. S. 1954. The significance of the histochemical localization of quinones in the differentiation of plant tissues. Phytomorphology *4*, 300–310.

VARNER, J. E. 1965. Death. *In* Plant Biochemistry. J. Bonner and J. E. Varner (Editors). Acad. Press, N.Y.

VENDRELL, M. 1969. Reversion of senescence: effect of 2,4-D and indoleacetic acid on respiration, ethylene production, ripening of banana fruit slices. Aust. J. Biol. Sci. *22*, 601–610.

WHITAKER, J. R. 1972. Catalase and peroxidase. *In* Principles of Enzymology for Food Sciences. J. R. Whitaker (Editor). Marcel Dekker, Inc., N.Y.

WOOLTORTON, C. S., JONES, J. D. and HULME, A. C. 1965. Genesis of ethylene in apples. Nature *207*, 999.

YAMAZAKI, I. 1974. Peroxidases. *In* Molecular Mechanisms of Oxygen Activation. O. Hayaishi (Editor). Acad. Press, N.Y.

ZELITCH, I. 1972. Comparison of the effectiveness of glycolic acid and xanthine as substrates for photorespiration. Plant Physiol. *50*, 109–113.

INDEX

Abscisic acid, 89–90
    determination, 89
    levels, 89–90
    role, 89–90
Absidia, 212
Acetaldehyde, 192
Acetate, 409
Acetic acid, 218, 235
Acid phosphatase, peanut, 379, 390
Acyl hydrolases, 402
Acyl-CoA reductase, 408
Acyl-CoA transacylase, 408
ADPG-pyrophosphorylase, 5
Affinity chromatography, 349, 356, 426
Aflatoxins, 216, 229–231, 375
    carcinogenicity, 231
    corn, 229–231
    incidence, 229–231
After-ripening, 18
Air classification, 255
Albumins, definition, 5–6
    in, breadmaking, 278–279
    cereal grains, 256
Alcohol dehydrogenase, peanut, 377–379, 390
    in flavor biogenesis, 409, 426–427
Aldehyde reductases, 409
Aleurone, barley, 27, 261
    rice, 251, 263
    wheat, 5, 260–261, 263
        description, 247
        separation from endosperm, 252, 254
Alkaline phosphatase, peanut, 379, 390
Alkanes, 333
Allinase, 337
Allium, 339
Alternaria, 176–177, 186, 189–192, 195, 197, 211–212. *See also* Stem end rot, *Alternaria*
Alternaria citri, 171, 195
Ammonia, corn drying, 235–236
Ammonium sulfate precipitation, 383, 426
Amylase, corn, 222
    inhibitors, 305

α-Amylase, in cereal grain processing, 28
    rice, 260
    sweet potato, 325
    wheat, 19, 259, 340
Amylases, cereal grain, 20
    wheat, 255, 259
α-Amylolysis, 259–260
Amyloplast, 321. *See also* Plastids
Anaerobiosis, 105–106
Anthracnose, 178, 180–181, 184, 187
Antimicrobial compounds, endogenous, 173
Apples, benzoic acid, 172–174, 180, 192
    internal atmosphere, 100–105
    scald, 106
Appressorium, 165, 178
Arachidonic acid, 331
Arachin, 373–375, 383, 390–392
Ardex No. *550*, 269
Aroyl-CoA reductases, 406
Arrhenius plots, in chilling injury, 148–152, 156
Ascorbic acid, in breadmaking, 270
    in wounded plant tissue, 116
    loss in modified atmosphere storage, 107–108
    reaction with isothiocyanates, 337
    soybean, 50, 53, 57
    stimulation of fungal germination, 172
Ascorbic acid oxidase, 105
Ascorbigen, 337
Aspergillus, 212, 225, 262
Aspergillus candidus, 219, 225
Aspergillus flavus, 212, 216, 219
    aflatoxin production, 229
    in peanuts, 376–379
Aspergillus fumigatus, 219
Aspergillus glaucus, 212, 215, 219
Aspergillus nidulans, 191
Aspergillus niger, 225
Aspergillus ochraceus, 231
Aspergillus oryzae, in peanuts, 379–380
Aspergillus parasiticus, in peanuts, 376, 379